America's Kingdom

Stanford Studies in Middle Eastern and Islamic Societies and Cultures

America's Kingdom

MYTHMAKING ON THE SAUDI OIL FRONTIER

Robert Vitalis

Stanford University Press
Stanford, California 2007

Stanford University Press
Stanford, California

Printed in the United States of America on acid-free, archival-quality paper

Library of Congress Cataloging-in-Publication Data

Vitalis, Robert, 1955-

America's kingdom : mythmaking on the Saudi oil frontier / Robert Vitalis.

   p. cm.

Includes bibliographical references and index.

ISBN-13: 978-0-8047-5446-0 (cloth : alk. paper)

1. Arabian American Oil Company--History. 2. Saudi Aramco--History. 3. Petroleum industry and trade--Saudi Arabia--History. 4. United States--Foreign relations--Saudi Arabia. 5. Saudi Arabia--Foreign relations--United States. I. Title.

HD9576.S334A738 2007

338.7'665509538--dc22

                           2006018541

Designed by Bruce Lundquist

Typeset at Stanford University Press in 10/14 Minion

We don't have a special relationship with them,
we have a deal.

Tom Friedman

We are the sons of the Indians who sold Manhattan.
We want to change the deal.

Abdallah Tariki

CONTENTS

Map and photographs follow page 26

FOREWORD

Americans have a pastime known as "reenacting." The journalist Tony Horowitz writes about Civil War reenactors in *Confederates in the Attic*. Every year, people meet across the country to re-create famous battles of the War Between the States. Americans reenact other wars too, from the War of Independence to the Wars of Empire: the American-Indian Wars, the War with Mexico, and the Spanish-American War. Battles draw the largest numbers, but many men and women also meet regularly for Old West reenactments, playing miners, settlers, trappers, gunslingers, cowboys, and Indians. Others play too. There are Old West reenactors in the Czech Republic, Vietnam War reenactors in Russia, and World War I and II reenactors everywhere, including in the United States, Australia, and England and elsewhere in Europe.

On February 14, 2005, the Saudi prince Abd al-Aziz Ibn Abdallah Ibn Abd al-Aziz took part in a reenactment. He played his own grandfather in a replay of a meeting that had taken place sixty years earlier between Abd al-Aziz, the first king of Saudi Arabia, known in America as Ibn Saud, and Franklin Delano Roosevelt, known as FDR. The president's grandson Hall Delano Roosevelt, a former city councilman in Long Beach, played FDR. The summit sixty years earlier had taken place aboard a destroyer, the USS *Quincy*, in the middle of the Suez Canal. It was during World War II. The two descendents met on a stage at the Ritz Hotel in Coconut Grove, Florida. It might be fair to add, during the U.S. Global War on Terror.

A second American, Condit Eddy, joined the other two reenactors on stage. Eddy was playing his uncle Colonel William Alfred Eddy, the translator on board the *Quincy*. Colonel Eddy had fought in World War I, ran the English

Department at the American University in Cairo, where he wrote the first Arabic rulebook for basketball, and served as president of Hobart College in New York before giving up academic administration for spying. At age forty-five, he dusted off his commission and joined the new U.S. intelligence organization, the Office of Strategic Services (OSS). As navel attaché in Tangier he ran resistance groups during the North African campaign. In 1944 FDR named Eddy America's first minister plenipotentiary to Saudi Arabia. Eddy spent the rest of his life involved in U.S.-Saudi affairs for the OSS's successor, the Central Intelligence Agency (CIA). His CIA cover was consultant to the Arabian American Oil Company (ARAMCO). Eddy had indeed translated for the two heads of state—he had learned Arabic as a child of missionaries in Lebanon—but "translator" is also an over-modest description of the historical figure Condit played that day.

Hard-core reenactors elevate verisimilitude above all else. They wear period uniforms, eat the foods that real soldiers carried with them into battle, and so on, but the sponsor, a recently founded organization named the Friends of Saudi Arabia, had other objectives, and verisimilitude was sacrificed that day. Neil Bush, the brother of President George W. Bush, and Anthony Kennedy Shriver, a nephew of President John F. Kennedy, climbed on stage with the others. You may ask yourself, why? Americans are much more apt to commemorate fifty-year rather than sixty-year milestones, and what would have been the golden anniversary of the *Quincy* meeting had passed without much comment beyond an ad or two taken out by the Saudi embassy in the *New York Times* and the *Washington Post*. Certainly no one had thought to organize a reenactment. Something extraordinary had happened in the interim to which the stiff piece of dinner theater was a response.

On September 11, 2001, at 8:46 AM, American Airlines Flight 11 burst through the upper floors of the North Tower of the World Trade Center in Manhattan. Seventeen minutes later United Flight 175 hit the South Tower. One hour later, a third plane, American Airlines Flight 77, tore into the Pentagon. The Towers collapsed about the time that United Airlines Flight 93 crashed in the Pennsylvania countryside. The best guess is that this fourth plane was aimed at some target in Washington too, but a group of passengers had stopped the hijackers. More than three thousand died that morning. Nineteen members of an organization called Al Qaeda had turned the airliners into weapons of mass destruction. Fifteen of the hijackers were Saudi subjects, warriors in a battle that is unlikely to be reenacted anytime soon. They served under another

Saudi-in-hiding somewhere in Afghanistan, Osama Bin Laden, the leader of Al Qaeda.

Americans channeled their outrage in various ways. Some enlisted in President George W. Bush's wars. The United States attacked and occupied Afghanistan in 2002 and Iraq in 2003. Filmmaker Michael Moore released what became the largest-grossing documentary film ever, *Fahrenheit 9/11*. Publishers rushed new books into print: biographies of Bin Laden, and studies of terror networks and of Saudi Arabia. Some without footnotes, others with. Some more sober, some less. Books with titles like *Hatred's Kingdom* and *Sleeping with the Devil*. The men and women who wrote in this vein viewed themselves as the nation's conscience, intent on exposing the ways in which interests and organizations beholden to the Al Saud (the "house of Saud"), as the ruling family is known, were undermining America's security. Not everyone agreed.

For the many defenders of the special relationship in Washington and New York—career Saudi watchers, ex-ambassadors, bankers, contractors, oil industry consultants, and geo-strategists—the days and months following 9/11 were dark and increasingly ominous. Different, they liked to imagine, from the days when Ibn Saud and FDR met on the USS *Quincy*. They feared that the tragedy of September 11, 2001, was in danger of being hijacked by ranks of misguided visionaries, neoconservatives, pro-Israel politicians, Christian fundamentalists, and other assorted "Saudi bashers" who were exploiting popular outrage by trying to steer U.S. policy in ways that were doing serious, perhaps fatal damage to the national interest. The unidentified board of directors of the Friends of Saudi Arabia may fit this category. The National Council on U.S.-Arab Relations most definitely does. Even older and more august organizations, such as the Middle East Institute in Washington and the Council of Foreign Relations in New York, weighed in, less ham-handedly, as is their style, through assisting the publication of books designed to instruct us anew on the history of U.S.-Saudi relations, *Inside the Mirage* and *Thicker Than Oil*.

America's Kingdom is not like all these other books, and is certainly not one rushed into print in response to the events of September 11. Professors generally operate on a time line different from those of commercial publishers, journalists, or think tanks. I first started thinking about a project on ARAMCO in 1989, when I was still finishing a first book on business in Egypt. I wrote the initial grants for the project in the early 1990s, and only began serious work on it in 1995. I have been working on it more or less steadily since.

Owned first by two and then four of the world's largest oil corporations, ARAMCO pioneered what was in the 1930s and 1940s a frontier of the world mineral market. *America's Kingdom* tells the story of its workers, assembled by the firm from a dozen countries in order to erect and operate the rigs and build the refinery, stores, and housing for the white American managers and their families. It is ARAMCO's massive investment in oil-producing technology, in the infrastructure to get the oil to market, and in its workers, beginning with the thousands of coolies, as the company's most famous geologist and future executive Tom Barger described the Saudis in his first letters home, that led the U.S. state to follow American capital to the kingdom, sending Colonel Eddy there in 1944.

The summer of '44 was the start of what the Friends call the special relationship and what others call, less reverently, "the deal," oil for security. Back then, the *New York Times* called it moving backward to the "old imperialism" and the era of "dollar diplomacy," as Eddy negotiated with Ibn Saud to build an airfield for the oil firm at taxpayers' expense. The *Times* came around about the time that the Harry Truman administration committed the United States to preserve the Al Saud in power. And when Truman's successors followed through on the pledge—Dwight Eisenhower agreed to train Ibn Saud's army, John Kennedy sent jets to defend the kingdom, and Lyndon Johnson sold missiles to the Saudis—the interagency correspondence returned routinely to reminding generals in the Joint Chiefs of Staff to remember oil and ARAMCO, "our largest single overseas private enterprise."

I am not the first to write on the U.S. relationship with the world's most famous family state, but, as we will see, the history of the oil settlement is little known and, as *America's Kingdom* shows, significantly, perhaps tragically, misunderstood. Two myths form the core of the standard account of the American experience in Saudi Arabia. The first is a familiar one, told about many places. ARAMCO is believed to have acted more generously and less exploitatively than enterprises of other nations, say, British plantations or mining firms in Africa. It is essentially the claim any firm that writes its history will pay to tell about itself. This book dismantles this idea piece by piece through careful use of ARAMCO's own records and its analysts' observations about competitors in the Persian Gulf and elsewhere.

Using the same method, it is even easier to expose to light the second myth about the kingdom, the idea that Saudi Arabia represents an unusual or exceptional case when compared to most other late-developing countries. The myth goes that market and state making there took place under conditions of relative international isolation. Oil company executives and other Americans engaged in all the most basic material and symbolic practices of nation-state formation, from drawing organizational charts and construction plans to building entire new cities, bureaucracies, and traditions.

Mythmakers will have to work harder on their stories about ARAMCO being magical, honorable, selfless, enlightened, and the like. The archival records tell a story that differs markedly from the drama of pioneering geologists and multinational managers in the wilderness, which, in the absence of serious and critical scholarship, has gradually come to be canonized in, for instance, novelist Wallace Stegner's *Discovery!*, the PBS documentary adapted from Daniel Yergin's *The Prize*, and now Thomas Lippman's *Inside the Mirage*.

ARAMCO's operations in the oil town, Dhahran, rested on a set of exclusionary practices and norms that were themselves legacies of earlier mining booms and market formation in the American West and Southwest. This was a system of privilege and inequality, which we know as Jim Crow in the United States, as Apartheid in South Africa, and as racism more generally. The "laws" that ARAMCO officials imposed on its employees against crossing the color line in its segregated Jim Crow compounds, forbidding Saudis from living with their families, and deporting Americans who pursued contacts with nearby Arab families, the compound's model of justice, and its labor problems generally, have not been documented and analyzed anywhere to my knowledge. Nor has anyone tried to document the movement that emerged to challenge this hierarchy in the oil camps, beginning with the first strike by Saudi workers in 1945. Doing so means having to excavate through multiple strata of company public relations campaigns and two or three generations of scholarship that rest on these foundations.

Most researchers though have constructed their stories about the U.S.-Saudi relationship without ever considering the encounter between Saudis and Americans on the ground. They write as if the history of mining enterprises elsewhere has nothing to teach us, as if there is nothing to compare. Some, who know better, continue to dissemble. And there are those who ignore rather than wrestle with arguments that don't fit the story they want to believe. Still, there are problems beyond naïve rehearsals of a firm's propaganda in response to Saudi workers striking and the rise of a state-building class that sought to limit ARAMCO's power and extract a greater share of the rents from its monopoly—a word that the firm insisted didn't apply to its, well, monopoly. There are more systematic problems or blind spots that affect our understanding of ARAMCO's or any other firm's role in the long history of empire, of which the moment forms a part.

One is the problem of exceptionalism—a way of viewing or narrating or thinking about the American experience generally. American exceptionalism assumes the deep structural autonomy of that experience, that American history is unlike and unconnected with all others. Exceptionalism grounds, shapes, frames, all the varieties of accounts purporting to prove American enterprise to be anything but agents of empire, of America being empire's antithesis, about the United States acquiring an empire late, or in a fit of absentmindedness, "learning early" to be good citizens, and the like.

The second blind spot is, to be blunt, the rich tradition of racism in American life. Although gender, ethnicity, nationality, and even religion have served in the United States as grounds for exclusion at the polls, in the workplace, in schools, and in neighborhoods, no identity has mattered more than race in determining and justifying hierarchy. ARAMCO, like all other large U.S. mining firms, organized life inside the camps on the basis of the then-ruling ideas about the superiority of whites and the inferiority of all others. Critics of the elaborate hierarchies built on the basis of skin color or facial features and on the alleged inferior and superior abilities of these differently marked bodies coined a term in the 1930s to characterize such practices. They called it racism.

Racism is exceptionalism's Achilles' heel, the contradiction at the heart of the "storybook truth about America," as Louis Hartz once described the country he called "eternally different from everyone else."[1] The problem is the same for the storybook history of ARAMCO. Illiberal institutions inside the United States were a "puzzle" for Hartz and a "dilemma" for the much more widely known Swedish economist Gunnar Myrdal, whose study of race relations influenced idea makers in post–World War II America. Myrdal said racism was really just a kind of irrational prejudice found mainly in the traditions of lower-educated whites of the backward areas of the South, and as such it was destined to disappear with time.

The argument though had been made many times before, and Myrdal was no more successful than others in reconciling it either with the history and sociology of racial science, a product not of working class whites of course but of the country's leading scholars and public intellectuals, or with the reality of the hierarchies that states and, as I explore here, firms had built. In the 1960s, ARAMCO's managers would try to rewrite their own past using the same Myrdalian idea, imagining a dwindling cohort of Texans whose prejudices may have been a problem "in the past," and they were successful because it is the story that writers still tell today. If only the firm had devoted the same energy to dismantling the institutions of hierarchy.

I made my first and only trip to the kingdom in December 1996. To get there, I dissembled, which gained me a visa and a two-week, subsidized tour. My benefactor was the National Council on U.S.-Arab Relations, one of those organizations convinced that the national interest is best served by a continued close relationship with the Al Saud, and pretty clearly in the royals' pay. Individuals or organizations inside the kingdom ultimately underwrote the study tour. The dozen Fellows sponsored by the National Council, including me, flew business class, on Saudia, courtesy of the chamber of commerce in Riyadh, I was told.

We were mostly educators, including two professors I have known a long while, and administrators from small colleges across the country. Two other Fellows though were undercover military intelligence officers, one from the Army and the other from the Air Force. The National Council told us that Saudis liked to receive little gifts from the States. A coffee mug from our college, say. It was the custom. I did not bring any. I hoped I paid back my hosts at the chamber by letting a friend there know that the National Council was providing this little extra service to the government. I also tried while I was there to look for a sponsor—the only way to gain a visa to the kingdom—that would allow me to continue my research on the workers who had organized the strikes that ultimately forced the dismantling of at least some of the Jim Crow institutions that the Americans had brought to the kingdom. Friends in Riyadh and Dhahran just laughed, as if it were a joke. The archivist at Saudi ARAMCO, which, as the name change indicates, is now a wholly Saudi-owned company, cancelled my appointment after she heard what I was after. We did meet Saudi journalists, academics, businessmen, and government officials, including the foreign minister, another grandson of Ibn Saud, and a son of the man who did most to crush the movement for reform in the 1960s. The Fellows from the National Council also were treated to briefings at the U.S. embassy in Riyadh and the consulate in Jidda. The best by far was by one of the embassy's economic officers.[2]

Nearly ten years later, in 2004, I participated in a Saudi succession-crisis game run by a firm known as Centra Technologies, which had been founded by ex-CIA agents. Centra does contract work mostly for the agency or at least that is what others who knew the business better had told me. We were a mix of academics, business people, oil consultants, retired diplomats, and men and women in the agency itself, the latter identifiable by tags listing first names only. Gaming the future is not like reenacting, although some of us were playing men who were young boys at the time of FDR's meeting with Ibn Saud. A man named Bob headed my team, the senior princes. He played Abdallah, then the crown prince and now the king. I played Salman, governor of Riyadh. Bob looked familiar. He was the economic officer in Riyadh, his cover during a short tour, he admitted. It is a small world.

America's Kingdom has taken a long time to write. Part of that time is accounted for by the vicissitudes of life, the classroom, love and love's loss, illness. I had heart surgery. I went through a divorce. I have tried to keep in mind that all the men and women in the book are human too, even those in the house of Saud who have been so cavalier about the people they have had killed and the lives they have destroyed in the course of hanging on to power. More of that time though is accounted for by my doing what was necessary to get the story right, which is about the best that we can expect from social science. Professors have a fancy word for what we do when we try to get the story right: method.

I did two things differently from those who worked on ARAMCO before me. First, I found and used more sources than the others typically have. In the early 1990s Georgetown University acquired the papers of William Mulligan, a career employee of ARAMCO, who took copies of many records from the Government Relations Organization (the company's Arabists and diplomats, in effect) with him to New Hampshire when he retired. It is an amazing collection. After all, firms are not like states in that they do not ordinarily preserve their records or, where they do, they do not ordinarily let just any researcher into the archive (if you don't believe me, then try). The Mulligan papers were not available when I began this project (unless you knew Mulligan), but since then I have read virtually every document in them. Journalists also conventionally insist on two independent sources to confirm any controversial claim before it can be reported as fact. If you check, though, you will find that the convention has generally gone by the wayside, and many treat Mulligan's

ARAMCO papers a bit too authoritatively. Other sources—the fifty-year-old memories of retirees, the published and, it turns out, redacted letters of the legendary executive Tom Barger—are treated the same way, unfortunately.

I believe I am the only researcher working on ARAMCO to have read, in addition, every page of the declassified State Department records for the same period—over 3,000 pages. Work like this takes time and money. So, although Lippman writes about the "lost" manuscript by the great Western American novelist, biographer, and historian-for-hire Wallace Stegner accepting an ARAMCO functionary's account of what its faults were, I read the original, as well as the writer's correspondence with the company, which tell a markedly different story from the one the firm invented and that the journalist simply reports as true. There are many such examples to be found in the next few hundred pages.

It takes time to resolve the many incommensurable truth claims and to go beyond pointing out the failures of logic and investigative powers by one or another writer. It is necessary—again, the norms of social science matter here—to offer a better, more convincing account than one that insists the Americans built a segregated enclave because the Saudis ordered them to or that American oilmen picked up their bad habits from British officials in Bahrain or that Crown Prince Faisal was a modernizer whose nonexistent reforms somehow saved the kingdom.

The second difference between this work and others about ARAMCO is its commitment to taking the problem of exceptionalism seriously. It takes time to develop the expertise necessary to write a better history, one that crosses all those boundaries that create the effect of a country unconnected with others and a past that doesn't matter. While it would be wrong to argue that the organization of knowledge in the American academy and the juggernaut of academic specialization are consequences of exceptionalism, it is easy to see how disciplines and specialization reinforce the effect.

I read American Western historians. I raised funds to retrain in African American studies in the course of writing this book. I began to teach on W. E. B. Du Bois. And I also taught courses on America's global frontiers. As a result, *America's Kingdom* brings together the history of America and the Middle East but not as if it is a story that begins in February 1945, with the meeting between Ibn Saud and FDR, as it does in too many accounts. *America's Kingdom* shows why it is imperative that we tear down the wall between the 1940s and all that came before, and that we topple that other, more formidable wall, once understood as dividing races and now as dividing nations or cultures, which protects the myth of an isolated and autonomous history of the United States of America.

I did not start out to write a book against exceptionalism. I had no argument to begin with, only the knowledge that the ancestors hadn't treated the history of ARAMCO very seriously, and that I had developed some usable skills with archives while writing my first book. When I started this project, I described myself as a political economist, and I thought of political economy as a kind of excavation project of material lying beneath the surface of ideology and culture. Now, at the end of a decade-long endeavor, I tend to use a different metaphor, and think of my work more in terms of reverse engineering of particular processes of mythmaking.

I have written a book that is eight chapters long, and have written it in an accessible style, but one that nonetheless advances the kind of historical synthesis and "strong, even heretical personal judgments" that Bernard DeVoto, a life-long friend of Stegner's, said "professionals had abandoned."[3] I have tried to tell a story. It is a complex one. There is no way around that fact. I have included a cast of characters to help readers keep all the actors straight. And I have chosen to do without the parenthetical or in-text references that somehow have crept into professors' writing since I joined the profession, which we use to convince ourselves that what we do is science and not art.

Those impatient for the argument, as I teach in my classes, and those who think the only question worth asking about a work is how it matters for something called theory, which is the question we ask constantly in graduate seminars, academic job talks, and, sadly, even after we've left the seminar room and reconvened at the White Dog or the Standard Tap, prepare for a rough time. The first chapter, "Captive Narratives," is written with Yuengling-drinking graduate students and Pinot Noir–drinking professors in mind, but the rest of the book is going to make your hangovers worse. The story unfolds at a relaxed pace. I am hoping that readers take this book with them to the shore.

That said, the structure of the book is a simple one. I have divided the story into two parts. Part 1 documents the setting up of the Jim Crow system in Dhahran in the 1930s and 1940s. The history though doesn't begin on the day the Chevron lawyer arrived in Riyadh to cut a deal with the king or when the first geologists put ashore near Jubail, the way all the other histories of ARAMCO begin. Unlike these other works, the goal is not to bury the parts that firms and their friends prefer that you not see. So Chapter 2 steals a device from ARAMCO's feature-length film, *Island of Allah* (1955), and any number of other films before and since. Think of the chapter as a flashback. Imagine yourself sitting around a desert campfire, listening to a tale of the rise of a large-scale mining enterprise and of work and life inside the various camps for whites or Anglos, for Mexicans, Chinese, and others. It opens with an image of Dhahran and what was then known as American Camp on a day in 1947, but then moves backward to the Western territories of North America and northern Mexico in the late nineteenth century through World War I, and from there to Venezuela and Colombia in the 1920s, and then on to Bahrain in the 1930s, before eventually reaching the east coast of Saudi Arabia once more. You will see why soon enough. Time runs one way only for the rest of the book, and Part 1 ends with the firm piecing together a response to the first workers' strikes in Dhahran.

Part 2 tells the story of the workers' challenge to the hierarchy of the ARAMCO camps in the 1950s and the parallel challenge by a small cadre of progressive Saudis in government in the late 1950s and early 1960s to the hierarchy of the international oil market. It was an extraordinary moment, and its details are mostly unknown or at least unpublished, but it was also short-lived. The final chapters trace the defeat of these forces by 1962–63 and the consolidation of America's kingdom under the third king Faisal and those of his brothers known as the Al al-Fahd or house of Fahd, the royal faction that rules even now.

If you can't wait, and need to know in advance what it all means, then read out of order. Start with the last subsection of Chapter 1, where you will find the argument summarized. Then read the subsection at the very end of Chapter 4 titled "Reprise: Dusk at Dawn," and section three of Chapter 8, through the subsection titled "Ayyam Jamila (Beautiful Days)." It is where you will find compact summaries of Parts 1 and 2. Maybe, like graduate students, you won't have much more time to spare than it will take to read those fifteen or so pages. If you do have the time and are willing to wrestle with a little bit of abstraction in trade for some background information, then read the book from front to back. And if you just want the story, skip the first chapter and begin with Part 1, "The Nearest Faraway Place," a title lifted from a now out of print biography of the Beach Boys.

Needless to say, I have left myself open for some criticism. I look forward to it, as anyone should who has come to care passionately about a work or about what the work tries to accomplish. Let me try to save some time and energy by answering two obvious and, sorry, not very compelling criticisms in advance. Maybe we can start from some higher ground.

For you who want a comprehensive history of ARAMCO, treating issues of technology, finance, administration, and above all American triumph over adversity, *America's Kingdom* is not the book you are looking for, not least because it is not intended as a history of the firm. Let me be clear once more. It is most basically an account of the building of a Jim Crow enclave on the eastern shore of Saudi Arabia at the end of World War II, an explanation for, if not the final word on, why the labor process was organized in this way, and what explains the start of its demise. The challenge for future researchers is to test and if necessary revise my claim about ARAMCO lagging behind its competitors in Iran and Iraq and its owners' own venture in Venezuela, in areas such as education, training, and promotion to management positions. The task for the future will be to falsify my theory of what accounts for the lag.

Then there will be those who will be more or less untroubled by what I have to say about ARAMCO but will worry about my failure to capture the Saudi experience in enough depth or to understand the so-called authentic character of the Saudis or to include enough and varied Saudi voices. These kinds of criticism miss the mark. I look forward to the possibility of a different and better world to come where some who live in the kingdom might finally start to do the kind of research and publishing that many have long wanted to see. My book, though, is not about Saudi Arabia. It is about America. It just takes place and tells you more than others have about somewhere else.

CAST OF CHARACTERS

LEADING MEMBERS OF THE HOUSE OF SAUD (THE AL SAUD)

Abd al-Aziz Al Saud or Ibn Saud (c. 1880–1953). Founder of the Kingdom of Saudi Arabia. Born in Riyadh. Lived in exile in Kuwait from 1891 to 1902, when he led the conquest of Riyadh. Extended conquests in 1910s and 1920s to lands that make up the kingdom.

Fahd Ibn Abd al-Aziz (1921–2005). Eldest of the "House of Fahd" brothers, sons of the king who were supporters of Faisal in the struggle with Saud. Ascended to the throne in 1982 upon the death of his brother Khalid. Served as Faisal's longtime minister of the interior and deputy prime minister. In 1995 was incapacitated by a massive stroke but remained the nominal ruler until his death.

Faisal Ibn Abd al-Aziz (1904–75). Governor of Hijaz and foreign minister during the early years of the kingdom. Became crown prince after the succession of his brother Saud in 1953. Overthrew Saud in 1964. Assassinated in 1975.

Saud Ibn Abd al-Aziz (1902–69). Succeeded his father Ibn Saud as ruler of the kingdom but lost in a long struggle with his brother Faisal for control of the state. Deposed in 1964. Died in exile in Greece.

Talal Ibn Abd al-Aziz (b. 1931). Leading "liberal prince," exiled from the kingdom in 1962 for his criticisms of Faisal, support for Nasser in Egypt, and calls for a constitution. Allowed to return in 1964 but never given an official position again until his appointment in 2005 as an advisor to the present King Abdullah.

SAUDI NOTABLES AND OFFICIALS

Kamal Adham (d. 1999). King Faisal's brother-in-law and head of Saudi Intelligence from 1963 to 1979. A main liaison with the CIA in the same period. Entered plea agreement in New York in 1992 for his role in banking fraud.

Anwar Ali. International Monetary Fund economist and head of its Middle East Division. Appointed governor of the Saudi Arabian Monetary Agency in 1958, a post he retained until 1974.

Muhammad Alireza. From influential merchant family of Jidda. Founded Jidda Chamber of Commerce. Ally of Faisal; joined forces briefly with Faisal's brother King Saud in 1960. After reconciliation in 1963, appointed Saudi Arabia's emissary to Egypt.

Muhammad bin Laden (d. 1968). Yemeni immigrant to Saudi Arabia in 1931 who became a wealthy building contractor and ally of Ibn Saud and his son King Saud. Father of Osama bin Laden.

Abd al-Aziz Ibn Muammar (1919–84). Technocrat, appointed counselor to King Saud in 1958. By 1960 the leading architect of a short-lived political liberalization. Imprisoned by Faisal for twelve years after the latter wrestled control of the kingdom from Saud.

Abdelrahman Munif (1933–2004). Writer. Born in Jordan to a family of Saudi origin. Never resided for an extended period in the kingdom. Obtained a Ph.D. in Belgrade in 1961 and worked in the oil industry in Iraq. Stripped of his Saudi citizenship in 1963 for criticizing the royal family. Author of *Cities of Salt* (*Mudun al-milh*), a quintet of novels about oil and state building in the fictional desert kingdom of Mooran.

Hisham Nazer (b. 1932). Protégé of Abdallah Tariki. Succeeded Zaki Yamani as minister of petroleum, 1986–95. Appointed first chairman of the nationalized Saudi ARAMCO in 1989 while he retained his government post.

Rashad Pharaon. Moved to Saudi Arabia from Syria in 1936 as chief physician to King Ibn Saud. Served as loyal political advisor to both King Faisal and King Khalid.

Nasser al-Said (d. 1979). Labor activist. Fled the country in 1956 at the time of the last major oil workers' strike and settled in Cairo, where he continued his campaign against the Saudi regime. After expulsion from Egypt, lived in Yemen and Beirut. Killed by unidentified assassins.

Umar al-Saqqaf. Staunch Faisal loyalist. First appointed deputy foreign minister under King Saud and later promoted to minister under Faisal.

Ahmad Shukairy (1908–80). Palestinian lawyer and diplomat. Saudi representative to the UN from 1957 to 1962, where he often criticized American policies. In 1964 he became the first president of the new Palestine Liberation Organization (PLO).

Abdallah Tariki (1917–97). First Saudi to obtain U.S. advanced degree in oil geology. Appointed director general of Saudi Arabia's Office of Petroleum and Mineral Affairs in 1954 and first minister of petroleum in 1960. Exiled under Faisal's administration and eventually settled in Lebanon.

Mustafa Wahba. Economist and close friend of Abdallah Tariki. Appointed deputy minister of finance in 1960. Son of Shaykh Hafiz Wahba, King Ibn Saud's Egyptian advisor and ambassador to England.

Ahmed Zaki Yamani (b. 1930). Appointed by Faisal to replace Tariki as minister of petroleum in 1962. Served as secretary general of the Organization for Petroleum Exporting Countries (OPEC). Famous for implementing the 1973 oil embargo. Dismissed by King Fahd in 1986.

Yusuf Yassin (d. 1962). Syrian who served as Ibn Saud's chief negotiator and de facto foreign minister during the early years of American-Saudi diplomacy. Remained closest advisor to King Saud until his death.

OTHER MIDDLE EASTERN LEADERS

King Hussein (1935–99). Long-ruling third king of Jordan, from the age of sixteen. From the family that ruled the short-lived independent kingdom of Hijaz in what is now western Saudi Arabia.

Mohammad Mossadegh (1882–1967). Elected prime minister of Iran in 1951. Nationalized Iran's oil industry and forced the shah to relinquish control of the military to the parliament. Ousted from power in 1953 in a coup engineered by the CIA in cooperation with British intelligence.

Gamal Abdel Nasser (1918–70). Leader of the "Free Officers" that staged a coup against the Egyptian monarchy in 1952. Served as premier and then president from 1954 to 1970. Following 1956 Suez War, emerged as the Arab world's most influential leader and supporter of pan-Arabism in the form often referred to now as Nasserism.

Abd al-Karim Qassim (1914–63). Iraqi officer and leader of a coup in 1958. Governed as the revolution's first prime minister. A staunch Iraqi nationalist, he ended a short-lived union with Jordan, refused to bring Iraq into the Egyptian-Syrian union, and opposed U.S. intervention in the region. Executed by members of Baath Party in 1963.

Abdallah al-Sallal (1917–94). Led coup against the imam in Yemen in 1962. Served as president of the new republic for the next five years. While Egypt supported Sallal's coup and provided diplomatic and military support in a long civil war, Saudi Arabia supported the royalists. Ousted in a coup in 1967.

ARAMCO OFFICIALS

Tom Barger (1909–86). Geologist. Joined Chevron in Saudi Arabia in 1937. Headed ARAMCO's Government Relations Organization in the 1950s. Named president in 1959. Retired as chairman in 1969.

James Terry Duce (1893–1965). Geologist. President of Texaco's Colombia subsidiary. Joined ARAMCO as vice president in 1939. Company's top Washington operative. Supported founding of the Middle East Institute in Washington and Harvard's Middle East Center. Retired in 1959.

William Eddy (1896–1962). Syrian-born American Marine, Arabist, OSS agent, and first U.S. minister plenipotentiary to Saudi Arabia during World War II. Joined ARAMCO as a consultant in 1947 while serving simultaneously as an agent in the newly created CIA.

William Mulligan (1918–92). Joined ARAMCO's new Government Relations Organization in 1946 after having served in the U.S. Air Force during World War II. Was unofficial historian of ARAMCO. Contributed extensively to *ARAMCO World*, the company magazine, and to the *Arabian Sun*, its newspaper.

George Rentz (d. 1988). Berkeley-trained Arabist. Directed ARAMCO's Arabian Research and Translation Department (1946–63). Moved to California as curator of the Middle East Section of the Hoover Institution.

Harry Snyder (1906–88). U.S. Air Force officer who headed the first Saudi training program at the Dhahran Air Base. Hired by ARAMCO in 1949 as coordinator of educational services.

Max Steineke (d. 1952). ARAMCO's Stanford-trained geologist known for his exploration and discovery of the first producing wells in the kingdom in the 1930s.

THE PHELPS DODGE DYNASTY

Bayard Dodge (1888–1972). Prominent educator. Son of Cleveland Hoadley Dodge. Joined the faculty of American University of Beirut (AUB) in 1913 and served as its president from 1922 until 1948.

Cleveland Hoadley Dodge (1859–1926). Phelps Dodge executive who financed Woodrow Wilson's campaigns for president. Son of William Earle Dodge, Jr.

David Dodge (b. 1928). Son of Bayard Dodge. Joined ARAMCO in 1949. Transferred to TAPLINE in 1954 and eventually became its vice president. After retirement, took over as president of AUB.

William Earl Dodge (1805–83). Merchant and philanthropist. Founding partner of Phelps, Dodge and Company. Founder of Young Men's Christian Association (YMCA). Original funder of what would eventually become AUB.

William Earle Dodge, Jr. (1832–1903). Led Phelps Dodge expansion in Western copper mining. Son of William Earle Dodge and brother of David Stuart Dodge, the founder of the Syrian Protestant College (later AUB).

James Douglas (1837–1918). Canadian mining engineer employed by Phelps Dodge in 1881 to examine properties in Arizona, leading to the acquisition of mines at Bisbee and Morenci. Chairman of the board from 1916 until his death.

Walter Douglas (1871–1946). Son of James Douglas. President of Phelps Dodge Corporation from 1918 to 1930. Led the company's campaign against the Western Federation of Miners during the 1915 labor strike at Clifton Morenci, Arizona.

Anson Phelps (1781–1853). Founded Phelps Dodge in 1834 along with son-in-law William Dodge.

OILMEN AND INVESTORS

Edward L. Doheny (1856–1935). California oil pioneer. Invested in Mexican oil industry. Was tried but acquitted on charges of bribery in the Teapot Dome scandal.

Juan Pérez Alfonso (1903–79). Venezuela's oil minister under President Rómulo Betancourt. Established the principle of fifty–fifty profit-sharing agreements with foreign firms operating in Venezuela, which became a world standard in the 1950s. Architect of Organization of Petroleum Exporting Countries (OPEC) in 1960 with Abdallah Tariki.

Nelson Rockefeller (1908–79). Investor and statesmen. Assistant secretary of state for Latin American affairs in World War II and director of Creole Petroleum, the Venezuelan affiliate of Exxon. Elected vice president of the United States in 1974.

Winthrop Rockefeller (1912–73). Politician and philanthropist. Brother of Nelson Rockefeller. Managed his family's Socony-Vacuum Oil Company, which was later merged into Mobil. Resigned in 1951.

AMERICAN BUSINESSMEN AND GOVERNMENT OFFICIALS

Charles Crane (1858–1931). Chicago industrialist and philanthropist. Among the first Americans to initiate contact with Ibn Saud. Offered assistance for surveys of water and mineral resources.

Allen Dulles (1893–1969). Diplomat and lawyer. In 1953 became first civilian to head the CIA. Credited with the removal of foreign leaders who posed threat to U.S. interests, including Mossadegh of Iran and Arbenz of Guatemala. Severely criticized for the Bay of Pigs disaster in Cuba and dismissed in 1961.

Parker Hart (1910–97). Career officer and early Arabist in the U.S. Foreign Service. Opened the first consulate in Dhahran in 1944. Served as ambassador to Saudi Arabia, 1961–65.

Harold Hoskins (1895–1977). Born in Beirut. Textile executive. Joined the OSS in World War II. Consultant to Exxon in the 1940s. Contract employee of the CIA in the early 1950s. Director of the Foreign Service Institute (1955–61). Member of the Board of Trustees of AUB.

Harold Ickes (1874–1952). Appointed Secretary of the Interior in 1933 by Franklin D. Roosevelt. Headed the Public Works Administration. Administered energy resources in the United States during World War II.

Issa Sabbagh. Palestinian educated in Britain who worked for the BBC. Hired by the U.S. State Department to start the Arabic service of the Voice of America in 1950. Joined ARAMCO in 1950s but returned to government service in the United States Information Agency.

Walt W. Rostow (1916–2003). Economic historian at MIT. Prominent "modernization" theorist. Deputy advisor for security affairs in the Kennedy administration. Appointed national security advisor by President Lyndon Johnson in 1966 and played crucial role in formulating U.S. policy during the Vietnam War.

OTHERS

W. E. B. Du Bois (1868–1963). Educator, activist, and prolific author. Founded the National Association for the Advancement of Colored People (NAACP) in 1909. Championed the cause of African American freedom before becoming disillusioned with the U.S. and emigrating to Ghana.

Harry St. John Bridger Philby (1885–1960). Writer, explorer, and one-time official of the British Foreign Service. Sent to Saudi Arabia in 1917. In 1930 resigned from the service, converted to Islam, and took the name of Hajj Abdullah. Remained in Riyadh as advisor to King Ibn Saud for thirty years.

Wallace Stegner (1909–93). Award-winning (Pulitzer Prize, National Book Award) American writer who founded the creative writing program at Stanford University. Commissioned by ARAMCO in 1955 to write a history of the company's early years, which it then suppressed.

Karl Twitchell (1885–1968). Mining engineer sent by Charles Crane to Saudi Arabia in 1931 to perform a geological survey of the area. His findings resulted in Chevron's eventual investment in the country.

Booker T. Washington (1856–1915). Born into slavery, and became one of the most influential African American leaders of his time. Founded what is now Tuskegee University in 1881. He was criticized by his contemporaries such as W. E. B. Du Bois for his refusal to fight segregation.

Minoru Yamasaki (1912–86). American architect who designed the Dhahran air terminal, the headquarters of the Saudi Arabian Monetary Agency in Riyadh, and the World Trade Center in Manhattan, among many other commissions.

America's Kingdom

1 CAPTIVE NARRATIVES:
A BRIEF AND UNEXCEPTIONAL INTRODUCTION
TO THE HISTORY OF FIRMS AND STATES

> *Doheny didn't overlook the visual arts. He commissioned a European*
> *painter, Detleff Sammann, to create a mural on the walls of his study*
> *at 8 Chester Place. Doheny proposed that the whole work depict the*
> *history of the United States—from the arrival of the pilgrims to the*
> *discovery of oil, the latter scene showing Doheny and his late partner,*
> *Charles Canfield, discovering oil in old Los Angeles in 1892.*
>
> **Margaret Leslie Davis, *Dark Side of Fortune***

INTRODUCTION

Al-Tih or *The Wasteland*, the first book in Abdelrahman Munif's quintet, *Cities of Salt*, tells the story of oil development and the emergence of a Saudi workers' movement in the 1940s and 1950s. Munif calls the kingdom Mooran, but we know he is writing about Saudi Arabia, the place that disowned him in 1963. A young Munif had joined his voice to those of workers, intellectuals, and other dissidents in calling on the ruling family, the Al Saud, to accept constitutional limits on its arbitrary and absolute power. That moment though ended in defeat for him and his comrades. Thereafter Munif lived the rest of his life in exile. Others were less lucky. When he died in Damascus in 2004, a new generation of Saudi dissidents was once again taking up the call for political reform and was once again being imprisoned for challenging the despotism of a still-intact Saudi *oil*garchy.

In conjuring the past, a world before oil, Munif turns to one of the most familiar tropes in western literature: the garden paradise that exists as if outside of time.

Wadi al-Uyoun: An outpouring of green amid the harsh, obdurate desert, as if it had burst from within the earth or fallen from the sky. It was nothing like its surroundings, or rather had no connection with them, dazzling you with curiosity and wonder.

1

But the wonder vanished gradually, giving way to a mysterious respect and contemplation. It was one of those rare cases of nature expressing its genius and willfulness, in defiance of any explanation.

The coming of the Americans to Mooran and its oasis town Wadi al-Uyoun sets history in motion in *Cities of Salt*, and history sweeps away the townsmen and their families. The oil drillers destroy the lush oasis and the oasis dwellers are made to abandon their homes, although some of them will—it seems are destined to—build a new form of solidarity as workers in the oil camps on the Gulf coast.

The destruction of Eden has served countless writers and, in our own time, filmmakers who continue to tell stories of nature, natural forms of solidarity, whole peoples and ways of life destroyed by forces coming from the outside—the White Man, western technology, capitalism, the railroads, or the oil industry. Consider, just in the case of the American West, Pulitzer Prize–winner Wallace Stegner's early novels. Leaf through one of the lavish 1960s photo essays by conservationist groups like the Friends of the Earth or the Sierra Club. Watch *Dances with Wolves* (1990). Many of us still cling so tightly to the belief in the existence of something like a pristine unspoiled nature or unspoiled people living in harmony with nature that even where scholars have tried taking apart these myths, it hardly makes a difference.

There is another myth though that may be even more powerful and productive today in the imagined histories of people than the one about paradise lost. There are many versions of a myth about a special or chosen people that is somehow able to escape a history that has trapped all those less-fortunate nations, tribes, and races before them. One version began to be told in the 1930s by men in the pay of the court in Riyadh and an oil company in San Francisco—Amin Rihani, Hafiz Wahba, Khayr al-Din al-Zirkali, Karl Twitchell, St. John (later Abdallah) Philby, and George Rentz. They said that it was the destiny of the Al Saud to unite the lands and tribes of the Arabian Peninsula and form a modern nation-state unlike any other, the Kingdom of Saudi Arabia.

Saudi Arabian Exceptionalism

I participated in a conference in February 2004 on the future of U.S.-Saudi relations where a prince in attendance argued, "sometimes I think we Saudis were dropped out of the sky and landed on the earth, like we are extraterrestrials, because no models or theories fit our special circumstances. We are always contradicting them all."[1] The first thing I thought was shades of Munif and Wadi al-Uyoun fallen from the sky. The second was that here was an expression in plain language of what professors call exceptionalism, the belief that some general rule or law or normal pattern governs the unfolding of history (the prince's "models" and "theories") in all cases but one, the putative exception. The most common form the argument takes in the case of the kingdom is that the Al Saud alone managed against all odds to keep imperialism at bay.[2] Other Arab and for that matter non-Arab countries were not so lucky, their leaders not so gifted.

It is hard to exaggerate the sway of this idea about the Saudi state's exceptional origins in a Middle East made by colonialism. Consider, for example, the full-page advertisement placed in the *Washington Post* on July 4, 1994, by Bandar Ibn Sultan Ibn Abd al-Aziz Al Saud, the prince who was appointed ambassador to the United States in 1983 and remained in that post for the next twenty-two years. "The Saudi state has been around in one form or another since about 1744, a quarter of a century before the American Revolution." This is a far cry from the more usual story of state after state in the Middle East coming into existence through the machinations of the British and French overlords after World War I, even if comparison with America's independence day presents some real problems.

The idea that something called the Saudi state is somehow older than the American republic is rather strange in fact given that the official date of independence of the kingdom is 1932, that is, about two hundred years later than the date noted in the advertisement. His excellency left out a detail or two. At some but not all points over the two earlier centuries Bandar's ancestors ruled one of the many tiny amirates—a town or two and some of the tribes that claimed grazing and raiding rights over the nearby country—that had risen in central Najd. The British explorer and Arabist St. John Philby, a convert to Islam, counselor to Ibn Saud, and father of England's most famous spy, Kim Philby, once likened the system of amirates to the Middle Ages in Europe when hundreds of tiny principalities existed in place of the relative handful of nation-states we know now. In 1744, the Al Saud tried to extend the frontiers of its domain in the name of a new and puritanical Islamic reform movement, which westerners call Wahhabism. It is the beginning of this expansionist episode that Bandar is thinking of when he imagines a Saudi state "in one form or another." He just doesn't tell you that one of those forms is imaginary or a mirage or a dream of the future.

Bandar might also have noted that like so many of those little principalities in Europe, his family's amirate was obliterated. This key fact is what makes the comparison with American independence in 1776 so fantastically misguided. That date, 1776, commemorates the successful defense of a sovereignty claim by the American colonies through victory in war with a European great power. The norms governing relations among the emerging European state system dictated recognition of the new state's strong empirical assertion of sovereignty. The same would have been true for an enlarged Arabian amirate—a Saudi state in other words—only if Bandar's ancestors had succeeded and a state had actually been built. When the Saudi-led, Wahhabi-inspired army crossed Najd to sack the holy cities on the Red Sea, the dominant imperial power, the Ottomans, launching an expedition from Egypt that crushed the movement and razed its towns.

The lands, towns, and scattered peoples of Najd remained nominal possessions of the Ottomans until 1915. They just didn't matter much to the sultan, however, in comparison either to the rich oases and small ports to the east in al-Hasa, along the Persian Gulf shores, or the western towns of the Hijaz, along the Red Sea, where the holy cities of Mecca and Medina were found. Warring amirates fought for regional dominance and control in Najd and were more or less left alone by the imperial center.

New generations of the Al Saud tried twice to reclaim and extend the family's domains, once in the 1840s, when the clan again failed and was driven into exile, and in the early 1900s. The outcome of the last long wars of conquest in the first decades of the twentieth century, when the towns and most of the tribes of Najd were ruled by the Rashids in Qasim, and an independent Arab kingdom existed in Hijaz, was hardly determined. Where the Saudis and their retainers like to portray the outcome as destiny, most professors prefer a more complex and contingent story about how the Saudis went from the conquest of Riyadh in 1905 to ruling over the peninsula in the 1920s, and we would underscore the decisive fact of intervention by the then great power Great Britain, which let one troublesome client in Hijaz go down in defeat at the hands of a new and more reliable one, Ibn Saud.

The twist in the story that everyone seems to forget is that Ibn Saud signed a treaty in 1915 with Great Britain that conceded sovereignty rights for protection. Today we refer to "security" rather than protection. Nonetheless, the new Saudi amirate-in-formation in Najd represented one of the last pieces of Middle East real estate brought formally into the empire. Great Britain declared a protectorate over Egypt the previous year, formalizing its two-decade-old occupation. In 1919 the League of Nations assigned the "mandates" of Iraq, Transjordan, and Palestine to Britain. These latter three countries were, basically, protectorates too in all but name. Protectorates represented the old imperialism. Mandates were intended to represent something slightly new. A key difference with both Egypt and Iraq, where the British were forced to put down rebellions and advance the date of independence, is that the Najdi amir (or "ruler"), Abd al-Aziz, *chose* formally to pledge allegiance to the British empire at a moment when Egyptians and other colonized peoples were advancing claims to the right to national self-determination. One cannot really blame Ibn Saud, of course, given the prospect that without British protection the lands he had conquered would likely be lost to bigger, more powerful nearby states. International recognition of the rival independent kingdom of Hijaz the next year, in 1916, under the banner of the Arab nationalist movement, though, created a serious dilemma for the ruler of the new Najdi protectorate, which intensified when Egypt gained its independence in 1922. The Saudis seemed to be moving in the opposite direction.

There are two points to make about the recognition a decade later (1927) in a second British treaty of Ibn Saud's "complete and absolute" independent rule of the twin kingdoms of Hijaz and of Najd and their dependencies, renamed the single or amalgamated Kingdom of Saudi Arabia in 1932. One is that the juridical status accorded his newly independent countries did little to alter the reality of the king's dependence on the British or the fact that the British state had more influence over its client, Ibn Saud, than Ibn Saud did over his own dependencies, al-Hasa and Asir. Most crucial is that Ibn Saud never attempted war against his British-allied rivals in Jordan and Iraq for the territories he claimed were his own. Instead, he had to resign himself, as later would his sons, to borders and frontiers imposed by others.

The second and more critical point is that Saudi Arabian independence illustrates the transformation in the norms of the interstate system ushered in by World War I and the Treaty of Versailles. The world of the twentieth century differed in a key respect from the world two centuries earlier when, as we saw in the case of the United States, sovereignty depended on a state's ability to defend itself.[3] Again, court and Arabian American Oil Company historians sometimes leave the impression that Ibn Saud's statesmanship somehow kept Saudi Arabia "freer" than Egypt and Iraq. Believing this requires one not look too closely or think too hard. Prince Bandar would have been on stronger grounds had he argued that the Saudi state had existed in one form or another only since 1927, and, as he knows better than most, on the basis of a broad accord or international protectorate regime, anchored first by British and after 1945 by American power. Chapter 3 recounts that moment when the Truman administration replaced the British as sole protectors of the Al Saud, taking over a role that dated back to the days of the Ottoman empire and its subventions to the amirs of Hijaz and Najd. As I have said, Ibn Saud had little choice, given the position of his kingdom in the regional and global orders. His independence was limited to bargaining on the terms of the new U.S. air base in Dhahran, the town founded by ARAMCO, and a few postwar loans and Export-Import Bank credits. The Americans had no problems, for instance, refusing the king's pleas for a treaty tying the kingdom even more closely to the United States and making American defense of the Al Saud more reliable.

As we shift the focus in the rest of the book from Saudi sovereignty claims to more mundane realms such as work, education, technology, finance, health, industry, and so on, we will see that the idea that the domains of the Al Saud escaped imperial domination or "external influence" is indefensible. On the eve of World War II the vast new lands of the kingdom were essentially still little more than a federation of tribes and towns. The conquest of Hijaz had provided the Al Saud with the rudimentary technology of administration that their agents sought fitfully to master. The transformation of the landscape that became Saudi Arabia was wrought in great part by foreigners, arriving in increasing numbers in the 1940s and 1950s, financed by foreign investment, foreign private and public aid, and large loans secured by future oil royalties. The king and his handful of agents had little capacity to direct or to even oversee these changes, and "government" of the economies (it is impossible to think of the domains as an integrated network of markets and administrative rules in any meaningful way) was essentially limited to keeping the Al Saud solvent. Given the king's generosity, as his American fans put it, one cannot help but admire Abdallah Sulayman, the state's most important economic agent for decades and an extraordinary rent-seeker in his own right. The consequences of the kingdom's chronic insolvency included the arrival in 1951 of the first of many "missions" to guide the creation of national-level policy agencies and practices.

Saudi Arabia in this sense was not that different from Panama and El Salvador in the same era, and, like these two American dependencies, lacked the capacity or will of other nearby states to challenge the prevailing order. After the death of the loyal American client Ibn Saud, his successors actually tried taking this treacherous path before giving up and accepting the less exalted but doubtless safer status of client of an emerging global "liberal" hegemon that some say is like no other.[4]

The Most Exceptional of Exceptionalisms

There is no small irony in the fact that the term exceptionalism, the one that professional historians and social scientists use to describe the most common way of writing the history of the United States, is owed to Joseph Stalin, or so Princeton Professor Daniel Rodgers argues in the best short account I know of the history of American exceptionalism.[5] The full force of that irony comes with the further realization that American exceptionalism is a product of the Cold War. The era invited if it did not require accounts of the course that led to the American century, that solved a putative puzzle about escaping the fate of Europe, and explained what seemed like liberalism's rapid progress and imminent triumph over the race, class, and ethnic divisions that had scarred the late nineteenth and early twentieth centuries. It was a time when some of the most brilliant thinkers insisted on a new, so-called realism in foreign affairs, defined against old, hopelessly idealist and utopian visions about ending war or transcending capitalism, and, without a hint of irony, also insisted that a color-blind society was within easy reach. Finally, it was the moment when the long course of American empire was forgiven and forgotten.[6]

Rodgers's essay makes three key contributions to our understanding of the idea of American exceptionalism. First, it underscores the misapprehension in a skeptic's standard response to hearing the claim that some people's past is exceptional. The skeptic's argument goes that all nations or states are unique or special in their own way. This view mistakes exceptionalist history writing for an argument about difference. "Exceptionalism differs from difference," as Rodgers puts it, so perfectly and succinctly. Where difference "feeds on polarities and diversity," exceptionalism contrasts "one's own nation's distinctiveness to every other people's sameness—to general laws and conditions governing everything but the special case at hand." The next quote is my favorite part of this argument. "When difference is put in exceptionalist terms, in short, the referent is universalized. Different from what? Different from the universal tendencies of history, the 'normal' fate of nations, the laws of historical mechanics itself."[7]

Rodgers's second crucial point is that the American historical profession did not always write the same kind of history about the United States.

There was a time before exceptionalism's triumph. Consider an account of the sources of American institutions that was popular in the late nineteenth and early twentieth centuries. Men like the historian Herbert Baxter Adams and his student, the political scientist and future president of the United States Woodrow Wilson, argued that American democracy was the result of the natural evolution of the Teutonic "germs" that produced the Anglo-Saxon race. Keep in mind that this was a time when the concept of nation was not clearly distinguished from that of race. Rather, most would have seen the one as an indissoluble part of the other. America was not so much exceptional, different from all other nations, as it was the highest stage of a common whiteness. Frederick Jackson Turner, perhaps the most famous historian of the twentieth century, offered an exceptionalist twist on the argument, through which he would account for America's uniqueness despite the common Teutonic germs carried by the settler colonialists. Turner argued that the successive frontiers of the eighteenth and nineteenth centuries remade Anglo-Saxons into Americans, interrupting or confounding the normal laws of social evolution.

In anticipation of an objection—surely, one or another strand of postwar American exceptionalism is a better, more compelling way of thinking about the historical process than the racist alternatives that were once written and taught—these are not the only choices before us. While exceptionalism may have seduced many in the profession after World War II, there were other, progressive historians, materialists, as we once liked to say, writing around the time of World War I who insisted on seeing America in terms of global economic processes and forces. Rodgers singles out two. "As for the giants of early twentieth-century American history-writing—Charles Beard and W. E. B. Du Bois—they had absorbed too much of class and economic analysis and instinctually thought on too large, world canvasses to be exceptionalists."[8] In turn we can trace the influence of these giants and their students on the undercurrents, dissident histories, and more ironic readings that were being produced at the edges of the exceptionalist consensus. These alternatives would emerge as a full-blown challenge by the 1970s and 1980s.[9] As Rodgers notes, though, "nothing showed the massive presence of exceptionalist assumptions more than the difficulty critical historians of the 1960s and 1970s had in surmounting them." His main example is the effort to bring American empire back in ("an empire which had been all but suppressed from political memory") by the great William Appleman Williams and his students in the Wisconsin school. As they and the many who are indebted to them argued (and still argue) America was an empire, all right, but a different kind of empire, rooted in the country's unique experience and values.[10]

Rodgers's third key point is a succinct statement and endorsement of the move that defines the alternative history writing of the past two decades, what he calls a postexceptionalist history. "The question of difference was, from the beginning, false and tautological. The antithesis at the core of exceptionalist history was never that between difference and sameness but between autonomy and connection." Think of all those histories that never embed America in larger circuits of peoples, ideas, and resources, and that stop "at the water's edge." The reality is that the places and peoples across what we now call the United States have always been linked in complex ways to other places and people beyond the shifting borders of the republic. This is not just true only for the nineteenth century or only after 1945. Nor is it reducible to the idea of common "external" shocks or "transnational forces" that spring a national elite (or a tribal chief) into action, who then steer the state toward some unique end or who forge a distinct path to the same one, modernity. Rather capital, nomads, copper, oil, workers, ideas, intellectuals, missionaries, terrorists, soldiers, slaves, and spies, moving constantly across the so-called frontiers, borders, and seas, are all part of one, ongoing story. It is an extremely hard one to tell.

The Exceptional Relationship

The logic, not to mention the irony, ought now to be apparent in the twinned Cold War stories of the United States and Saudi Arabia, "the special relationship." On the one side is the global hegemon and on the other is its client, arguably America's most important client. President Kennedy's national security advisor, McGeorge Bundy, told the U.S. Joint Chiefs of Staff in 1963 that Saudi Arabia was the location of America's single largest private overseas investment anywhere, namely, the Arabian American Oil Company or ARAMCO. The company that operated the Saudi concession had been founded by two of the world's oil giants, Chevron and Texaco, before two even bigger ones, Exxon and Mobil, bought a forty percent share of the business after World War II.[11] ARAMCO in turn was the mechanism through which the United States secured what the policy makers liked to imagine was control over the world's single largest untapped source of a vital resource, oil.[12]

We would expect precisely what we find in the scholarly literature, let alone in the pronouncements of the various heads of states and their functionaries (and, as we discuss below, the firm and its functionaries). ARAMCO is conventionally portrayed as operating under a principle of enlightened self-interest that ostensibly distinguishes it, although historians of British Petroleum might object, from its rivals in the Gulf. American exceptionalism had come to dominate accounts of the United States when ARAMCO was operating full blast in Saudi Arabia, when it hired Wallace Stegner, himself a dedicated exceptionalist, to write its first history, when it made films, took journalists on tours, and subsidized scholars and Middle East centers, and when the first generation of radical critics of American Gulf policy were writing.

American exceptionalism and its account of the U.S.-Saudi encounter help make possible Saudi exceptionalism itself, where the rhetoric and substance of imperialism are rejected as a way of telling the story of U.S.-Saudi relations and of Saudi state formation. The latter story thus always begins with an account of the distant and "natural," "indigenous," "internal," "pristine," and so on foundations of the state. The seemingly much later story of the dawn of the new American era is structured by rote comparison with other, earlier national imperial styles and epochs. The capacities of clients and dynasts in Saudi Arabia and elsewhere are romanticized, emphasizing their ability to manipulate great powers for their own objectives, seemingly undaunted by systemic, global power asymmetries.

Saudi Arabian historiography works to put maximum possible distance between the kingdom and America. A narrative devoted to the autonomy (using Daniel Rodgers's term) of the kingdom's path through time has the effect of adding resonance and depth to the new state's sovereignty claim. The narrative was invented in the 1950s and 1960s as a response to those in Nasser's Egypt and Baathist Iraq who disparaged the Al Saud's right to rule and called for the family's overthrow. The arguments of the kingdom's opponents are familiar ones. The kingdom was, and some still insist is, a puppet state, a country that imperialism simply invented (Saudi propaganda would say the same about both Egypt and Iraq), a set of tribes with a flag, manifestly not like other imagined real or authentic nations.

Exceptionalism is thus both probably inevitable and to an extent useful given such powerful prejudices and conceits. We are then faced, however, with a dense shroud of myths, as if the identities that states assert are liable otherwise to disintegrate at first light. And, as we will see, a lot of money was spent on mythmaking. A lot still is being spent. We have a valuable resource of our own, however. The archive.

WHY WRITERS HAVE GOTTEN THE ARAMCO STORY WRONG

Until now, the debate about the adequacy of exceptionalism as a way of writing history has been one about the history and identity of *nation states* (or *peoples* more popularly). The debate is one that is also generally confined to the world of professors, although on occasion the brawl may spill out into the wider public arena, as when the Smithsonian Institution's National Museum of Art mounted the exhibit "The West as America" in 1991. The show was a relatively tame rehearsal of the historical revisionism ("reinterpreting") going on in the academy. Patricia Limerick, a best-selling historian of the West who teaches at the University of Colorado, defended the multifaceted exhibit, sensibly to my mind, when she said that the "conquest of the West is a story of success and failure, heroism and betrayal, capitalist triumph and labor exploitation. Colorado miners and Mexico peons are as much a part of the story as pioneers on the Oregon Trail." Not all agreed, however. The *New Republic*'s Charles Krauthammer wrote that the exhibit had "a crude half-baked Marxist meaning" that would have been at home in Moscow "thirty years ago." It showed "contempt for every achievement of Western expansion."[13] Expansion is the word that Americans use in place of another, more familiar one, imperialism, that somehow does not fit the story of America.

Business (History) as (Un)usual

Firms are a lot like states when it comes to telling their own stories, and they are implicated in the stories that Patricia Limerick would tell instead. Company histories read like American exceptionalist fables. Look again at the quotation that opens this chapter. Legendary investor Edward Doheny, whose vast holdings in California and Mexico earned him the title Emperor of Oil, would see his drilling of the first Los Angeles producing wells as the providential end of a journey begun when the Pilgrims set sail aboard the *Mayflower* in 1620.

A firm, the Phelps Dodge Company, say, or Doheny's Pan American Petroleum Company, or the Arabian American Oil Company, ARAMCO, reliably overcomes immense obstacles to become a giant that then marshals its tremendous resources on behalf of its own people, but also on behalf of the countless others who are beneficiaries of its largesse. The "uniquely progressive and socially capacious character" that two or more generations of American historians have ascribed to American political development at home and the exercise of its power abroad is, it turns out, basic to the identity of those same firms that claim to have opened up the American West or the Middle East or Asia or Africa and that have forged a partnership in progress and development with Mexicans, Venezuelans, Indonesians, Iranians, Saudis, Libyans, Nigerians, and so on.[14]

ARAMCO performed better than its competitors, the fans say, paid higher wages, served the surrounding communities responsibly, and recognized a kind of compact with the host society to uplift the race, as it once was put. Services, benefits, education, housing programs and the like all expanded continuously. When labor unrest occurred, infrequently, it reflected politics outside the work place, nationalism, and xenophobia. The regular upgrades in wages, benefits, and uplift had little if anything to do with those same unions, strikes, or populist parties. The causal arrow runs in some other direction. And since the story turns out to be the same for every one of ARAMCO's competitors, it is probably a good thing that analysts don't read one another's accounts.

The point about a firm's history taking a particular form is only made more obvious if we consider why a firm commissions its history in the first place. ARAMCO did so in response to what its managers viewed as profound anti-company biases first and foremost in the United States and secondarily in the Arab world. It viewed the company's history in other words as a weapon in the public relations arsenal—magazines, movies, stories planted in newspapers, and the like. It may be that not all firms that hire historians do so for this reason. We cannot be sure because there are no histories of company history-writing. We do have histories of company public relations programs, however, which show this very logic at work in the invention of company magazines, such as *ARAMCO World*.

An obstacle in the way of our ability to generalize about motives is that in a key respect firms are not like states. As Gregory Nowell notes, "corporations see historical records as a cost to maintain, and at times as a threat to the firm—they do not systematically preserve their political records the way nations do."[15] Even firms that preserve their records are not like states, or at least not like liberal democratic states. Companies are more like authoritarian countries. They keep records hidden. They pay for extravagant hagiographies. They open their archives only to those they hire. They insist on the right to approve what is written. They pay for histories and then file them away in drawers. There are no sunshine laws and no Freedom of Information Acts against corporate privilege.

Hegemonic Decline?

Many other factors besides a penchant for secrecy have operated to preserve the hegemony of the company-eyed view of the world in the case of ARAMCO, although those who once worked for the company would reject this characterization as unfair, since, in the view of ARAMCO's executives, virtually everyone had it in for the giant oil firms, and even the writers and scholars they hired could not be trusted to get it right. One factor is timing. The 1970s and 1980s were the heyday of debate on the multinational corporation and U.S. foreign policy, but also the decades of the great OPEC price rises and the nationalization of foreign oil firms across the Middle East. It was hard to argue that Saudi Arabia in 1980 looked much like Mexico, say, in 1910 or Iran in 1920 (but to be clear, nor did Mexico or Iran in the 1970s in the 1980s look the same), where caudillos and shaykhs could be bought and armies hired, and oilmen could still imagine their governments possibly sending gunboats to effect regime change in a pinch. Those were the old days when firms operated more like states within a state. In response to those who argued nothing had changed in the meantime, analysts like Helen Lackner and Fred Halliday worked overtime to educate a generation of radicals about how the world had changed, and how simple notions of the Al Saud as American "puppets" were wrong.[16]

A more generalized form of the argument emerged roughly at the same time.[17] The basic point is that as firms increased their fixed investments in oil rigs, refineries, and so on in places where sovereign, increasingly populist governments rather than colonial consuls or occupation authorities ruled—the fundamental change in the world order in the twentieth century—these investments left them increasingly vulnerable to the threat of nationalization. The mere existence of a giant firm operating somewhere did not automatically translate into loss of sovereignty for the host country or into bargains that invariably cost the host more than firms or gained hosts less. Outcomes varied across places and over time, according to a number of factors, but the trend away from extraterritorial privilege to domestication was clear.

At the same time, though, the transition from one era to another was messy, and the archives were not (and in some cases are still not) available for anyone interested in writing about the post–World War II era. The U.S. state had intervened in Mexico in 1910, but the Roosevelt administration refused to do so in 1938 in response to the Cardenas government's takeover of American firms' assets. Harry Truman expressed the change in view more plainly still, against his own State Department and the big oil firms that the

Department of State was backing. He wasn't going to return to supporting "the old Doheny interests and people who robbed the Mexicans before."[18] In 1951 a populist prime minister in Tehran, Muhammad Mossadegh, nationalized the oil facilities of British Petroleum when it refused to rewrite its concession agreement to reflect the better bargain that Venezuela had obtained from owning firms there, with prodding from the U.S. State Department. Truman was no longer president, however. The Eisenhower administration approved a Central Intelligence Agency (CIA) operation to overthrow Mossadegh on the grounds that he would otherwise surrender Iran to Moscow. The last covert interventions by U.S. agencies in oil-producing states during the era when a handful of companies still managed worldwide price and supply stability, an era that was coming to an end, were under the Kennedy administration in Iraq in 1963 and the Johnson administration in Indonesia in 1965. The history of these interventions is still mostly unknown.

The key point though is that by about 1972 the tide had turned.[19] In the Middle East, Algeria, Iraq, and Libya all moved fastest toward the full nationalization of the firms' wells, pipelines, and refineries. The decline in prices and the cutting off of the American market, which had spurred the formation of OPEC, pushed producing states to overthrow the western-administered international oil regime. It was the objective argued for first and most vociferously by Saudi Arabia's first oil minister, Abdallah Tariki, in 1963 right after he was forced out of office and into exile. A decade later his radical principle had become the new norm in the industry, in and outside the Middle East, and what would distinguish so-called radicals from moderates would be by what means—through decree or negotiations—and over what period of time the oil-producing states would take over the assets of the companies.

The explosion of scholarship on ARAMCO in the 1970s and early 1980s, including the first histories to draw on newly declassified State Department and White House records, thus reflected two contradictory strands of that energy-crisis-filled, post-Vietnam, and post-Watergate era, the years of the gasoline lines and the calls in *Commentary* magazine to occupy the Saudi oil fields. The firms confronted the most serious challenge in Washington since the Truman administration's investigation of the international oil cartel, and a new generation joined in the great twentieth-century debate about the relationship between private and public power. In 1973, a special subcommittee of the Senate Foreign Relations Committee, headed by Idaho senator Frank Church, launched a multiyear investigation of multinational corporations and U.S. foreign policy. The Church committee took on the oil companies, along with airlines, agrobusiness, ITT in Chile, and, not least, the CIA for its role in the assassination of Chilean leader Salvador Allende. For some, it was the zenith of countervailing congressional power in the aftermath of the Watergate crisis, for others the collapse of the corporatist oil policy that had stabilized world prices and supply for twenty years or more.

At the same time, and hard as it might be to believe now, many argued that both U.S. industrial capacity and its power to lead the world economy were in sharp and irreversible decline. While these arguments proved false, the declinists got a subsidiary argument correct, as I have suggested. The firms faced a challenge not only from the U.S. Congress but also from Saudi Arabia and other oil-producing states. The problem is that then some analysts tended to write as if these dramatic changes were prefigured in ARAMCO's original negotiations with Ibn Saud, and the presentists (those who see the present in the past) were most impressed with those moments when ARAMCO or any other firm anywhere was bested in some negotiation with a ruler or his agents. Some imagined ARAMCO having actually pioneered a new course for other firms to follow. The competitors that were not wise enough to do so, such as, allegedly, British Petroleum (BP) in Iran and BP's American partners in Iraq, would pay the price of a hostile takeover.[20]

The details of Saudi institution building got lost in the debate between those who viewed the companies as agents of so-called neoimperialism (like those who today argue the same about firms and globalization) and those who insisted that states in the Third World had turned the tables on the multinationals. There was apparently little interest and, in the case of the Middle East at least, ability to write about the oil companies' polices on the ground. Irvine Anderson's 1981 study of ARAMCO was the first and really only account of "the field personnel" as he put it, by which he meant the managers, and of what he saw as the tension between them and the executives in San Francisco.[21] Only two Saudis mattered: the king, who was constantly beating the Americans at their game, and the infinitely resourceful finance minister, Abdallah Sulayman. Even the new leftists-turned-graduate students downplayed the question of labor (hard as it is to believe) in the history of oil. As Petter Nore put it, in a book that remains an artifact of that era, *Oil and Class Struggle*, "the history of oil workers' struggle has yet to be written," which remains true twenty-five years later. Nore meanwhile guessed "the high organic composition of capital and the high rent element" in the industry made it easy "for the companies to 'buyoff' oilworkers with high salaries and create a type of aristocracy of labor."[22]

The Norm against Noticing

In her remarkable essay "Black Matters," the writer and Princeton professor Toni Morrison discusses the turn after World War II in the United States to ignoring race.[23] She calls it "a graceful, even generous liberal gesture." Postwar generations have been conditioned not to notice, she says. Morrison was writing about postwar literary critics and their silence about racism in the history of letters, the construction of literary canons, and the criticisms worth making about the canonical texts. We can, however, easily extend the argument beyond the English departments of the universities to the wider state and society.[24]

Abd al-Aziz Abu Sunayd, one of the seven Saudi workers who led a 1953 strike against ARAMCO's Jim Crow system, one of a series that had begun in 1945 and ended in 1956, after which the state made strikes illegal, had first sought to bring his complaints directly to management. Rational Saudi actors thought twice before taking such a drastic step as a strike, not because they were aristocrats but because others had been beaten, jailed, and killed for doing so. Sunayd addressed a letter to ARAMCO's president asking why he was being treated in his own country the way he had been treated in Washington, D.C., where he and a few of his comrades, who had been sent to the States to teach Americans a little Arabic before heading to Dhahran, had been banned from a movie theater due to the color line. It was the same color line, he said, that the firm had drawn in Dhahran.

The norm against noticing is thus another factor that has worked to preserve the company-eyed view of things. It was impossible to find a scholar's account of the 1953 or any other strike save a one-page-long version in a book written in 1960 (and hardly cited since). The existence of Sunayd's letter notwithstanding, the book, by Berkeley's George Lenczowski, said absolutely nothing about the Jim Crow system that the firm had exported to Dhahran as the grounds cited by the leadership for striking. Perhaps the letter by Sunayd had not been publicized. As we will see, however, the leaders of every strike for a decade denounced ARAMCO's racism and discrimination. The attack on racism played a role not just in the work stoppages by Saudis but also in those led by Italians, Pakistanis, and others.[25] It may be that Lenczowski left the fact out of the account to please the ARAMCO partner that subsidized his research yet went unacknowledged in a long list of grantors that he thanked in *Oil and State in the Middle East*, but Morrison's explanation is a better one, applicable as much to the conservative Orientalist-turned-oil-company-consultant as to left wingers who thought that firms had bought off the working class and created an aristocracy of labor. The strategy was an old and somewhat different one: to divide labor by race as a mechanism of social control. The aristocracy-of-labor idea might better be understood as a hierarchy, with white workers on top and, in the kingdom's case, Saudis, whom Americans imagined as "black" and as "coolies," on the bottom.

The American military officials who were training Saudis in the late 1950s and early 1960s reported that two questions flustered American interlocutors most, requiring special attention and better propaganda: America's Palestine policy and the pervasiveness of racial discrimination in the United States. Recovering moments such as these, and there are a wealth of such moments, needless to say, forces us to reckon with another of Morrison's key arguments, about how "certain absences are so stressed, so ornate, so planned, they call attention to themselves."[26] A company buries records of its efforts to preserve the color line intact. In Washington, ARAMCO's vice president tries to convince a skeptical State Department official that protests against racial discrimination in Dhahran are really another case of following the communist line, which is what Southern politicians were saying about African American demonstrators in Montgomery at the same moment. Decades later a son excises the racial slurs from his father's letters before publishing them as a testimony to enduring U.S.-Saudi friendship and cooperation. And in 2004 the ex-*Washington Post* journalist Thomas Lippman would simply ignore the inconvenient facts of Saudi protests against American discrimination rather than wrestle honestly with the argument.[27]

Across the long second half of the twentieth century, intellectuals have adopted one of four positions in trying to reconcile their beliefs about the exceptionalism of firms and states with America's rich racist heritage. The first and most common, as Toni Morrison shows, is not to notice or pretend it is not visible, say, in the layout of the oil camps. The second, well-known one, the Myrdalian position, is to argue that racism is a kind of cultural atavism, a holdover from some earlier time, an import from some other place, with a shorter rather than longer half-life. ARAMCO's agents argued along these lines in the 1960s. The third position, seemingly more forthright, acknowledges the reality of racism at a particular moment as a prevailing social norm.

Thus if a company deployed racism as part of its arsenal of labor-control mechanisms, it is hardly the fault of its owners and managers, who are, after all, products of their time. The problem, however, is then reconciling the prevailing norms-of-society idea with exceptionalism, particularly when exceptionalism is defined, nonetheless, by the firm's putative commitment to uplift, reform, development, welfare, and so on for its workers.

A fourth position, the one I have come to, is to reckon with the institution of racism honestly. This means, first and foremost, acknowledging the long, unbroken legacy of hierarchy across the world's mineral frontiers. It means developing an account of the agency of firms in its building and maintenance—one of the great, usually unheralded feats of social engineering by American multinational or, we now say, transnational oil companies. Finally, it means thinking harder about, and looking beyond the firm for, a causally adequate explanation of the changes in the regime that race built.[28]

THE ARGUMENT

In the first decade or two after the Civil War, in the period known in the South as Reconstruction, white southerners began to institutionalize a system of legal separation of the so-called races. The law was a primary tool in the assertion of white supremacy, but the Jim Crow system, as America's version of apartheid came to be known, involved a panoply of mechanisms, including economic coercion and organized extralegal violence, which defined, divided, and determined the life chances of the dominant and subordinate social castes for most of the next century. "Free" blacks in the south could not enter public spaces reserved for whites, from parks to saloons and swimming pools. They were denied access to education and health care, or were taught in inferior schools. Blacks earned less than whites for the same work. And they lived in homes and neighborhoods with inferior, often nonexistent, public services.

A great deal of intellectual work went into naturalizing the hierarchy. Races were real, the professors showed, and the reality of the differently marked and abled bodies explained why segregation and discrimination were necessary. The truth is, however, that it was the legal and extralegal institutions that worked to create and stabilize an arbitrary racial binary. This is what W. E. B. Du Bois meant when he said that ultimately a black person was someone made to ride in a Jim Crow car. We now say the same thing with less bite. Race is a social construction.

The mining industry was not prominent in the export agriculture-based economy of the South. The coal mines of West Virginia and Alabama are the exceptions. As large-scale mining enterprises emerged elsewhere in the United States in the 1870s—coal in Pennsylvania, Ohio, Kentucky, and Wyoming, copper in the Southwestern territories, and oil in Pennsylvania, Ohio, and the West—mining firms organized production in the way that the post-Reconstruction South organized society. Most basically, norms of separate and unequal rights and privileges governed life in their rigidly segregated mines, camps, and company towns. The precise mix of races differed—Mexicans, Chinese, Serbs, Greeks, Irish—across the mineral frontiers, but, as in the South, it was possible for some to become white and thus escape the subordinate status to which the inferior races were consigned.

Dividing labor by race was a strategy to inhibit the organization of unions in the mines, and unions were key for those who sought to end the racial wage system. All firms paid miners, drillers, and other skilled and unskilled labor different wages according to race. And ending the racial wage became the issue that pitted the subordinate races against not only the white owners and managers but also the privileged caste of workers in strike after strike across the nineteenth and early twentieth centuries. The full panoply of Jim Crow institutions, from segregated housing and differential access to services to the degradation and humiliation of white supremacist thought, worked to buttress the labor-control regime.

Large firms also deployed a second mechanism, paternalism, known formally in U.S. industrial relations as welfare work, in their efforts to defeat union building in the mines and the oil fields.[29] Owners would begin to offer benefits and build housing, churches, and recreation facilities for some, ideally small segment of employees as a means to secure loyalty and thus stability. Thus, even while racism's ethics governed the hierarchical distribution of these benefits, a firm's beneficence ostensibly demonstrated how so-called outsiders and third parties, including the local state—Arizona, for example—were second-best (or worse) options for securing a decent life. Needless to say, all the large firms brought the model with them when they began producing oil beyond the U.S. borders.

Franklin Roosevelt's New Deal is celebrated by some as the eclipse of firm-based welfare work paternalism and criticized by others as little more than welfare work by other means. The Saudi Arabian case offers no insight into resolving the debate. While it is true that increased regulation was one of the factors that drove labor-intensive industries, mining included, to invest in places beyond the reach of federal welfare state regulation, oil is a less labor-intensive industry than copper and coal mining. The factors driving firms to increase investment in oil exploration abroad differed in part from those driving the overseas investment strategies of copper mining firms. What was common across the different industries and mineral frontiers was reliance on the mix of racism and welfare work paternalism.

Racism drove down labor costs and divided workers among themselves. Organizing under such circumstances was as likely to deepen the racial fault lines as it was to produce class solidarity. All other things being equal, the more workers were kept divided the less there was to fear of unions or of strikes succeeding. Using racism's ethics in distributing welfare work provisions reduced the costs to the firms, obviously, while reinforcing the hierarchy that labor organizers either believed in or else found daunting to overcome. Racial science, including the body of knowledge that came to be known as "race development," showed why blacks, Mexicans, and other biologically and culturally inferior peoples required long and slow tutelage and uplift rather than the higher wages, better housing, schools, and the like that were reserved for whites.[30]

Keep in mind that the idea of uplift was contested in America, seen as dangerous even. Some experts, perhaps most, insisted that any resources expended on the unrecuperable races were wasted. At the dawn of the twentieth century in the United States it would be left to visionaries like Du Bois and organizations like the National Association for the Advancement of Colored People (the NAACP) and the Pan-African Congress to think and agitate for the unthinkable: immediate freedom and equality, expanded access to universities, and the right to run the commanding heights of the economy rather than to slave away in the cotton fields and coal mines. Nationalists in Mexico, Venezuela, Iran, Iraq, and Saudi Arabia, among many other places, would make the same arguments. Finally, and mostly late in the game, where firms sought to replace white labor, owners might use paternalism to buy the loyalty of the lesser races. They would position themselves as defenders of blacks against white unions, the day-to-day abuses of lower-level managers, and the hostility of the surrounding community.

Firms supported and used a third institution in the war against unions: violence. The history of labor organizing and strikes against the racial wage and other aspects of the separate and unequal world-mining frontier is replete with examples of companies deploying private police or, preferably, gaining the support of the local state or its equivalent to deploy against threats to order. The crushing of the strike wave in southern Arizona in the 1910s is an example. Doheny's alliance with the caudillo Manuel Peláez during the long Mexican Revolution is the stuff of legend. And it is conventional to recognize the role of post–World War II American administrations in arming and training the Saudi king's warriors-turned-army. What is never discussed is the role of ARMACO and the U.S. embassy in arming the amir of the oil province, Ibn Jiluwi.

Generalizing about outcomes in the multi-front, decades-long war between business and labor is difficult, given that comparative study hardly exists. Certainly, in many cases union-building efforts were solidly defeated, and a firm might celebrate—were its future historians brazen enough—its ability to preserve the racial order in its camps intact for its entire life, before, as happened in one case, copper prices plummeted, and the owners shut down the mines and shifted production to Northern Rhodesia (now Zambia). In other cases, populist politicians allied with unions, forcing firms to work harder to preserve the privileged enclave and resist demands to end segregation, while conceding "peripheral advantages." The term is from Michael L. Conniff's remarkable study of the Panama Canal Zone, the American colony that Raymond Leslie Buell, who was then America's leading specialist in problems of colonialism, said was a model for the evolving relationship between Britain and Egypt in the 1920s.[31] In the Canal Zone the labor system was organized around a two-tier system of wages and benefits, which divided white Americans from the vast pool of West Indian migrants imported from Barbados, Jamaica, Martinique, and elsewhere to build the canal and ancillary works.[32] Labor was assigned to different gold (U.S. coin) and silver (Panamanian peso) rolls. The payroll became the means, according to a member of the canal commission staff, "for the Government to draw the color line—a practice it could not openly attempt under the Constitution of the United States."[33] The commissioner meant the U.S. government, of course, which built and operated the segregated enclave that resembled the one ARAMCO founded in Dhahran three decades later.

Although the U.S. oil industry remained one of the most segregated sectors of the economy through the end of World War II, the strategy of race-based labor control and discrimination began to break down over time. I have found no good, synthetic account of the process, unfortunately. Gary Gerstle's general account of the intertwined course of labor and civil rights in *American Crucible* will have to suffice. First, Gerstle argues that the New Deal state was indeed a setback for the antiunion, welfare work alternative in American industrial relations. Unions grew more powerful and legitimate as the federal state backed the right of workers to organize. Second, parts of the left moved further than the Roosevelt administration did, seeking to "break through" the limits of exclusionism and build cross-race alliances. The Congress of Industrial Organizations campaigned against discrimination. Nonetheless, supremacist currents continued to dominate politics in the South and the West, where the New Deal "did little to challenge" disfranchisement. The third and most significant force in its weakening was the mass movement from below that drove the civil rights revolution of the 1960s.

Gerstle argues, unusually for an American social historian but on good authority, that of the many factors involved in propelling the civil rights movement the most important was "the collapse of the European empires in Africa and Asia" and the mobilization of colonized peoples' movements for independence.[34] A forgotten but telling illustration is the presence of the twenty-eight-year-old Southern preacher and organizer of the Montgomery bus boycott, Martin Luther King, Jr., in Accra on the night in 1957 that Ghana, the first new state in sub-Saharan Africa, gained independence.

The significance of the point is that successful challenges to the oil firms' organization of work by race might have occurred first and advanced faster outside rather than inside the United States, in Mexico and Venezuela, say, but certainly it was not a matter of America having shown others the way. I am confident, too, about two matters. Texaco, Chevron, Exxon, and Mobil—ARAMCO's owners—accumulated decades of experience in dozens of locales: Beaumont, Bakersfield, Coalinga, Maracaibo, Oilville, and Tampico. And they laid out each field and camp everywhere the same way, decade after decade, with the labor force divided, segregated, and paid different wages according to race. Where unions successfully gained ground or where local politicians' and the firms' interests aligned in defeating unionization, the firms would eventually begin to make concessions in living conditions for the majority, agreeing finally to build the schools that were promised ten years earlier, say, or authorizing a wage increase, and so on. The same concessions had to be wrested, piece by piece, by workers in California, Oklahoma, Colombia, Venezuela, Indonesia, and Saudi Arabia. It was not a matter of learning by firms. The incontrovertible fact is that it was a purposeful strategy deployed consistently and unaltered across most of a century. What the firms learned, in other words, was that it was rational to continue to deploy the tried and true strategies for as long as possible. As late as the 1950s, even as ARAMCO's owners were forced to make concessions to its Arab workforce, a new firm, Getty, begin production in the Saudi Arabia–Kuwait neutral zone using the strategy of divide by race and rule.

In addition, equality between white workers and others remained everywhere an unfulfilled ideal. In part, nationalization ended the responsibility of American firms for organization or reform of industrial relations. In part, too, firms began disguising the supremacist origins and resonances of their labor regimes during the Cold War, consistent with the general account of the pressures facing the postwar U.S. state and society. So, ARAMCO would decree name changes for its camps, start to crack down on the petty abuses to which the white workers subjected Saudis, and re-inscribe the grounds for its continuing commitment to separate and unequal work and living spaces. It was not a question of racism but of skill levels, the firm would begin to insist in its ramped up public relations campaigns. Markets, not hierarchy, dictated that some workers received their pay in dollars, others in riyals. Needless to say, Saudi (renamed by the firm "general") labor and other less privileged ("intermediate") castes were not fooled.

The ideal of equality went unfulfilled as well because where nationalist and populist governments alone or in combination with growing working caste action compelled firms to agree to provide equivalent goods to the majority, the enormity of the costs in doing so at times drove the companies to offer less-expensive alternatives. We will see this sequence in the case of educating the children of the Saudi workers as opposed to those of the white managers and labor, as well as in the case of housing. Firms provided whites with free housing. As labor struggles in Maracaibo, Jakarta, and Dhahran won recognition of the rights of workers to equal housing benefits, companies stopped building houses for workers (save white workers) and began to offer loans to them to build their own homes, saving tens of millions of dollars at each camp site. At such moments there is always an earnest and fair-minded assessor ready to judge the step a "clever and generous solution" by the firm.[35] Generous is a euphemism. And the solution is generous only by comparison with an original decision, never discussed, to provide decent housing and services to a tiny, privileged caste while others were provided squalid quarters and not much more. Saudis continued to protest the unfairness. A firm's protection of the bottom line, the unreconstructed paternalism of its white executives, and the segregation of the campsites themselves all trumped equality, apparently.

Finally, company records show that its competitors in Iraq and Iran (joint British-U.S.-owned firms) moved faster and further on all the dimensions that Saudi workers began to mobilize to change and that the small Saudi state-building class pressed ARAMCO to honor. This is true about wages, housing, training for employees, the education of employees' children, and the advancement of Arabs and Persians into management positions. The explanation for this is simple, known to be true at the time and since forgotten. While all three countries were monarchies, only Iran and Iraq had functioning parliaments, parties, and unions. Populist politics, which emphasized inclusion and redistribution, gained ground in both places after World War II, culminating in the famous nationalization of the Anglo-Iranian Oil Company in 1951 (only partially reversed after the successful CIA coup two years later) and revolution in Iraq in 1958. Conditions for the nascent Saudi labor movement and the relative handful of officials who sought to move Saudi Arabia in a more inclusive and redistributive direction were, to understate the obstacle of absolute rule, inauspicious, and the firm there had a freer hand to deflect, ignore, and counter demands for fairness and human capital development.

While a number of sources support this analysis, the most severe criticisms of the firm's labor regime came from the consultants and advisors to the Rockefeller interests, the single biggest shareholders in Exxon, who had negotiated to purchase a 30 percent holding in the Texaco-Chevron operation in Dhahran and toured the facilities at different points in the late 1940s. We have the reports of the consultants' trips to the camps and some evidence of the faction's efforts to overthrow the unreconstructed racists who were running ARAMCO in Dhahran, fresh from stints in Colombia and Venezuela.

This intervention by the Rockefeller group, undocumented anywhere else, may represent an exception to my general statement about oil firms preferring and continuing to build hierarchy anew. Recent work on Nelson Rockefeller's personal involvement in reform of the labor policies of the Mobil subsidiary, Creole Petroleum, in Venezuela after World War II is evidence for an argument about learning. Texaco and Chevron ran their operations, well, more consistently with racist custom, earning them the distinction at one point of being called "a disgrace to American enterprise" by the State Department official reporting on labor troubles in the oil camps, as we will see. Mobil people were brought into the Dhahran operation and the ARAMCO planners visited Creole operations to study labor relations issues. ARAMCO's vaunted home ownership plan was in fact an innovation adopted from Creole. As I have said, this reform was itself consistent with preserving hierarchy, not least the preference for keeping ARAMCO's American Camp itself white, although this custom too would give way, over the longer term, with nationalization. Conflict among the owner factions, the relative autonomy of the managers in Dhahran, but, above all, the weakness of the worker movement and its progressive allies in the 1940s and 1950s gained the firm exemption from the concessions that populists were pressing for and winning elsewhere.

ARAMCO might have chosen differently at any point in the 1950s, as laid out in one of its planning sessions and discussed in Chapter 8. One option it had was to align with Saudi nationalist and developmentalist forces, end segregation, provide serious resources for building hospitals, schools, and roads, bring Saudis into management as was done with the local population in Iraq and Iran, and, as the strategists argued, to "get the people behind us" and prepare for the possibility of regime change, "drastic or otherwise." It did not do so. Instead it bet, correctly, on the house of Saud, securing its survival more or less intact for another decade.

The outcome of the story is rather sad, and not only because it reveals the hollowness in one or another account of ARAMCO's dedication to preparing "Saudis as quickly and as soundly as possible to operate the Saudi oil industry," as one retiree wants desperately to believe. Nor is Lippman able to explain why "by 1949, despite all the training programs," Saudis weren't available for promotion to management positions.[36] The answer is because there were no training programs in place at the time designed to prepare Saudis for advancement. It was a moment when the firm refused to hire Saudis or other Arabs who were qualified, as we will see, because of the racism that organized life in the camps. Some others may be saddened more by having to give up the myth of Good Prince Faisal as the ruler who saved Saudi Arabia, arresting the decay of nascent institutions and putting the kingdom on the path of development and reform.

The saddest part of the story, though, is the fact that so many sons of the Arabian Peninsula like Abdallah Tariki, OPEC's founder, were driven into exile. Others were tortured and, like another visionary, Abd al-Aziz Ibn Muammar, who for a brief moment led a kind of administrative revolution in the kingdom, were made to endure years in the dungeons of Riyadh. Still others gave up their lives in the course of seeking to change the terms under which ARAMCO operated in the kingdom and to check the power of the family that was and is still beholden to the United States for its survival. The significance of the story for theory building, as we might ask now, fifty years later, is a question to be entertained in its appropriately diminished context.

It should no longer be possible to leave labor out of the familiar (and not so familiar) stories of how firms and states transformed the twentieth-century world oil economy. My book complicates one such story I learned when I began to study political economy in the 1970s and 1980s, and that remains the conventional wisdom thirty years later. Firms were hardly the agents of the changes referenced by ideas about post-imperialism, the doctrine of domicile, the end of gunboat diplomacy, and the rest. Although oil companies were undeniably more capital intensive than other mining industries, which my mentor in political economy, Thomas Ferguson, argues is a key to the coalition that emerged in the United States to support the New Deal, my book suggests how in another specific sense this structural factor hardly mattered. ARAMCO, a subsidiary of four of the world's largest and richest firms, would have to be pulled into the twentieth century by its Saudi workers.

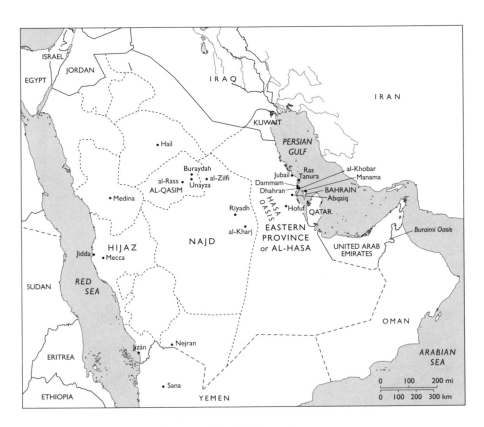

Map 1. Saudi Arabia and Surrounding Region

Figure 1. Seventy oilmen, diplomats, bankers, and Arabists at ARAMCO's banquet in honor of Crown Prince Saud at Manhattan's Waldorf Astoria, 1947. Courtesy of the Department of Special Collections, Princeton University Libraries.

Figure 2. Barastis (palm-frond huts) provided to the Saudi workers in Dhahran stand in sharp contrast to the California-style ranch houses for the white drillers and managers. Courtesy of Archives and Special Collections, Georgetown University Library.

Figure 3. "Eddy of Arabia." ARAMCO's "consultant" and undercover CIA agent Bill Eddy dressed for an audience with King Ibn Saud in Riyadh.

Courtesy of the Department of Special Collections, Princeton University Libraries.

Figure 4. Prince Faisal and his admirer Bill Eddy meet again.
Courtesy of the Department of Special Collections, Princeton University Libraries.

Figure 5. Abdallah Tariki, Saudi Arabia's first minister of petroleum, OPEC founding father, and ARAMCO's bête noir.

Courtesy of Archives and Special Collections, Georgetown University Library.

Figure 6. The Saudi visionary Abd al-Aziz Ibn Muammar. He led an aborted administrative revolution in the kingdom and suffered Prince Faisal's wrath. Courtesy of Haifaa Muammar.

Figure 7. ARAMCO-Americans relaxing at home. A Saudi flag hangs behind the bar.
Courtesy of Archives and Special Collections, Georgetown University Library.

Part 1 INTRODUCTION
The Nearest Faraway Place

David Dodge joined the Arabian American Oil Company, ARAMCO, in 1949. A Beirut-born and Princeton-bred patrician, heir to an immense American fortune, Dodge was an ideal addition to ARAMCO's Government Relations Organization, its combined diplomatic relations and intelligence operation. He grew up speaking Arabic. His mother, Mary Bliss, was the granddaughter of American missionaries who built schools in nineteenth-century Syria. The biggest was the Syrian Protestant College, which was renamed the American University of Beirut (AUB) in 1920. The Dodges were college trustees. His father, Bayard, became president of AUB in 1923, the year after his son was born.[1] David grew up in Lebanon, spending fourteen years there before leaving to complete his schooling in the States. During World War II he joined the Office of Strategic Services, operating in Egypt and the Levant. He then returned to Princeton to finish his senior thesis and study for an MA in the Department of Oriental Studies. On his ARAMCO job application, when asked to list any hobbies, sports, and avocations, he wrote "tennis, swimming, horseback riding, military intelligence."[2]

While it appears he was given the chance to sign with the Central Intelligence Agency (CIA), the successor organization to the Office of Strategic Services (OSS), Dodge says he did not do so. James Terry Duce, ARAMCO's vice president in Washington and the firm's primary liaison with the CIA, instructed William Eddy, the legendary ex-OSS-officer-turned-CIA-agent working undercover in ARAMCO, to invite Dodge at the time of his application to consider "training as a Special Employee-Trainee before going to the field."[3] They were guarded in their correspondence, but Eddy makes clear that Dodge had turned the offer down. Instead he headed straight for Arabia, working in ARAMCO for five years before transferring to ARAMCO's affiliate, the Trans-Arabia Pipeline Company (TAPLINE).

Dodge spent the next twenty years with TAPLINE in Beirut, rising to vice president and head of government relations. He retired in 1976, the year the Lebanese civil war began, and turned to shepherding his family's varied Middle East philanthropic projects, from the Near East Foundation to the American University of Beirut. He took over as acting president of the AUB in July 1981. One year later, as Israelis laid siege to the city and their Lebanese Phalange allies began kidnapping Iranians, he was taken by Shiite militants and secreted to Tehran, where he was held for exactly one year. David Dodge had become America's first hostage in Lebanon.

Some will see in this story a "metaphor for the United States in the Middle East," a condensation of the American experience as many have imagined it.[4] Missionaries and private philanthropists paved the way for the oil companies that, as Colonel Eddy, another Syrian missionary's son, would say, were themselves like giant missions, "involved . . . in programs of education, health, imparting skills for vocations, and extending goodwill for the U.S.A."[5] American businessmen and diplomats were not like the old British and French overlords. Yet the United States would nonetheless be confronted by "emotional nationalists," "hostile propagandists," and "emergent unrest" each decade after World War II.[6] Dodge's kidnapping in fact marks the beginning of a period of unprecedented violence against Americans in the region. Kidnappings, assassinations, the attack on U.S. forces in Beirut in 1983, the bombings in Riyadh in 1995 and Dhahran in 1996, the attack on the USS *Cole* in Aden in 2000, another round of bombings in Riyadh in 2003: it is as if Dodge's—and America's—generosity has been repeatedly and ungratefully spurned.

This story about Dodge or about the United States in the Middle East is, however, far too simple. It leaves too much out. It reads, and works, in fact more like ritual than like history: reducing "complex events into 'typical scenes,' based on the formulas of popular literary mythology."[7] And those who are convinced by it or want us to believe it have a difficult time resolving its many inconsistencies. To this day, Dodge believes TAPLINE did not provide cover for CIA personnel in Lebanon, and he claimed ignorance about his mentor, Eddy's, own involvement with the agency.[8] The missionaries-turned-Arabists who are among the founders and funders of our Middle East institutes and centers have also tended to reach for ritual in explaining why resistance to American hegemony continues. Dodge blames the Israelis for his ordeal.[9] Eddy and others have blamed Israel for America's troubles for a long time now. As Colonel Eddy's passionate and generous-hearted wife, Mary Garvin Eddy, once told the Saudi ambassador, it is the Jews ("Yahudis") who have a stranglehold on the Congress, producing policies counter to America's *real* interests and turning the people of the region against us.

We begin a different, more complicated story, ranging across places that are more typically viewed as having separate histories and that are the appropriate preserves of distinct disciplines and expertise. The new story has the Dodge family in it too.

2 ARABIAN FRONTIERS

This was Arabia as a romantic imagination might have created it;
nights so mellow that they lay out under the scatter of dry bright
stars, and heard the silence beyond their fire as if the whole desert
hung listening. Physically, it might have been Arizona or New
Mexico, with its flat crestlines, its dry clarity of air, its silence. But
it felt more mysterious than that; and the faces of soldiers and
guide and interpreter, dark, bearded, gleaming in teeth and eye as
they spoke or laughed, corroborated Hamilton's sense that this was
authentic Arabia, hardly touched by the West.

Wallace Stegner, *Discovery!*

On a warm, windy day late in January 1947, three thousand visitors gathered at the new United States Air Force base in Dhahran to await the arrival of King Abd al-Aziz Ibn Saud. The American ambassador, J. Rives Childs, had come from Jidda to head the committee welcoming the king to the "American oil colony on his country's eastern shore."[1] The visit was only the king's second since two American geologists had splashed ashore fourteen years earlier to begin a search for oil. Once, in 1939, on the eve of World War II, Abd al-Aziz had traveled by motorcade from his palace in Riyadh, a three-day journey, to witness the loading of the first tanker of Saudi crude. This time the king traveled by plane, a gift of the late American president Franklin Roosevelt.

A special committee convened in Dhahran to decide on an appropriate inscription for the gold key inside a gold box that would be given to the king to commemorate the visit.[2] They debated "The key was the key to our hearts" and "We, as a company, had provided him with a key to open new horizons of advancement for his kingdom." Then the Berkeley-trained Orientalist George Rentz, head of ARAMCO's newly created Arabian Affairs Division, an in-house intelligence and research department, earned his salary that day with a line from the Quran, surat al-Mala'ika xxxv, 2, "The mercy that God openeth for mankind none can withhold."[3]

The same week, six thousand miles to the west of Dhahran, the king's son and heir to the throne, Saud Ibn Abd al-Aziz, stepped from a train onto the platform at New York City's Pennsylvania Station. He had arrived from Washington, D.C., and a meeting with President Truman, the first stop on a month-long tour of the United States. ARAMCO handled the arrangements, hauling the crown prince and his companions across the country, showcasing Air Force bases, dams, farms, the Douglas Company factory in Santa Monica, where the planes for the kingdom's air force were built, and Oriental studies at Princeton, which the oil company was supporting. A job of Floyd Ohliger, the Dhahran-based ARAMCO executive who accompanied Saud, was to keep the entourage's more libidinous escapades out of the newspapers.[4]

On Monday evening, January 20, ARAMCO hosted a dinner for seventy-five at the Waldorf Astoria in Prince Saud's honor. Invitees comprised the elite of the oil world, the owner companies, the king's New York bankers, State Department Arab hands, and the Bechtel Company's intelligence-agent-turned-oil-consultant, C. Stribling Snodgrass. On Saud's right sat H. D. Collier, president of Chevron (or Socal as it was known then). On the prince's left was W. S. S. "Star" Rodgers, president of Texaco. Sitting next to Rodgers was Colonel William Eddy, who was invited to serve as translator, reprising the role he had played for FDR and Ibn Saud aboard the USS *Quincy* in 1945. Eddy called the banquet "an incredible affair, the first such I have ever attended, since the official government parties for royalty are, by contrast, modest and unpretentious, but I suppose, with the oil company making millions out of Arabian oil, they can afford to go Hollywood as hosts."[5]

The guests assembled around a single, vast, oval table, a half acre in size. Richard Sanger, the State Department desk officer for Saudi Arabia who was assigned to the entourage for the duration, described it. The center

> was laid out to represent a New England village in winter time. A river flowed down through the center between snow-covered hills, past an old mill and under a bridge. A church, a school and houses dotted the hillsides and there was a Texaco Filling Station along one side of the main highway. Winding through this scene in a figure eight was a miniature electric railway complete in every detail. The controls were located at the Prince's plate and he enjoyed himself running the train, blowing the whistle, producing puffs of smoke and loading the coal cars.[6]

The Waldorf gala, commemorated by a handsome white calfskin menu and twenty-page engraved souvenir book in Arabic and English, represented the Saudi kingdom's debut on the western world stage, orchestrated with great panache by the American oil giant, ARAMCO. Push the coming-out date back slightly, if you'd like, say, to the arrival of Saud's brother and foreign minister, Faisal, to head the Saudi delegation to the new United Nations, the costs again borne by ARAMCO. There is still no denying the significance of the moment, given that the kingdom, as Eddy described it in 1945, was a "country which has at present very rudimentary foreign relations, no experience in international law or world trade, no corporation law, and no banking system" and "not more than ten officials who have any knowledge at all of the basis of commercial and political relations between modern nations."[7]

The events that led to a crown prince running a toy train in a hotel ballroom in midtown Manhattan are now well known. In 1933 a California oil company loaned an amir-turned-sultan-turned king £50,000 and promised an annual rent of £5,000 (about $2.3 million in 2005 prices), all in gold, for the right to prospect for oil. The sums enabled Abd al-Aziz, the new king of Saudi Arabia and badly in debt as a result of the tail end of the world depression, to begin to secure his new, nominally unified domains. The consequences of his choice to let in the Americans were immense because the discovery of oil in 1938 increased the prospect that nearby, more powerful rulers would want the resources for themselves. A major objective of Saud's trip to the United States in 1947 was in fact to obtain aid and weapons with which to defend what was now understood to be a vast reservoir of a vital strategic commodity in an increasingly polarized Cold War world.

The bargain struck by Abd al-Aziz was more fateful still because it meant the start of a kind of invasion of his domains that he and his clan would then have to defend as best as they could. It is no insult to note that in a contest with actors possessing superior knowledge of "the basis of commercial and political relations between modern nations," as Eddy put it, and with far greater resources than those commanded by the conqueror of Hijaz and al-Hasa, the odds were hardly in the Al Saud's favor. Few expected the rulers to do as well as they did, in fact. Perhaps Ibn Saud thought that if the siege grew too threatening he could hunker down behind the mud brick walls of his new Murabba Palace. In the early days, the oilmen were allowed into Riyadh only with advanced permission and when they dressed as Arabs. The invaders though kept multiplying—ideas, men, women, books, wealth, goods, weapons, technologies, radio broadcasts, intoxicants—like one of those late twentieth-century video games that has the player facing ever larger numbers of attackers, striking from all sides. Imagine a game called incorporation into the world market or frontier capitalism or one called oil or simply ARAMCO.

No other outsider observed ARAMCO's early operations in the kingdom more carefully or more keenly than did J. Rives Childs, America's first ambassador to Saudi Arabia (1946–50). He had arrived in the midst of a vast inflow of soldiers and civilians to Dhahran, which the company would start to claim was the largest single postwar concentration of Americans abroad. ARAMCO lore portrays the moment as the end of the first pioneering phase in its history and of the "passing of the frontier." The postwar boom had led to the opening of a U.S. consulate in the new oil territories. Childs's staff could monitor developments more closely than was possible from Jidda, on the opposite coast, but the small, overworked consular staff also depended on ARAMCO's superior organization and expertise for reporting on developments in Dhahran and its hinterlands. It even depended on ARAMCO for housing.

In a key dispatch to Washington in March 1947, just weeks after the spreading of the welcome mats for King Abd al-Aziz in Dhahran and Crown Prince Saud in Manhattan, Childs reached for the metaphor that the turn-of-the-century western novelist Frank Norris had made famous in depicting "the huge conglomerate West" of the railroads, ranching capital, and mineral firms. Childs warned that ARAMCO was an *octopus*. Its tentacles extended from American Camp in Dhahran, as its main settlement was then known, "into almost every domain and phase of the economic life of Saudi Arabia."

> We can, of course, make a fetish of the free enterprise system and in its name avoid any attempt to exercise a control over the octopus represented by ARAMCO. The longer we delay, however, the deeper its tentacles will be spread and in the end the policy of the Government of the United States in Saudi Arabia and in the Middle East may be dominated and perhaps even dictated by that private commercial company.[8]

At State, Richard Sanger said the same in *Arabian Frontiers*, the book he wrote in 1948–49 while on a study leave. ARAMCO was an empire expanding outward from its capital in Dhahran. A company executive advised that he look for another term. "While this is an apt word in a very general sense, it has a connotation which will give fuel to those critics who accuse us of being 'imperialists'—a criticism which applies to the American Government as well as ourselves. I don't exactly know what word to suggest as an alternative—perhaps arena."⁹ Sanger, who like others at State and elsewhere in Washington, New York, California, Harvard, Stanford, and Princeton depended on the company for data, photos, and more, made the change and also removed a nice description of how the American-Saudi special relationship had begun, "with a King who thinks like an oil company and an oil company that thinks like a King."[10]

ARAMCO would ultimately invest millions of dollars to sway the public toward its own particular reading of history. After World War II, it began to cast its fifteen-year-old investment in Saudi production and, as a subsidiary interest, in the stability and security of the Al Saud, as an unprecedented "partnership in oil and progress." In 1946, Chevron's monthly magazine, the *Standard Oil of California Bulletin*, published its first such account of a land called "Sah-oo-dee Ah-*ray*-bee-ah," where "part of America has been set down amid the rock and sand" and where the new "quarters being constructed for Arab employees will be among the best in the Middle East." Workers there had just gone on strike in dispute of such claims. In 1947, during Saud's tour of the States, ARAMCO screened its first documentary film on the subject for Americans. The crown prince found the account of the firm as the kingdom's salvation insulting, so the sponsors stopped screening it for the duration of the tour. By 1948, ARAMCO executives were portraying the Saudi investment as a "junior Marshall Plan," and after 1949, once the Truman administration launched its new aid effort in Greece and Turkey, it had become a "private Point Four program" that was "free for the American people." The same year, the company began to publish a magazine of its own, *ARAMCO World*, from its new headquarters at 505 Park Avenue in Manhattan, to spread the good news.[11]

Ambassador Childs and Richard Sanger, neither of them muckrakers, may have had trouble squaring various facts on the ground with ARAMCO's carefully crafted image. Neither of them, though, provided a theory of institutional origins, as an early twenty-first-century political scientist might say, although there were plenty of clues. If comparisons to emperors and kings made the firm's principals cringe, no one in the company raised an eyebrow at the idea of Arabia as frontier or at comparisons of ARAMCO's venture to the opening of the nineteenth-century trans-Mississippi West.

It was a time when the company's geologists could still sit around the campfire and recount for their Arab guides tales of the war for Kentucky. They might lament the costs that state building exacted both on the tribes in North

America and the Arabian Peninsula. And top officials of the firm would choose, of all people, Western writer Wallace Stegner to tell ARAMCO's early history. Stegner had just published his biography of that intrepid one-armed adventurer-scientist, Major John Wesley Powell, who during the late 1800s had charted the forbidding expanse of the canyon lands, the Great Basin Desert, and the Colorado Plateau. Powell was America's Theiseger or Philby, renowned explorers of the peninsula, and ARAMCO's first geologists and amateur Arabists imagined themselves as following in their footsteps.

No one, though, would be looking backward much longer or talking about two moments of incorporation into the world market let alone thinking very hard about the origins of the institutions that ARAMCO brought to Dhahran. Twenty years of the Cold War and the spread of the powerful norm against noticing would produce a new understanding of historical space and time, not least by academic historians for whom the kingdom and its encounter with the oil firm seemed "*quite unlike anything in previous American experience.*"[12]

The reason for our moving backward in time before moving forward in the story is simple. Doing so is one of the best ways I now know to neutralize exceptionalism's power. Part of the problem is simply that the history of mining firms was just not that well known—by ARAMCO's own employees and those who observed its operations at the time, in the 1950s and 1960s say, let alone by all those who would turn to write about ARAMCO in Saudi Arabia thirty or forty years after the commencement of oil production there. If you doubt that lack of knowledge was or is a problem, then stop and think for a second about how much you know in detail about most areas of American history in 1925. The reality is that many very knowledgeable people, experts on Saudi Arabia or oil firms in the Middle East or Middle East history, will be able to talk about the nineteenth-century American South or West only in the most general terms or in clichés and images from Hollywood. This isn't a criticism, by the way. Nor will merely increasing the stock of knowledge about the past solve the problem.

We also have to overcome the sense of a gulf or divide between the 1940s and after and the nineteenth century, the sense that people have that the past is another world that has been obliterated, transformed, overcome. So bear with me. For the rest of the chapter at least entertain the possibility that by the time of World War II or the Cold War the corner had not (yet) been turned on the Trail of Tears, for example, or Reconstruction or Ludlow or the Spanish American War.

There is a great deal to learn about the institutions in place on Sanger's *Arabian Frontiers* in 1949 by looking at the practice of white supremacy and tutelage in the segregated world of the mines and, later, oil fields of the newly conquered territories of the American Southwest. The war for northern Mexico, the Treaty of Hidalgo (1848), and the Gadsden Purchase (1853) altered the political boundaries of North America, opening up new vast territories to a continent-spanning project. Eastern capital, subordinated Mexican labor, waves of Anglo settlers, and the Native American resistance movements in the gold-, copper-, and oil-bearing lands of the territories are the actors in a complex story of domination and dependency.[13]

We begin with the copper mining ventures of Phelps, Dodge and Company in the newly conquered regions, among the most important of the mid-nineteenth century, before moving on to oil. As the firm's senior partner, William Earl Dodge, took the company into the territories, he also plowed a tiny part of his family's vast fortune into founding a new missionary school in 1866, the Syrian Protestant College, now known as the American University of Beirut (AUB). Dodge's latter investment has long served for some as a main proof that Americans in the Middle East, at least in the nineteenth century, had nothing to do with empire. They were wrong.

A TRIP TO BAGDAD AND SOME OTHER PLACES NOT IN IRAQ

The natural wealth of the nineteenth-century West included timber and cattle but it is the waves of mineral resource booms and busts that most indelibly mark its past: first gold, and then copper, and, by the 1920s, oil.[14] Hard as it might seem to believe now, oil production was once feasible and profitable with relatively small levels of investment.[15] Copper and other mining production, by contrast, required quantities of capital that local San Franciscan investors could not supply on their own, and entrepreneurs turned to East Coast and European investors to develop the "new" Mexican territories, which included "some of the richest copper deposits in the United States," and to build the railroads necessary to bring the metals to market.[16] Phelps, Dodge and Company was one of the largest and best known of these ventures.[17]

The founder of the firm, Anson G. Phelps, a Connecticut Yankee in New York, had grown rich at the turn of the nineteenth century by financing Southern planters, trading cotton in London for tinplate, and running ships up and down the East Coast of America. In the 1830s, he reorganized the business, forming partnerships with his two sons-in-law, one based in Liverpool and the other in New York. Phelps oversaw new brass and copper manufacturing ventures in Connecticut, while his New York partner, William E. Dodge, from another transplanted New England family, led the firm's entry into timber lands, first in Pennsylvania and, as those resources were consumed, then in the forest lands of Canada, the Great Lakes states, Georgia, and Texas. He also turned the firm from shippers to railway builders by bankrolling, among others, the New York & Erie, New Jersey Central, Houston and Texas Central, and the Union Pacific lines. As the railroads snaked westward into the periphery, so did the firm's holdings in mines beyond Pennsylvania to Michigan and, ultimately, across the Rio Grande River to Apache territory or what we now think of as Arizona and New Mexico.[18]

Race Development

Two great obstacles stood in the way of security of investment and efficient exploitation of the mineral wealth of the Western territories, however. One was the existing Native American settlements and traveling bands, both those indigenous to the area and the relocated Eastern tribes, including the Cherokee, Choctaw, Creek, Seminole, and others that had been forced on the Trail of Tears as the white man took their lands. The other problem was the attempt to extend the Southern slave system to the conquered lands, an issue that would ultimately lead to war between North and South. The owners of Phelps Dodge played leading roles in the now infamous campaigns to resolve these issues. Both David Phelps and William Dodge served terms as presidents of the American Colonization Society, the leading force behind the effort early in the nineteenth century to rid the North of its free black population and solve the slavery issue through voluntary repatriation of the entire race to Africa. The first Liberian project had been a failure—few black men and women were ready to trade their less than full citizenship rights it seems. Nonetheless, white interest in the Colonization Society grew in the decade or so before the Civil War in tandem with the increasing threat that slaves might soon be freed en masse. The Colonization Society opposed their emancipation and revived instead its ill-fated idea, finally abandoned in the 1870s, to expel the entire black population to Liberia.[19]

As a member of the Board of Indian Commissioners, the federal body that oversaw the expulsion of Native Americans from their lands and secluded them on reservations, Dodge played a leading role in another, much more successful effort designed "to separate whites and Indians, clear the path for westward expansion, and force Indians into a settled lifestyle that would allow their eventual 'civilization.'"[20] Missionaries would direct the civilizing part of the policy, as it evolved, by introducing the benighted race to the industrial arts and the rudiments of reading and writing, the better to "settle down . . . , cultivate the land, and learn to support yourselves, and become part of the American people and children of the Great Father," as Dodge put it in person to a council of tribes at the Quaker-run Camp Wichita in Indian Territory in the summer of 1869.[21]

After the failure of the back-to-Liberia plan, the Indian mission school model of uplift would be extended to the freed and then subsequently and rapidly disfranchised population of African Americans. Once again, both Phelps and Dodge, together with their descendents, served on and for decades aided the American Board of Commissioners for Foreign Missions and the American Home Missionary Society, two of the most important agencies of assimilation and uplift in the West and in the South during Reconstruction.

In a landmark speech at Atlanta in 1895, the "wizard of Tuskegee," Booker T. Washington, the preeminent black educator in America and spokesman for the industrial education and uplift model, acknowledged support by Northern business and philanthropy, William Dodge's son Will Junior among them. Maybe just as famously now, at least among those who study African American history, the intellectual giant of the twentieth century, W. E. B. Du Bois, first came to prominence through his brilliant attack in *Souls of Black Folk* (1903) on the Northern-backed Tuskegee machine and Washington's acquiescence to African American's disfranchisement, demobilization, loss of civil rights, and the establishment of the Jim Crow system.[22]

Two points matter to our story. One is that Du Bois sought to build a social movement that would press for the restoration of citizenship rights and expanded access by African Americans to all-white ruling institutions of power, culture, and the economy.[23] A critical part of that program was, he insisted, a radically different vision of the education of black people: the creation of black universities and the nurturing of a national elite or "talented tenth" of teachers, publicists, writers, scholars, lawyers, and the like. The second point is that the struggle over the nature and objectives of education for subordinate races as they were imagined would be repeated across the colonized world in the following decades, and Du Bois himself would begin to link the struggles at home with incipient anticolonial movements and ideologies abroad. When the Hampton Institute trustee Thomas Jesse Jones released the massive report and blueprint for reform in British dominions, *Education in*

Africa: Recommendations of the African Education Committee (1926), based on the success (!) of America's vocational education model, Du Bois viciously tore it apart in the magazine he edited called the *Crisis*. The Jones report blamed higher education for unrest in India, devalued the rudimentary college preparatory schools in Nigeria and South Africa, argued for redeploying resources and for adopting the Hampton-Tuskegee model. Du Bois called it a program designed to make Africa "safe for white folks." More crucially, he denounced its sponsors, the Phelps-Stokes Fund, a charitable foundation that was the longtime backer of the Tuskegee machine, representing the fortune accumulated by the third partner in Phelps Dodge.[24]

There is a third point. Across virtually the entire twentieth-century oil frontiers, firms faced with the demand for building schools and educating an emerging elite to run the countries and the companies where oil was found—Mexico, Iran, Venezuela, Colombia, Nigeria, and Saudi Arabia—would respond both by deflecting these demands and breathing new life into the nineteenth-century idea that the native peoples were best off with long, slow tutelage in the ways of work. They tried keeping it alive for much too long a time.

The Copper Kingdom

With the deaths of the founders, it would fall on a new generation to turn Phelps Dodge into one of the world's leading copper mining firms. William Earle Dodge's son, Will Jr., whose brother David Stuart had gone to Beirut to teach, would oversee the Western copper holdings. Will also became the leading patron of the Syrian Protestant College. Will's cousin Daniel Willis James was co-organizer of the expansion effort. And two more generations of Dodges would eventually follow Will Jr. west to Arizona and New Mexico, including Cleveland Hoadley Dodge, whose fortune financed Woodrow Wilson's campaigns for president. One of Cleveland's twin sons, Cleveland Earl, went to Bisbee while the other, Bayard, went to Beirut.

Under William Dodge, Jr., the firm sunk capital in 1880–81 into the Detroit Copper Company's mine at Clifton and smelter at Morenci, two towns along the Gila River, about five miles apart and near the Arizona–New Mexico border. The firm then gambled even more on another claim at Bisbee, up in the Mule Mountains, six miles from Mexico, ultimately taking over the Copper Queen mine there. The Copper Queen would turn out to be one of the richest veins in North America, and by 1909 Arizona would lead the world in copper production.

All these Phelps Dodge properties though were on lands claimed by the Apache, and this people had not yet given up the war with the colonizers when the firm started buying mines. This was barren country, part desert, and one of the least populated parts of North America. The first claims had only been filed in 1877 by three U.S. cavalrymen who had taken time out from their warring with the Chiricahua Apache bands. The last recorded Apache raids on the properties and ore trains had taken place in 1882. Geronimo, the chief who was born where Clifton now stands, had surrendered in 1884 but two years later fled captivity. Before the settling and civilizing could take place, the resistance would have to be broken once and for all. The U.S. military sent 5,000 soldiers to hunt Geronimo and his thirty-five men and their families down, capturing him again in 1886. The renegades received sentences of hard labor, and were deported and imprisoned two thousand miles away in Florida.[25]

As Phelps Dodge began copper mining, its investments encompassed more than the production technologies for extracting and refining ore. The firm ultimately built one thousand miles of railroad, which knitted together its investment properties, including an additional mine across the Mexican border at Nacozari, near Douglas, Arizona, where it built its new smelter, and joined the mining towns to the Southern Pacific's hubs east at El Paso and west at Tucson.[26] The owners also had to invest in the settling or modernizing of towns for its managers and workers. Copper was found in remote terrain, and the firm had no choice but to assume the added costs of building homes, schools, stores, and the like.[27] Phelps Dodge was routinely heralded for generous and benevolent stewardship of its Arizona and New Mexico domains.[28]

There is no real dispute among contemporary historians about the objectives and timing of hospital- and school-building campaigns, the provision of health and accident insurance, contributions to social clubs, the YMCA, and the like. The goal was to forestall the union organizing and labor unrest that might compel owners to provide higher wages and a shorter working day and week, and end different rates for different races, the system used by Phelps Dodge and all other firms to organize work.[29] Rather, what would allow us to distinguish among scholars today is the degree of dissent from an interpretive framework in which the company must confront and defeat "the forces of darkness"—Apaches, bandits, Italians, revolutionists, half-breeds, corrupt governors, and Mexicans—as one insider historian put it in 1940.[30]

Despite its best efforts, however, Phelps Dodge failed to prevent the Western Federation of Miners (WFM) and the Industrial Workers of the World

(IWW)—known as the Wobblies—from gaining ground in the Southwest. Major strike waves rocked the company's Copper Kingdom in 1903–4, 1907, 1914, 1915–16, and, most infamously, 1917, leading to the mass deportation of over a thousand miners. Democracy had proved unreliable. Populist tendencies in the Arizona territorial legislature were strong, and politicians had passed an eight-hour-day law in 1903 directed explicitly at Phelps Dodge. These same forces then led the territory into statehood as the forty-eighth and most labor-oriented state in the union in 1912, intensifying the fears of the mining elite.

Worried that developments in the Bisbee and Clifton Morenci districts were shifting against the firm, the viscerally antiunion boss James Douglas prepared the ground for a new model of development across the border in even more isolated western New Mexico. The new settlement in Tyrone would give the company greater control over both the environment and its inhabitants. "Unlike Bisbee and other towns where Phelps-Dodge operated, from the beginning the company completely controlled Tyrone's planning and development and the colony was owned by the company."[31] The firm built an idealized Mexican colonial town nestled in the Burro Mountains, with a wide central plaza, arcades, a movie theater, a company club and high school, and plans for a massive Catholic church for its Mexican workers, never built, and it operated the police and fire department, jail, "sanitary squad," and so on. To reach Tyrone, one used the company's new railway line.

For our purposes, the Tyrone settlement is akin to a natural experiment or as close as we can get to one for understanding the choices made by the firm's principals, renowned reformers motivated by their religious convictions and commitment to improving the lives of their employees and organizing work and life as they thought best. No other firms or agencies shared power or responsibility for the shaping of Tyrone. Its own workers had little say, certainly, which we can gauge not only by what has been claimed generally about companies like Phelps Dodge but by how easily the Presbyterian owners ignored "prevailing biases of American society" in many respects. So the owners set up a town in which residents had no say in governance—no town council or the like. They banned labor unions in a country that recognized freedom of association. They decided that residents would not be allowed to build and own their own homes. Prostitution was banned inside Tyrone, although presumably many of its male residents would have chosen differently, to judge from nearby towns. Doubtless workers would have chosen differently when it came to the banning of saloons and the consumption of alcohol.

In precisely the same way, the owners made the choice that Jim Crow would rule there. In Tyrone, unlike Bisbee, which union activists successfully kept as a preserve of white miners, the owners counted on using Mexican labor (in the firm's payrolls this category included American-born, Spanish-speaking citizens, Yaqui Indians, and so on) underground in the mines. For this reason, it was critically important to preserve the racial wage differential that drove the process of displacing white American labor in the first place. At the same time, the firm insisted on separate and decidedly unequal housing and other services in Tyrone for all non-white residents, following the custom at all its other camps.[32]

The company's great social experiment at Tyrone survived for a few years only. The crushing of the labor movement in the Arizona copper belt, on the one hand, and the plummeting world market price for the metal at the end of World War I, on the other, led Phelps Dodge to close down its New Mexico operations in 1921. The largely Mexican work force drifted away. Tyrone turned into a ghost town.[33]

All the White Labor the Law Allows

The first mine union organizing in Arizona in the 1890s appears to have been triggered by white miners seeking to keep out Mexican labor. A spontaneously organized union at the Old Dominion mine in Globe, which forced management to restore wage cuts and fire its newly hired replacement workers, invited in the newly organized Western Federation of Miners, intent that Globe remain another "white man's camp" in a place where white and American were virtual synonyms.[34] The Globe local in turn aided union organizing in nearby mines.

Union strategies varied in different places and moments. Organizers at times sought to bring Mexican workers into the movement, and white miners often split into inclusionist and exclusionist groups. Inclusion provided a means to build union strength and could rob companies of a main strike-breaking tactic through the threat of importing "foreign" labor. The limits to inclusion were also clear. In no case did white miners and unions take up the cause of equality between races.[35] Thus, so-called Mexican (and Italian, Greek, Serb, et al.) workers might make common cause with white miners in defense, say, of contracted wage rates if firms were underpaying them, but the battles to end the racial or "dual wage" system and the racial division of labor were ones that would be waged in Arizona by the Mexican and Mexican American workers for decades.

Squeezed by both sides, the foreign workers, including U.S.-born Hispanics, launched the first big strike in Arizona in June 1903, which was the first mass action anywhere in the United States led by Mexican and Mexican American labor. The workers called for the end of high prices at the company store, of firings without cause, and of mandatory deductions for the company's welfare work benefits. The really critical demand, however, was for wage parity with white workers. But Phelps Dodge officials refused to negotiate any of the issues. The strikers succeeded in shutting down the mines for twelve days, before a combined force of local police, the Arizona Rangers, National Guardsmen, and a U.S. cavalry unit dispatched by President Theodore Roosevelt crushed the rebellion. The leaders were imprisoned for years on charges of rioting, and the company preserved its dual wage and hours system while imposing the pay cut on its underground miners that had triggered the conflict.

Through the rest of the decade, workers on both sides of the U.S.-Mexican border continued unsuccessfully to resist the reign of the racial wage system. The copper boom of the early 1900s had transformed the Sonoran villages of Cananea and Pilares de Nacozari into booming enclaves and the Arizpe and Moctezuma districts into the fastest-growing districts in the state, closely bound to the United States. Arizpe was the location of the largest firm, the Cananea Consolidated Copper Company, and Moctezuma the home of Phelps Dodge's mines, linked by rail to its smelter across the border in Douglas, Arizona. Workers in the Sonoran mining camps "inhabited a segregated universe designed by the Yankee."[36] American families lived in elegant houses, had clubs, lavish hotel dining rooms, and tennis and golf courses for their exclusive enjoyment, and sent their children to company-built, segregated schools, while the Mexican workers lived apart, in shacks, which they owned themselves or rented from the companies.

In the run up to statehood, during an era when progressive politics dominated at the federal level, Arizona's labor grew into a formidable political force, and the great cattle and copper interests found their grip on the state slipping. Thus, in 1908, the Democrats, who gained control of the legislature after pledging support for labor's objectives, disbanded the Arizona Rangers. Progressives also dominated the constitutional convention in 1910 and, in the first years of statehood, women gained the vote and labor won the right to picket and to better, safer working conditions. The power wielded by a coalition of agrarian populists, small businessmen, and labor reached a peak in 1915–16.[37]

Populism was hardly liberal and only partially progressive as manifested in a project, beginning at the Arizona constitutional convention, to rid the state-in-formation of its Mexican population. Anglo delegates called for restricting non-English-speaking employees at any mine to 20 percent of the workforce. Defeated in 1910, the "Americanization" measure was put forward by the legislature and passed by a referendum in 1914. The new Alien Labor Law triggered worldwide protest and a series of challenges in the courts. In 1915, the U.S. Supreme Court declared the law in contravention of the Fourteenth Amendment.[38] At the same time, Arizona's governor, George Hunt, a liberal Democrat, for the first time in the mining industry's history used the militia to *protect* the Phelps Dodge workers during the second big Clifton Morenci strike. Hunt feared the copper company might follow the course the Rockefeller interests took in their bloody confrontation with the coal workers at Ludlow the preceding year, and sought to prevent another massacre.[39] He traveled to Clifton to address the workers, where he called for "adjustments" by the company and, to the delight of the crowd, threatened to throw the mine managers "into the bullpen" if that is what it took to bring them to the negotiating table.[40]

The great Phelps Dodge boss Walter Douglas and the company-controlled Arizona newspapers attacked Hunt and portrayed the strike as a plot hatched across the border in revolutionary Mexico.[41] Douglas took on Hunt in the pages of New York's progressive *New Republic*. The company stalled. In the end, however, the workers got their wage hike, at the cost of agreeing to repudiate their affiliation with the Western Federation of Miners. Destruction of the WFM was an objective from which Phelps Dodge never wavered. Douglas, who also vowed never again to negotiate with a union, led the mining industry's successful campaign in 1916 to bring down the Hunt regime in Arizona.[42]

The company's real triumph came one year later, in 1917, when it delivered a death blow to union organizing in southeastern Arizona, abetted by the IWW's own ruthless campaign against a weakened rival, the WFM. The U.S. entry into the war in Europe in April 1917 had turned production of copper into a great patriotic duty, or so the companies would insist, and workers' quiescence into a litmus test of loyalty to the republic. This was at a time when copper prices (and, thus, profits) and workers' costs were skyrocketing. The companies prepared for war by organizing and arming vigilante groups in the vicinities of the camps.

Workers walked out for the first time at the Phelps Dodge camp in Bisbee on June 27, 1917. "Big Copper Strike Blamed on Germans," reported the *New York*

Times. Bisbee was where the Wobblies had concentrated their recent organizing among the white miners. On July 1 workers at Clifton Morenci struck, the first WFM action where a majority of the strikers were Mexican. In Bisbee, the company press had been warning of stringent reprisals against "any sign of disobedience," although there had been no violence up to that point, and by July 4 the *Los Angeles Times* was reporting that over half the miners were back at work and the strike had been broken. On July 11, the general managers of the town's three large firms met at the Phelps Dodge dispensary to finalize plans to destroy the union movement. Over 1,100 men were rounded up in twenty-three cattle cars, and railroaded across the desert to New Mexico, where they were left without provisions. The majority of the deportees were not Wobblies but men who either belonged to the WFM or strikers who did not belong to a union, and many were longtime residents with homes and families back in Bisbee.[43]

Wrapping themselves tighter than ever in the flag, mine owners in nearby Globe, Arizona, a stronghold of the WFM, promised to turn their operation into "an 'American camp'—no more foreigners with their violence and strikes and unAmerican practices and purposes."[44] Vigilantes turned to running "illegal" Mexicans out of the main Arizona mining towns. The Clifton Morenci strike, led by Mexican workers and arguably the most radical of the 1917 actions, was crushed. "During the course of the strike, a large group of Mexicans was arrested on the charge of rioting and planning a march on Clifton to attack and clean out the whites."[45] Seven hundred "American citizens" among the miners would later blame the "alien element" for fomenting the trouble at the Phelps Dodge camp. The companies had won, and collective bargaining was a right that continued to be denied to copper mine workers until the New Deal.[46]

THE EMPIRE OF OIL

The giants of the pioneering oil era in the American West loom larger than the copper barons in America's political, social, and cultural history, and none looms as large as Edward L. Doheny, founder of an oil empire with Los Angeles as its capital. Doheny drilled the first successful oil well in California, built the first major producing firm in Mexico, and became notorious during the Teapot Dome scandal of the 1920s, which exposed his payoff to the secretary of the interior and one of his oldest friends, Albert Fall, an ex-miner and senator from New Mexico, in return for cheap oil from U.S. Navy reserves. Fall went to jail for accepting the bribe. Doheny, one of the richest men in America, went free.

The history of oil in the West is entangled with copper and the other mining industries, and is not a world apart. Doheny started out as a gold and silver miner. He prospected in the Arizona and New Mexico territories unsuccessfully for twenty years before moving to California, where his luck would finally change.[47] The sad truth is that we know much less about the social history of the oil autocracy, as California workingmen called it, than we do about the copper kingdom.[48] Until now, we have little with which to balance the breathless firm-as-pioneer idea where the tycoon has exemplary relations with his men, pays higher wages than his competitors, builds better housing, and, while refusing to recognize workers' rights to organize, nevertheless "never reacted hysterically against unionism."[49]

The landscape of production that emerged in the first decades of the great California boom in the Los Angeles basin and the San Joaquin Valley was one that shared a great deal in common with other mining camps and little with the romance about oil that Hollywood would later help to invent. Most camps were isolated communities, with little or no separation between work and home. "Both literally and figuratively, oil pervaded almost every aspect of a worker's life." The smell of crude filled the air. Drilling went on day and night. Many of the oil field workers and their families both in and outside the camps lived in what were known as "rag rows" of tents and shacks, which could still be found in areas of Los Angeles into the 1920s.[50]

The World War I years in the Oklahoma and California fields were, as in the case of copper mining, a moment of sharpening conflict between the firms and a nascent workers' movement led in the mid-continent fields by the IWW. Much of the story thus follows a predictable course. The Wobblies had made Tulsa the center of Western radical oil unionism in 1915–16 and were then defeated in the same kind of repressive campaign waged by the copper mining industry in 1917, the year the United States entered the war.[51] The IWW did not make inroads into California until after the war, thus depriving companies of a main weapon against the increasingly politically oriented and effective oil workers' movement there. The leading firms, the Union Oil Company (Unocal), which was the biggest, and Standard Oil of California (Socal, later Chevron), responded with improvements in housing and working conditions in their camps through the early 1920s.[52] The moment gave rise as well to the first company publications, such as the *Standard Oil of California Bulletin*, where company men began to craft the identity of the firms as benevolent stewards of the land and its worker-consumers. We have seen this sequence before and we will witness it again.

A key difference with Western copper is that in California and elsewhere oil was a white man's preserve, and it remained one of the most segregated industries across the United States through World War II.[53] One reason might be the more limited demand for unskilled workers in the industry. Oil took less labor to produce than did copper. Labor demand was heaviest when production was just getting started, as drilling rigs had to be put up, camps built, pipelines laid, and so on. Many oil-field workers were in fact migrants who moved frequently from site to site. More permanent work, as in the case of drilling, depended on small, autonomous crews possessing specialized skills. While there was a hierarchy of labor in the camps, there was no real equivalent to copper's vast ranks of men toiling underground.[54]

There doubtless were exceptions to the industry's rule by whites, especially in the least desirable occupations, and the profile of the workforce in Oklahoma and Texas may have been different from California's. Employers in the mid-continent camps sometimes played "off the *few* black and Mexican workers against the whites."[55] Tulsa business and policemen in a secret Ku Klux Klan–like organization, the Knights of Liberty, led the war against the oil workers' movement there in 1917. The Oklahoma KKK was founded two years later, in 1919, and through the 1920s became the main vigilante force against three undesirable populations: immigrants, African Americans, *and* the radical unions in Tulsa, Oklahoma City, and the surrounding oil camps.[56] Since there were no formal apprenticeship programs or schools in the oil states to teach drilling skills until the mid-1930s, white drillers, who were expected to pass this knowledge to others on the job, did not have to don robes to preserve oil's color line in the West. Beyond those borders, however, the castes in oil and copper camps looked much more like one another.

Rule, Petroleum!

Although American oil-producing firms in the 1870s led the move to what would later come to be called the "multinational" or "transnational" enterprise, keep in mind that in those first few decades foreign investments were almost exclusively in refining facilities and marketing operations. U.S. firms, most famously, John D. Rockefeller's Standard Oil, then the world's largest corporation, built a vast infrastructure to sell Pennsylvania and Ohio oil and oil products in markets from Mexico to Europe, Russia, the East Indies, and China. The quest by America's oil giants and their rivals for new supplies of crude oil through control of foreign concessions really began in the 1910s and 1920s.[57]

A U.S. Supreme Court decision in 1911 had declared the Standard Oil trust a monopoly and it was cut up into smaller pieces—Standard Oil of New Jersey (later Exxon) and Standard Oil of New York (later Mobil), which are today a newly merged firm known as Exxon/Mobil, Standard Oil of California (later Chevron), and so on. The first two of these were marketing and refining firms more than oil crude producers and as such were suddenly in need of their own sources of supply. At the same time Standard's greatest foreign rival, Shell, and others were prospecting and discovering new sources. The conversion of naval forces to oil power in the run up to World War I also meant a vastly increased demand for what had suddenly turned into a vital strategic good.

This "late" entry of the big U.S. firms also meant that the new oil fields in Mexico, where more oil was pumped than anywhere else in the world in the decade between 1910 and 1920, including California, were the possessions of others at a key juncture. Two new firms had raised funds for prospecting the jungles around Tampico, risked all, and made fortunes. One was Edward Doheny's Pan American Petroleum. The other was his rival, the British contractor Wheetman Pearson (Lord Cowdray), the builder of New York City's Hudson River Tunnel and the owner of El Aguila S.A. or Mexican Eagle. Both men waged heroic campaigns to protect their interests, first, against the upheavals of a ten-year-long revolution in Mexico that began in 1910, and, second, against the larger and more powerful companies that wanted in on the Mexican oil boom. Doheny and Pearson succeeded in the first campaign. Mexican oil production continued uninterrupted through the decade. They failed in the second. Cowdray fended off the American giant Exxon before selling Mexican Eagle to a British-Dutch rival, Shell, in 1919. Doheny was bought out by Exxon six years later, though some argue that the connection to the world's giant had developed earlier, the outcome that Mexican nationalists had always feared.[58]

These campaigns are more intricate and infinitely more fascinating than I portray them here, of course, as they involved: rival revolutionary armies, German spies, wartime "high" Anglo-American diplomacy, two rounds of armed American intervention across the border and in the oil fields, the rise of the legendary Pancho Villa, and millions of dollars paid by the oil firms to Mexican protection rackets, coup plotters, private militias, and the like. The magnificence of a canvas of this size and complexity may explain why in two decades of remarkable scholarship on these matters, beginning with Friedrich Katz's *Secret War in Mexico*, there is still virtually nothing written on conditions inside the oil camps themselves. The difference with work on Mexico's copper mining enclaves is dramatic.[59]

We do know that, as in Arizona and Sonora, a racial hierarchy was visible and continuously reproduced in the layout of the camps, where Anglos were a privileged minority. The camps were segregated. Races were paid in different currencies—gold-backed for Americans and paper for Mexicans. The amenities companies provided to Mexican workers differed markedly from the Americans' quarters, hospitals, schools, movie theaters, and so on. Crucially, access to housing or medical care depended on the workers first demanding these as rights and then struggling to improve the terms that they were offered or else the companies introducing improvements in working and living conditions when the politics of the revolution threatened to engulf the oil region. Thus, Doheny's firm built its first primary school for Mexican children in 1918, almost two decades after his oil venture had begun, with limited instruction designed to train children to become gardeners. Paternalism by the oil firms was in general a strategy to deflect Mexicans' demands for redress of the most glaring inequality in the industry, the different wage rates for foreign and native workers.[60]

The familiar liturgies followed. There was of course a rich store of knowledge about the races to explain why Mexicans required less than Anglos or could not make good use of schooling or did not really mean it when they demanded more. It made no sense to train Mexicans for skilled positions. Such arguments notwithstanding, once American oil workers abandoned their jobs in fear of the revolution, Mexican workers took over many of the positions.

The superior analytical capacities of the oilmen also could detect the hand of various foreign agents manipulating the overly emotional spics, greasers, and peons, as Mexicans were known. These foreign agents included the IWW, the Germans, and by 1919 the Bolsheviks.[61] How else could one explain rebellion by workers who were in fact better paid than those outside the oil enclave? The retrospective judgments of historians like Alan Knight and Jonathan Brown agree that the oil field workers were privileged by comparison to those in the textile mills and on the farms, but that they were also bound in a relationship of subordination to Anglos. Refinery and oil field workers were no vanguard of the revolution, and the focus of their demands was on improved wages and living conditions in the foreign-dominated enclave, not on ending capitalism or taking over the companies.

Others did press for the revolution to take such a course, and as a result oilmen like Doheny found themselves pursuing a new kind of argument about their companies' contribution to the welfare of Mexico as a whole. Perhaps more accurately, firms and their agents added a new twist to older, more venerable strands of thought about bringing civilization via Christianity and about the uplifting of the lesser races. Confronted for the first time by what we in the early twenty-first century now consider to be defining characteristics of the twentieth century as a whole, the age of populist nationalism and state formation in the colonies and semi-colonies of the world economy, American oil investors in Mexico began to portray themselves as dedicated agents of nation building, unlike the plunders and exploiters of old.

Doheny's pioneering contributions to the industry's nationalism-management technology are indeed noteworthy. His company began producing a new monthly magazine in 1916, the *Pan American Record*, to promote the strategic value of Mexican oil in World War I and a new anthem for the postwar world order,

> The Muses, still with Freedom found,
> Shall to thy happy yachts repair;
> Blest oil! with matchless power crowned,
> To rule the sea and guard the air.
> Rule, Petroleum! Petroleum rule the waves!
> Fuel-Oil is not for slaves.[62]

He also founded one of the country's first private think tanks, the Doheny Research Foundation, in 1917. Its first and only project was a massive study of Mexican social and economic development, to form the basis for a sounder U.S. foreign policy. Doheny's foundation was in effect a shadow "Inquiry," as the Wilson administration's recently assembled stable of scholars planning the peace in Europe was known, and when the Inquiry expanded its focus to include Latin America, the two organizations shared information and personnel.[63] The foundation hired Robert Cleland, a young historian of the West at Occidental College, who wrote the massive report of the mining and petroleum industries. He is the same man who forty years later would write *A History of Phelps Dodge*, the first study of the firm that bore the name of his "friend and fellow-member of the class of 1909 at Princeton University," Cleveland E. Dodge, the brother of AUB head Bayard Dodge and the vice president and director of Phelps Dodge after 1926.

Doheny's magazines and stable of scholars were parts of a campaign to move the U.S. government to intervene late in the war to protect American investors when the new Mexican constitution had declared all oil-bearing lands property of the state. Private firms have long since accepted this kind of claim by states as a legitimate one. In 1917 it drove firms and its agents to lobby for what in some versions was a plan for the annexation of Mexico's mines and oil fields. Doheny instead pressed for a form of protectorate and by late 1918 he and other investors had organized a formal association that in turn had sponsored a highly publicized Senate investigation of the "outrages" against American business by a "Bolshevistic-Carranza government," as one of the lobby's pamphlets put it. And this kind of public relations exercise was only the visible part of a much larger, subterranean program of stockpiling arms and trying to foment rebellions to force the issue of intervention.[64]

The oil industry's size, scale, and notoriety, going back as far as the 1904 muckraking accounts of the Rockefeller "Trust" by Ida Tarbell, may in turn explain why students of the copper mining industry spend less time study-ing the industry's "grand strategy" and record of diplomacy and interven-tion in the same era. In seeking the truth about the companies' alleged role in funding the revolution and in arguing about the limits of investors' abili-ties and willingness to shape the political field, we have come to learn a great deal about Edward Doheny. This Los Angeles oil baron had bought himself a private Mexican army and a number of high Wilson administration figures to protect his investment. But Doheny's ostentatious displays of wealth and heavy-handed political style, for instance in bragging about his influence with Woodrow Wilson, also made him an inviting target.

The contrast between the Westerner, Catholic, and nouveau riche Doheny and a Presbyterian patrician like Cleveland Hoadley Dodge couldn't be greater. The Phelps Dodge head was a patron of the arts, religion, and edu-cation on three continents. A statue honoring his grandfather William Earl Dodge stands in New York's Bryant Park. Cleveland H. Dodge was also much closer to Wilson than Doheny could ever dream of being, although the latter's campaign contributions doubtless helped reduce the distance a little. Dodge was Wilson's lifelong friend, his patron during the years in New Jersey, and the single largest donor to his 1912 and 1916 presidential campaigns. When Wilson died, Cleveland Dodge was invited to help carry the president's casket to its final resting place, an honor the ailing friend and patron had to decline.

Phelps Dodge was no less heavily involved in lobbying the Wilson admin-istration than was Doheny's Pan American Petroleum, and Dodge's interests stretched beyond Mexico to the Near East. The family believes that it was

Cleveland who secured Wilson's agreement to avoid war with Germany's ally, Turkey, in 1917, where his sons and daughters lived among the American teacher-missionary community. And it was Dodge together with another Wilson backer, Chicago industrialist Charles Crane, who led the effort in 1915 to aid the survivors of the Armenian genocide. Nonetheless, historians have yet to tell us much about this investor's wartime activities in the Near East let alone his firm's response to revolution in Mexico, which was known at the time as America's Mesopotamia.[65]

Mexican petroleum output began to decline around 1921, and for the rest of the decade the midland fields in Oklahoma and Texas produced more oil than any other place on the world oil frontier. Doheny never built a strong marketing organization outside the Mexican borders and, additionally, guessed wrong in banking on fuel oil rather than on gasoline sales. His search for new sources of supply led to the U.S. Naval Reserves in Wyoming, to the infamous bribery charges, and ultimately his downfall. The continuing impasse in Mexico likewise drove other U.S. oil firms south to Colombia and Venezuela and east to Iraq, the real Mesopotamia, and the shores of the Persian Gulf in search of oil. Chevron was one of them.

Charles Crane, Dodge's partner in the Near East relief effort, paid a Princeton-trained engineer named Karl Twitchell to assist a struggling new state builder in the Gulf named Ibn Saud with some rudimentary economic modernization schemes. Twitchell reached Saudi Arabia to search for water and minerals. He banked on the old gold mines in Mahad al-Dhahab, in the western part of the kingdom, the Hijaz, and Twitchell set up a mining firm there that operated profitably for a few years. But Twitchell also thought that the geology of the east coast in a part of the kingdom known as al-Hasa resembled that of nearby Bahrain, an island where Chevron was already prospecting for oil. Lore has it that the cash-strapped king and his finance wazir were happy to take the money of a company that paid up front for something that probably did not exist.[66]

All that follows is typically portrayed as a new era in the world oil market, in U.S. international relations and, not least, in Saudi Arabia's history. In the most breathless accounts, for instance, the first history commissioned by ARAMCO, as I noted, something "magical" and unprecedented is said to have taken place.[67] The men who splashed ashore on the Hasa coast near Jubail in September 1933 are said, it is worth repeating, to have crossed the borders into a desert kingdom "unlike anything in previous American experience."

Perhaps, and yet a familiar part of the American experience was brought with the geologists and planted in the sands of Jabal Dhahran, the local name for the highest point in the area, where the whites billeted together and their two Chinese cooks were put in a tent of their own.[68]

BEGINNINGS

The founding myth about American Camp in Dhahran is that ARAMCO created its enclave on the coast at some distance from the oasis towns under orders from the king who wanted to protect his subjects from westoxification or, slightly more cynically, to avoid fanning the flames of religious opposition to the westerners encamped in Wahhabi Arabia, as St. John Philby once named it. One of Philby's biographers, Anthony Cave Brown, offers a variant of this idea in *Oil, God, and Gold.* "To protect the American community from the attentions of the religious police (among much else besides), the company established an enclave."[69] The implication is that had ARAMCO not been forced to defer to the monarch or to the *mutawwaa* ("religious police"), American Camp and its twins would have been configured differently.

A problem with this story is that, as we now know, virtually all oil and mining towns and camps built by ARAMCO's owners and other firms for the fifty years prior to the founding of the American colony in Dhahran took the same form, in places far from the reach of the mutawwaa and where caudillos and governors rather than amirs and kings ruled. It turns out, in fact, that the Saudi government's only representative and the director of customs prepared for the Americans to build their first camp not in some isolated spot but in the midst of the big oasis town of Hofuf where most of the sedentary population of the province resided, the easier to regulate and monitor the foreigners. The Americans chose instead to take the tried and true road, moving the center of their operations to the coast, starting with a small house in Jubail and setting up a tent camp in the vicinity of Dammam. No place called Dhahran yet existed. "A few mud huts and dirty natives mark the metropolis of Khobar. One of our Ford cars with large tires conveyed us some six kilometers inland to a temporary camp consisting of tents for the white men and straw-huts for the natives."[70] The men started setting up bunkhouses and the like at Dammam camp, near the site where the first drilling would commence in 1935, as the first drilling crews and their Chinese cooks arrived.[71]

Most of the first geologists, drilling crews, and camp bosses who worked in Saudi Arabia and who are celebrated as the ARAMCO pioneers learned their trade as wildcatters and contract employees elsewhere, notably, in the South American fields of Colombia, where ARAMCO's vice president James Terry Duce worked as president of the Texaco subsidiary before joining ARAMCO, and Venezuela, where Chevron had first ventured abroad in the early 1920s. The boom that had drawn some 20,000 mostly unskilled Venezuelans into the oil fields by 1929 had made the country the second largest producer of oil behind the United States. ARAMCO's Venezuela veterans included Floyd Ohliger, who first handled government relations in Dhahran and brought his hatred of unions with him, Bill Lenehan, a senior geologist who represented the firm in Jidda, and Max Steineke, the legendary discoverer of the first large producing wells. The first drilling teams too reached Dhahran from Maracaibo in the western Venezuelan state of Zulia by way of Bahrain.

Organization of work and life in the camps of all the major foreign producers in Venezuela, which included Shell, Gulf, and Exxon through its subsidiary Creole Petroleum, the largest single investment abroad by any American firm until eclipsed by the Saudi concession, followed the familiar pattern. The firms kept the mixed labor force of Anglo-Saxons, British West Indians, and Venezuelans divided. Camps were segregated with gross disparities in living quarters, wage scales, and access to services.[72] Barbed wire fences kept everyone but the house servants out of the Anglos' reservations. A letter home from a Johns Hopkins geologist on a one-year contract at Venezuelan Gulf's Iguana Camp in 1924 offers a window into the place and time.

> There is a "colored gentleman" from Trinidad working on the rig who is quite interesting to listen to. His English is absolutely faultless, none of our "darkies" dialect a'tall. On the contrary his pronunciation is most exact. They say all Trinidadians talk that way. For example he will say "Good afternoon, Mr. Douglas, it is quite warm this evening, is it not?"[73]

Oil field workers led two major strikes against conditions in the camps, a two-week work stoppage in June 1925, the first strike in Venezuelan history, and a second, bigger, month-long strike beginning in December 1936. The firms had counted on the country's dictator, Juan Vicente Gómez, to prevent the emergence of a workers' movement. With Gómez's death, the oil state of Zulia erupted in a popular uprising centered in the camps and towns, which were burned and looted. The military regained control, crushing the rebellion, but the new ruler, Eleazar López Contreras, accepted the return of limited political contestation. The Venezuelan congress passed a new labor law, the firms became targets of attacks in the left-wing and populist press, and new forms of regulation became the norm. The firms nonetheless continued to resist recognition of the petroleum worker unions, which led to the December 1936 strike.

While the army once again ended the strike, with the companies' help, leading to the deportation from Zulia of thirty labor leaders, the justice of the demands were conceded by none other than Nelson Rockefeller, whose investment in Creole brought him to Venezuela as a director in 1937. He was then just twenty-nine years old. Rockefeller condemned conditions in the Creole camps and began promoting creation of charitable foundations by the leading firms in the Venezuelan industry targeted at the oil regions. During the early post–World War II years, when Venezuela was led by a left-wing populist, Rómulo Betancourt, at the head of a radical junta that pioneered the famous fifty–fifty profit-sharing formula, which Saudi Arabia would force ARAMCO to adopt in 1950, the firms and foundations began investing millions of dollars in public works, hospitals, schools, and the like.[74]

Chevron had discovered oil in Bahrain in 1932, after disappointing results in South America, triggering the decision to seek the Saudi concession. Ed Skinner, an assistant manager in Maracaibo, was promoted to manger in Bahrain in 1931 and he took his favored crew with him, including Jack Schloesslin, Slim Williams, and another fabled pioneer, Bill Eltiste. Even the two Chinese cooks, Frank Dang and Chow Lee, had worked at Chevron's Venezuelan camps. The organization of the camps in Bahrain, which were being built up by Skinner and others in 1932 and 1933, and the town of Awali at the site of the original Jabal Camp, were organized in the time-tested fashion. The camps and town were of course divided into separate reservations, in this case, for Anglo-Saxons, Bahrainis, Iraqis, and Indians respectively. Whatever the relative state of primitiveness or luxury that American employees enjoyed, for instance, living originally in corrugated metal and cement huts, all non–Anglo-Saxon employees were to be given and expected to be content with less. Arabs and Indians in the camps lived in thatched, palm-frond and floorless huts, known as "barastis," remaining in them even after Americans moved into permanent housing, with air conditioning among other amenities. ARAMCO would use barastis for workers in Dhahran until the strikes began. Bahraini men also lived alone in the camps, traveling home to visit their families once every few months and, later, every few weeks, another custom that would be imposed on local labor in Dhahran and the satellite camps. Only Americans were permitted to have families live with them.

No one to my knowledge has argued that the system of separate and un-equal wages, rights, privileges, services, dwellings, and the like was imposed on Americans by Bahrain's ruler, Shaykh Isa bin Ali al-Khalifah, or that an American enclave took shape there as protection from religious police, since no such institution existed in Bahrain. In fact, Bahrain became, as it remains today, a kind of haven for those who sought a more relaxed and convivial at-mosphere than that found in the Eastern Province. Some men were drawn to the "Persian" and Somali prostitutes who worked on the outskirts of Manama, as one ARAMCO engineer reports in a privately printed memoir of life in the 1940s, which includes the words to a song he wrote (to the tune of "Managua, Nicaragua").

> Manama, Bahrain Island is a wonderful town
> You'll meet your heart's desire for a few rupees down.
> At evening time you think you've got a beautiful rose,
> But, come morning, there's a ring in her nose.[75]

One would-be historian argued, however, that the first ARAMCO Ameri-cans learned their racism at the feet of the Britons in the colonial Indian service who dominated expatriate life on Bahrain. William Mulligan, who began working for ARAMCO's Government Relations Organization in 1946 after serving in a U.S. Air Force communications squadron in Aden during World War II, had intended to write a history of the company. He lived in American Camp for over thirty years and produced many articles about the early days for the camp's long-running newspaper, the *Sun and Flare*, and for *ARAMCO World*. Mulligan believed that Americans "started import-ing some of these attitudes and the vocab, which was a problem." The two examples he gives are *coolie* and *Sahib*. "All the workmen were coolies. . . . Americans who went over there would come back and have the natives call-ing them Sahib—'You call me Sahib boy.' *This was very un-American and un-Arabian*."[76] The same argument is occasionally found elsewhere in chronicles of one or another oil firm abroad.[77]

Coolie is a word traced back to Urdu and Bengali (*Koli*, which identified one as part of a tribe in Gujerat) but one in widespread use in English in the 1600s to refer to hired Asian labor. In the United States it was used routinely in the gold rush era in the West. For example, in 1862 the governor of California, Leland Stanford, the railroad baron and university founder, somewhat cyni-cally—given employment patterns in his railroad building schemes—called on the legislature to block entry of Chinese workers because they threatened to degrade the superior race of Anglo-Saxons. The legislature followed by passing an Act to Protect Free White Labor against Competition with Chinese Coolie Labor, and to Discourage the Immigration of the Chinese into the State of California (April 1862). Decades later, the campaign continued in the building trades and in other redoubts of "white man's country" in California, such as the Japanese and Korean Exclusion League, threatened by "an unlimited Asiatic coolie invasion" until passage in the 1920s of new national quotas.[78]

One of the most revered of the early ARAMCO pioneers, Tom Barger, a geologist from Minnesota, had worked only in the West, in places like Butte and Great Bear Lake rather than in Maracaibo or Manama, before joining the firm. He landed in al-Hasa in December 1937 and rose eventually to become president of ARAMCO. His touching letters home to his parents and to his new wife, Kathleen, daughter of a North Dakota rancher, during his first two-year contract are among the most important sources for chroniclers of the pioneer days of oil exploration in Arabia. Barger had led ARAMCO's first 1938 expedition into the vast southeastern lands claimed by Abd al-Aziz and known as the Empty Quarter. The novelist Wallace Stegner, whose history of the frontier days is based in part on the letters, called Barger a "shirt-sleeve democrat fascinated by the Arabs and their rugged life" who struggled with the dialect of his Bedouin guides on the four-month expedition.

> I enjoy the poems even though I can't understand a word. The listeners all repeat the last two or three syllables at the end of each line. Khamis [ibn Rimthan] usually summarizes the story for us beforehand. Jerry [comember of the party Jerry Harriss] told them about the escape of Daniel Boone from the Indians, the "American Bedu," by throwing tobacco in their eyes and swinging across a river gorge on a grapevine. I stumbled through an account of Custer's Last Stand. Abdul Hadi [ibn Jithina] . . . wanted to know what tribes the American Bedu belonged to. Khamis said none that he (Hadi) would know, but Abdul Hadi said he had been up to the border and perhaps the American Bedu were related to some of the tribes in Iraq! After I had told about them cutting out Custer's heart and eating it to make them brave, Khamis said the Bedu did the same thing with a wolf's eye—they carried it with them when they did not want to be seen at night. Khamis is member of a once-powerful northern tribe, the Ajman, who rebelled against the King Ibn Saud, and were defeated in battle. . . .

Many fled to Iraq, had much of [their] land taken away, and their largest city is now falling into ruins.

Asked to finish the story of the American Bedu, Tom Barger drew two uneven circles in the sand. "Before the war their land was like this, after like this." Here was Barger's version of the story we just read a dozen pages back of mining capital's expansion into Apache and other Native American people's lands. Khamis concluded, "Yes . . . that is what happens to the Bedu in Arabia when they make war with the government."[79]

His deepening regard for his companions complicates but cannot erase the reality of the time and place, with Barger calling himself one of the first white men to set foot in the Quarter.

> Our guides are all soldiers. The army we have of ten men is furnished by the government and paid for by the company. In order to make it look as if we used them, we have to take four with us in the field at all times. The seismograph crew, who have a lot of Coolies to work with them, are going over country fully mapped and consequently don't care whether they get along with the soldiers or not.

Barger's son published a collection of Tom's letters to Kathleen in 2000, including a long one dated Camp Tarfa, Rub al-Khali, February 18, 1938, the same date Tom sent his parents the letter quoted above, found in Wallace Stegner's papers at the University of Utah.[80] The son has removed all references to white men, coolies, and the like from the originals. Like father, like son. Tom Barger had done more or less the same fifty years earlier when he had Stegner strike a passage from *Discovery!* describing the expansion "of the coolie camps at Khobar and Dammam" in 1937–38 in the transition from wildcatting operation to celebrated American Camp.[81] The record shows that the Americans, roughnecks and executives alike, continued to use the term coolie through the 1940s and early 1950s, until, as Mulligan put it, although he cannot be trusted to have gotten it right, Barger and presumably others "recognized that [the] effendi, coolie, raghead situation was wrong and started to do something about it."[82]

The development of the coolie camps came at the time of the first, brief period of oil production before the outbreak of World War II. The permanent if rudimentary headquarters nearby the first producing wells, four miles from the coast, on the highest point in the area, known as Jabal Dhahran, included new housing (air conditioned, two bedroom portables) for the Americans, some offices, a small hospital, mess hall, and storage facilities. Workers had constructed a simple pier six miles north at al-Khobar, a tiny fishing village. Presumably, the camp's first paymaster, Joe Carpenter, nicknamed Joe Khobar, had by then done away with laying his six gun on the table as silver coins were counted out and the coolies handed their pay. The camp was nonetheless beginning to be plagued by theft of building materials and other goods, a form of resistance to the imposition of the new and hardly legitimate authority. Americans too were resisting the local police-in-formation that was attempting to discover and test its power. Violence was inevitable "because no Arab policeman could take hold of an American without a row, and the Saudi police were not trained to gentleness." This dynamic of meeting Americans' often brutal sense of innate superiority with acts of violence or theft would repeat itself regularly over the coming decades, but meanwhile a fence went up around American Camp and company officials secured the right of the firm to police inside its perimeters.[83]

The first families, wives, and children of the senior managers began to arrive in the spring of 1937. Casoc Town, as Barger called it in December, had six married couples living there when Crown Prince Saud made his first trek from Riyadh to al-Khobar to visit the new oil colony. The king himself would follow one year later. Saud invited all six women and the four little girls to his outdoor *majlis* or public assembly. Tom Barger had just reached Saudi Arabia that day from Bahrain, then a three-hour journey "by a native boat with a crew of four variously colored natives." "Natives" is another word he later had Stegner strike from the story of the pioneer years. With the families came the first movies shown in camp and likely the first in Saudi Arabia. "Arabs were forbidden to attend, but they were about five deep all about the outside of the house peeking in the windows. The picture was Ann Harding in the *Gallant Lady*."[84] Later, also for the first time, there were home-cooked Christmas dinners, and also the first houseboys, mostly Indians and Arabs from Bahrain. The prejudices of the time typically held a local population to be incapable of and thus inappropriate for servant work. They were not trusted in these private and intimate spaces. The language barrier reinforced these judgments.[85] ARAMCO had Stegner cut out any discussion of houseboys for the oilmen, and today the facts may be rearranged to portray Americans in Dhahran as enlightened men and women who treated the Saudis more as equals than inferiors. Saudis "were not servants," Nicholas Lemann tells us.[86]

The press to begin commercial production in 1938–39 occupied the sixty or so American "staff" and their "army of young Arabs" full time. The men ran power, phone, fuel, and sewer lines. Inside camp they put up additional housing, a new staff hospital, and a recreation center. The camp's four-hole golf course came next. The firm supplied the police with living quarters away from the Americans and built a new jail. New permanent oiled roads crisscrossed camp and stretched beyond to the Dammam Dome field and to al-Khobar. A second camp was laid out at Ras Tanura and supplied with water and power in preparation for building a new pier on the coast to accommodate tankers.[87] A new landscape of power was taking shape on the coast, an oil town, which the *petroleros* from Colombia and Venezuela began to call Dhahran.

Barger returned in June 1939 after another six-month desert trek to find that "there were so many new buildings, I hardly know the new part." The invasion had begun. The idea of rapid change was fast on its way to becoming a cliché in letters home by returnees who first set eyes on the kingdom in the 1930s or during the war or in the first postwar years. One month earlier, in May 1939, King Abd al-Aziz and visiting dignitaries had been feted aboard the SS *Scofield*. It was the king's first visit to the new American colony. At the pier in Ras Tanura his highness turned the valve on the pipeline and sent off the first tanker full of Saudi crude to pay for all the roads, railways, ministries, prisons, pipelines, and palaces that the new California construction company Bechtel Brothers would ultimately build for the Al Saud.

3 AMERICAN CAMP

Vast investments and huge organizations have expanded the humble efforts of a few score, penniless missionaries, but the purpose remains largely unchanged. As consultant on Arab Affairs for the great Arabian American Oil Company, I have been involved as were my father and grandfather in programs of education, health, imparting skills for vocations, and extending goodwill for the U.S.A., which is the legacy of our pioneer missionaries.

William Alfred Eddy, "The American Invasion of the Near East"

OIL'S NEW DEAL

World War II started on September 1, 1939, with Germany's "lightning" invasion of Poland. Hitler's military had occupied Czechoslovakia earlier in the spring while the troops of his ally Mussolini had conquered Albania. Germany's foes in the First World War, Great Britain and France, tried desperately to head off a second one, turning a blind eye to these early conquests while allying with Poland—along with Turkey, Greece, and Romania—in the hope that Hitler would be deterred. He wasn't. By the end of the first year of fighting, the Axis powers had conquered most of Europe and had taken the war into North Africa. The United States began emergency aid to Britain in 1941, known as Lend Lease, not the least of which was oil, and joined the war formally in December after the third major Axis member, Japan, attacked the American base in Pearl Harbor, Hawaii.

Oil mattered critically to war fighting. The German military with its highly mechanized and airborne oil-fueled offensive forces was unlike anything seen before. Hitler's project in the 1930s for building a strong, war-capable state included conversion from a coal-based to petroleum economy. Germany even began to build a synthetic fuels industry. Hitler's turn on an ally of convenience, the Russians, in 1941, was driven in part by the need to secure the oil supplies of Rumania, the Ukraine, and the Caucasus. Among the basic Japanese motivations for war was the drive to free the country of its dependence on western petroleum supplies, which militarists there viewed as a means by which its enemies could prevent Japan from ever achieving great power status. Thus among the first objectives of the generals in the Pacific invasion was the takeover of the Indonesian oil fields, but Japan's inability to supply its navy with all the fuel it needed during the war remained its Achilles' heel.[1]

President Roosevelt signaled the importance of oil in the war for the United States by appointing Secretary of Interior Harold Ickes to a new position, petroleum coordinator for defense, and then once the United States entered the war, petroleum administrator for war. As a consequence of the creation of the new wartime energy agency, Ickes (pronounced like "icky"), an aggressive, skilled bureaucratic infighter who had headed the New Deal's vast Public Works Administration, would henceforth be known by a simpler title, oil czar. He had taken over at Interior a decade after the Teapot Dome scandal in an office still tainted by Doheny's influence peddling and Albert Fall's conviction on bribery charges. Ickes mistrusted the industry and had made enemies of many oilmen over the past decade. The task of coordinating oil production and supply in war, however, would require the industry's expertise and cooperation, and Ickes reached out to Fred Davies, a vice president of Chevron (or Socal then), the firm that had negotiated the Saudi concession, to join as his deputy and to run the day-to-day operations of the new federal agency, the Office of Petroleum Coordinator.[2] Many more oilmen would follow Davies to Washington.

To win the war the Allies would need to marshal their existing oil resources, especially those in the Western Hemisphere, where there was some spare capacity and supply was secure. The problem was getting the oil from the southwestern and Caribbean producing fields to the British, Russian, and ultimately U.S. forces in two widely dispersed theaters. Meeting the needs of war meant shifting consumption from civilian to military uses, diverting oil tankers from the East Coast to England, and countering the deadly German U-boat campaign in the Atlantic. One of the great engineering feats spurred by the war effort that we take for granted today is the building of the country's first two massive 24-inch (the Big Inch) and 20-inch (Little Inch) pipelines from Texas to New York and Philadelphia, each over one thousand miles long, one for crude oil and the other for gasoline. A political equivalent with even greater consequence was the U.S. government's push into the Middle East, driven first by the need to successfully prosecute the war, but by 1943–44, as the prospect of victory appeared much more certain, by plans for consolidating its power in the postwar era. FDR pushed first and most visibly in Saudi Arabia, where the United States became the new patron of the Al Saud and protector of its domains.

A great irony of the war is that although the Saudi fields represented a vast reservoir of a vital commodity—"the greatest single prize in all history" a State Department analyst called it in 1943—ARAMCO could not get the oil to market. Transport disruptions and the diversion of materials like steel to the war effort forced the firm instead to cap the wells and evacuate the families living in Casoc Town, as the Americans first referred to Dhahran. Employment had peaked in early 1940 when there were 370 American men along with 38 wives and 16 children in the various camps, with the men managing some 3,300 Indian, Bahraini, Iraqi, Saudi, and other non-American workers.[3] One year later, a skeleton crew of about 100 Americans remained along with 3,000 others, although it is remembered in ARAMCO lore as the "time of the Hundred Men" (the others didn't count apparently). Tom Barger left for the States to see his wife for the first time in two years.

The inability to keep production going in turn posed a serious problem for ARAMCO's owners. They approached the Roosevelt administration in April 1941 with the proposal that it loan the Saudi king $6 million as an emergency war measure. Why? When commercial sales of oil had begun in 1939, the firm began to pay royalties in advance of the payments due from buyers, but insisted on a kind of accounting discipline, limiting the advances to what the Saudis would actually earn from the quarter's sales. The war though cut off at once the cash-starved king's two main sources of revenue, his oil royalties and the taxes collected from pilgrims to Mecca. Travel was hardly possible for Muslims or anyone else at a time when virtually all commercial ships and planes had been requisitioned for use in the war. So Abd al-Aziz and his agents began to pressure the company to meet the shortfall by loaning the king money to be repaid out of future sales, only the future didn't look all that bright in 1941. The company's owners turned to the Roosevelt administration to absorb some, ideally the major, share of this risk. The ploy worked.

James Moffett, the chairman of another Chevron-Texaco joint venture, the Bahrain Petroleum Company, and a close friend and advisor to Roosevelt on oil matters by dint of his having raised millions for the president's election, made the pitch directly to FDR on ARAMCO's behalf.[4] Roosevelt agreed to the bailout, but his White House counsel feared a political backlash and ultimately had the British government pay the king $10 million in 1941–42 out of its own massive $425 million wartime loan.[5] While the strategy solved one of ARAMCO's problems it exacerbated another, because the American oilmen increasingly feared, as did at least some of their State Department allies, that the British government would seek to use the war as an excuse to gain control of the Saudi concession. The owners appealed to the administration once more to take over the subsidization of the Saudis, and in February 1943 FDR authorized direct Lend Lease aid to the kingdom on the grounds that it was vital to the course of the war! A first shipment of sixty trucks released from Allied stocks was airlifted with great fanfare to Jidda in American planes to make sure the British allies could not take credit for the delivery. The purpose of this aid, the *New York Times* would soon wonder, cynically, surely was not "to give us a toehold on all the petroleum fields in the world."[6] The U.S. government delivered $18 million in goods and services over the next two years as its share of a joint Anglo-American economic and military assistance program.

Leave aside the very real fact that the owners saved millions of dollars that they otherwise would have had to pay the king without the prospect of repayment in a month or two. It was a gift, sure, and they were banking on more of the same, as I show below. Ultimately the Dhahran Airport would be built for the company with taxpayers' money, and vital shipping space and materials would be made available for its new refinery at Ras Tanura. This latter deal was sweetened by long-term contracts for oil signed with Secretary of the Navy Forrestal, a one-time lawyer for Texaco, at prices 20 percent above those of an already elevated world market.[7] This was not all the firm had hoped for from the government during the war, but what massive, government-assisted corporation ever gets all it believes it deserves?

Discount a little for the padding in which any appeal for private gain in a democracy must come wrapped, such that policy makers may justify their support in the so-called national interest and defend themselves against the objections of many other equally privileged claimants. Then discount a little more for the sad but nonetheless real fact that, in the course of an expensive, six-year-long war during which many in both countries lost their lives, Americans and Britons grew to suspect each other of seeking future advantage at the other's expense. Little evidence exists that British officials planned somehow to dislodge the Americans from the region or that the Americans sought somehow to sabotage the British position there, but the systematic, mutual misapprehension of motives resulted in a kind of Anglo-American rivalry during the war, and many began to fear that the stupendous prize could be lost if the Americans let down their guard.

Alongside these somewhat suspect and exaggerated claims and the firm's finely tuned instinct for rent-seeking, ARAMCO's overtures to the U.S. government and to the ill-fated partnership with Ickes reflected additional and in retrospect well-grounded fears that its $30 million investment might be lost. One is that the outcome of the war itself was unknown, and in the period 1941–42 the prospect of an Axis advance on and defeat of Allied territory in the Middle East was quite real indeed. In the more nostalgic versions of the war years in Dhahran, it is told as a story of a moonless night in October 1940 when an Italian plane bombed the oil installations in Bahrain and, by accident, the eastern Saudi coast, which led to the evacuation of the remaining wives and nonessential personnel. Well, the story is a bit more complicated than that.

In March 1941 a pro-Axis military junta seized power in Iraq, a country that was at once the central node of British power in the Gulf and the most serious potential threat to Saudi control over al-Hasa.[8] By April, British and Jordanian troops (themselves led by British officers) had reoccupied the south and by May the invading forced had restored the Iraqi monarchy. It is during this increased instability in the nearby states that the remaining dependents were ordered to leave Dhahran. Among the women, only two nurses would remain. In the same period, British troops would also be used to secure more reliable governments in Iran in August 1941 and in Egypt in February 1942. Cairo was the center of Allied Middle East operations, including the intelligence-gathering expertise that ARAMCO would later import for its burgeoning Saudi operations. In the bleak days of the summer of 1942, the Germans were massed only a day's march from Alexandria, the British fleet had retreated to the Gulf, and westerners prepared to flee too before the British army's improbable victory against Rommel.

The war's effects on the region remained a problem for foreign investors and the "friendly" Allied occupation authorities—keeping in mind that for many Middle Easterners the Axis "threat" represented the prospect of liberation from colonialism—long after the invading forces themselves had been pushed back. The wartime inflation, the migration from the countryside to cities like Baghdad, Teheran, and Cairo, and the expansion of the urban workforce gave great impetus to the various Arab and Iranian trade union movements.[9] Nationalist politics shifted in more populist directions. All the oil companies operating in the region, including ARAMCO, were beginning to struggle with unions in the oil fields and refineries. Under the circumstances, the firm would have wanted some additional insurance in the form of American backing for the house of Saud against the parties—students, lawyers, and, above all, workers—that were beginning to challenge the monarchies that ruled over the neighboring, more powerful states.

Beyond all these various challenges confronting ARAMCO's owners, Chevron and Texaco, at a critical juncture, another factor drove the firm's negotiations with the consummate New Dealer, Ickes. It had to do with the structure of the world oil market and the owners' competitive positions vis-à-vis their rivals. The same prize that excited the strategists raised a stupendous problem for ARAMCO's parents. It was going to cost millions of dollars to get large quantities of Saudi oil to Europe, the biggest potential market, because they simply did not have sufficient refining and transport capacity, let alone a marketing network on the continent. Europe and most other markets had been parceled out and controlled beginning in the late 1920s by a cartel of the world's largest firms, Exxon, Mobil, Shell, British Petroleum, Total, and Gulf, to use their contemporary names. The same cartel controlled the price and supply from the Middle East–producing fields until ARAMCO began production on the eve of the war. At the same time, with the outbreak of the war, the companies' cartel agreements had come undone.

Through the war years and beyond, factions among the ARAMCO owners and executives fought over two possible courses for the company. The bolder one, pushed most heavily by some inside Chevron, then a second-tier firm, was a risky, go-it-alone path. ARAMCO would go up against the cartel, if it were to be reconstituted after the war, by investing in the necessary infrastructure and winning markets through a price war, using the fantastically cheap-to-produce Saudi oil. The second path open to the firm was to let one or more cartel members buy into ARAMCO in return for access to the world markets under their control. The war gave a momentary boost to those who wanted to use ARAMCO to build a new giant company or what the head of the Federal Trade Commission's investigation of the cartel would later call "the largest supplier of oil in the world."[10] The boost came when the fearless oil czar Ickes opened up negotiations with the owners in 1943 for the government to buy a share in ARAMCO in return for the capital to build a pipeline network and refinery. But opponents inside and outside the company mobilized to thwart the plan, and ARAMCO's owners would instead return to negotiations, interrupted by the war, with cartel members Exxon and Mobil, leading to the sale of 40 percent of ARAMCO to them by the end of 1946.

At any rate, we now have a context for understanding the emergence during the war of the institutions in Dhahran that, for the true believers, makes ARAMCO exceptional in its enlightened commitment to the welfare of the Saudi people, and in its generous and fair labor, housing, education policies, allegedly in contrast to its nearby British rivals. There are three institutions in particular on which these fables depend, which I look at in detail below. The first is ARAMCO's Government Relations Organization, which probably was unique. The second is what most misidentify as ARAMCO's precocious development role or agency on the Saudis' behalf, building model farms and railroads for the people, which was in reality a profit-making contracting (or construction and services) program on behalf of the king, his sons, and a handful of others, such as the finance minister, Abdallah Sulayman. The third is the campaign for the hearts and minds of the media, academics, and related cultural elites at home who ideally would spread the gospel about ARAMCO's "very special and unique relationship . . . with Abdul Aziz, Saudi Arabia's George Washington."[11]

THE BEGINNING OF GOVERNMENT RELATIONS

Tom Barger returned to Dhahran in May 1941 with a new position in a new department, called the Government Relations Organization (GRO). A year earlier he had refused the same offer. To give up geology because he could speak what was then known as Sahb (for Sahib) Arabic seemed the wrong career move for someone hoping to climb the Chevron ladder and eventually get out of Saudi Arabia. He weighed the renewed offer while in San Francisco in early 1941, where ARAMCO maintained its first headquarters until a move to New York after the war. Barger was pretty sure that he "didn't want to spend any great part of my life in Arabia." He would in fact spend the rest of his career there, taking the position after wooing by Vice President Terry Duce, who was on his way to join Ickes' staff in Washington, and by Duce's replacement as head of ARAMCO's political affairs, Roy Lebkicher. The idea of a government relations bureau had emerged in an ad hoc way as men in Dhahran and Jidda were given liaison assignments with the king and his agents. Lebkicher intended to turn it into "a sort of diplomatic service. Picked men, good pay, and good conditions to attract them."[12] Leaving Kathleen in Hawaii and with a baby daughter on the way, Barger set sail for Arabia once more, making it back just as the last dependents were leaving Dhahran.

Government Relations became ARAMCO's equivalent of the State Department, charged with many of the tasks an embassy routinely performs. If it is not stretching the metaphor too far, the organization was even run out of Washington, D.C., where Terry Duce spent an increasing amount of time from the war years until retirement in 1959. If the Government Relations Organization was like State, then the GRO's Arabian Affairs Division, set up in 1946, was like the Office of Strategic Services (OSS), literally so, because it was modeled on the wartime intelligence service's Cairo branch. The connection was preserved as Duce and Barger began to open up positions in Arabian Affairs to newly minted spies from Washington. Bill Mulligan, who spent his career in Arabian Affairs, called Duce and Barger the two "darlings" of the postwar Central Intelligence Agency (CIA). These institutions were only closed down as the phased nationalization of the firm began in the 1970s and as Saudis gradually replaced Americans because, as an internal company memorandum notes, "it is no longer necessary to have a group of people to interpret local customs."[13]

The analogy matters ultimately in answering the question of why the firm took the particular path of constituting a formal diplomatic organization, complete with a research and analysis branch. Other firms negotiate with states and collect political intelligence, but most, perhaps all, do not create their own foreign ministries, at least not since the time of the East India Company. But most firms that operated elsewhere on the world oil frontiers followed the flag to places where their countries' embassies and residencies already functioned. In the case of ARAMCO, the situation was the reverse.

As the doctrine of ARAMCO exceptionalism began to take shape in the waning days of the war and early postwar era, its creators and disseminators would actually downplay what was truly unique about the firm's operations in Saudi Arabia, that is its "relatively elaborate Government Relations organization," and hold up as unique "development" efforts that were found everywhere on the world oil frontiers.[14] As we saw in Chapter 2, virtually all big mining and oil concessionaires undertake a range of ancillary operations as a necessary cost of production. This added cost is an accident of geography. The world oil frontiers are typically located far from urban centers. ARAMCO, like its parents and competitors operating in Oklahoma, Venezuela, or in southwestern Iran, hung power lines, laid roads, and built housing in order to produce oil.[15]

The reason for downplaying the role of the GRO in its company histories' public relations campaigns is simple. While it is possible to find occasional descriptions of one or another ARAMCO official's abiding close, personal relationship with some Saudi principal—for instance, Floyd Ohliger with the king and crown prince, William Burleigh and later Harry McDonald with Ibn Jiluwi, the amir of the oil province, or Barger with Saleh Islam, the Saudi government's representative in al-Hasa—the political ramifications of much of GRO's routine work made it difficult to acknowledge let alone analyze in the frank way that Childs would in his 1947 memo, when he called ARAMCO an octopus.

Like Being the Garden of Eden's First Lady

Many of the services that ARAMCO personnel performed for the royal family during the war, for instance, installing air conditioners in the growing number of palaces in Riyadh and Dammam and the extra-wide running board for hunting fitted on the royal Cadillac, are not what comes to mind when we speak of development. Little wonder that those intent on emphasizing ARAMCO's dedication to Saudi nation building reference instead the experimental farms at al-Kharj, about fifty miles southeast of Riyadh. Not that white elephants like these ever mattered much save to establish the precedent of sinking vast sums into uneconomical projects. The farms were started by the finance minister Abdallah Sulayman, who had imported Egyptian and Iraqi émigrés in the 1930s to run them for the king. With each telling of the story, ARAMCO's minor role grows more expansive. By the 1950s the project was being depicted as ARAMCO's own "private Point Four program for Arabia, long before there was a Point Four program in Washington." Fifty years later, in 2002, the *New Yorker*, legendary for its elaborate fact-checking system, called it ARAMCO's futile effort "to introduce agriculture to the kingdom."[16] Nothing could be further from the truth.

Until the advent of the oil era, most Saudis would have depended either on camel-herding or farming (or a combination of the two) for their livelihoods while a smaller number of urban dwellers engaged in trade. Dates were the famed export crop of al-Hasa and had financed the conquest of Najd and Hijaz by Ibn Saud and his band, although the richest farmland is in the southwestern Asir province, annexed in 1930. One of the best-known events in the creation of the Saudi state was the settlement of the *Ikhwan*, the fighting force used in the conquest of the kingdom, in dozens of new agricultural colonies, *hijrat*, in the 1910s and 1920s. Most farmers practiced small-scale, simple production.

What was likely new in the 1930s was the amassing of large tracts in places like al-Kharj and Khafs Dagrah near Riyadh and in the Hasa oasis and Qatif in the eastern oil province by members of the royal family and their trusted allies. In a remarkable piece in the Hijazi semiweekly newspaper, *al-Bilad al-Saudiya*, in 1949, Abdallah al-Malhuq attacked the problem of what he called "agricultural feudalism." Malhuq worked as private secretary to Muhsin Ibn Jiluwi, amir of Dammam district and among the biggest of the new landowners. The Americans had begun to assist this same small class in Najd, where the king, his finance minister, Sulayman, and the king's brother owned the "bulk of the arable land in the al-Kharj oasis."[17] Foreign Arab technicians had been brought in there originally to install and service pumps to improve irrigation.

The importance of the royal farms to the court increased because the war had simultaneously raised the costs of food and all other vital imports, disrupted shipping lanes, and, as has already been noted, curtailed the ready resources at the kingdom's disposal. The farms might have produced substitutes for the imported foods for the court, but the Egyptian technicians had all but abandoned the project, and the government was pressuring ARAMCO to take over and pay for the operation. Barger and a colleague had surveyed the al-Kharj properties and produced a report about them, which becomes the fact that is recounted to establish an ARAMCO connection, however misleadingly "from the very beginning." What gets lost is the more important fact that the company refused the finance minister's request to take it over. "Ohliger [the ranking ARAMCO official in Dhahran during the war] stated that the company's position in the country is that of a commercial oil organization with a vast amount of work to do along those lines. The organization did not have the men available to properly handle its own work, particularly that of an engineering nature."[18]

Richard Sanger, the State Department Saudi hand, produced a good account of the work at al-Kharj up to about 1949. He makes clear that ARAMCO played a modest role at best, billing any support it lent to Riyadh, which was ultimately paid for by American taxpayers through 1946 and by oil rents thereafter. He documents the primary role played by two U.S. government agricultural missions (never mentioned in ARAMCO-centric accounts) in planning and implementing the model farms project. The first, under that intrepid leader of missions to the kingdom since the 1930s, geologist Karl Twitchell, headed a White House project that brought experts on irrigation and reclamation from the Department of Agriculture and Bureau of Indian Affairs in

1942 for a seven-month study of the country. The two developed the first real plans for large-scale irrigation and reclamation work at al-Kharj and in other parts of the country.

In response to the finance minister's continued entreaties, Ohliger and Government Relations installed a few pumps at one site, Bijidiyah, and supervised the digging of a canal. The firm charged for this work against royalties and taxes as it did all such services, for example, wiring the palaces or farming out a steward to work as the king's chef, although preparing food may not be as noble as growing it, so José Arnold, the cook sent to Ibn Saud, is not usually counted among the new model missionaries. The humbleness of ARAMCO's actual contribution "from the start" is set into relief by the dramatic food shortages of the next year and the emergency dispatch of a second U.S. mission to al-Kharj in December 1944, this time by the State Department rather than directly by the White House. For the next two years Americans working for the regional Lend Lease unit in Cairo took charge of building the showcase royal farms—vast wheat fields, alfalfa for the king's 500 horses, fruits and vegetables for "the people" of Riyadh, and the melons sent by the planeload as gifts by Abd al-Aziz to the rulers of Egypt and Kuwait—using a labor force of some thirteen hundred Saudis.[19]

A State Department assessment in 1947 by Nils Lind, the retiring cultural attaché at Jidda, stressed the great political value that had accrued to the United States by supporting the al-Kharj farms, although his account of the lineup of opposing sides is not the conventional Anglo-versus-American or emerging Soviet-versus-American axes.

> The oil company did not abdicate its dominant position easily. It strove to keep alive its influence on the king through one means or another; but it could not supply him with food. Indeed, the company behaved as if its own government were unwarrantably interfering in an exclusively company area. But no argument was as convincing as food and so it was forced to accept the inevitable, the necessity of recognizing the leadership of its own government in all future negotiations with the king.[20]

Meanwhile, the American who headed the wartime planning organization, the Foreign Economic Administration, in Cairo, had denounced the entire project as a waste of scarce resources, and when the Truman administration declined to continue to pay for the king's "kitchen garden," Abd al-Aziz finally had to pay for his little piece of Eden himself.[21]

Tensions inside the tiny American compound—at its most luxurious in the early 1950s it comprised a single adobe structure with one four-room apartment for the manager and his family and single rooms for the sixteen or so staffers—drove out most Americans in 1948, and ARAMCO's Government Relations Organization recruited outside replacement staff. Two years later, the oil company accepted formal responsibility for the running of the farms, and they became the showcase for its newly created Arab Industrial Development Department under Bill Eltiste.

The claim by ARAMCO to a share in development of the farms in al-Kharj and its offshoots—dairies for the crown prince's son in Riyadh and for finance wazir Abdullah Sulayman in Qatif, another property for Sulayman near Jidda—are thus more defensible if limited to a few years in the early 1950s. ARAMCO's internal audit showed, however, that it was running the farms at a loss. Pricing of goods was arbitrary, as in socialist and planned economies, and if "real prices" (or what economists later called shadow prices) were used the farms "would not be able to show a profit." The audit also revealed that production figures were no more reliable than prices.[22] In this respect the resemblance to other development projects of the era is noteworthy. In 1954, a year after the death of Ibn Saud, the firm basically gave the farms back to the new king to run, while continuing to supply technical assistance for a fee.

As a romance, the story is better—let alone more appropriately—told from the point of view not of peripheral figures like Barger or Eltiste but of the Sonora, Texas, county agent Sam Logan and his wife, Mildred Montgomery Logan, the "first white woman to live in the heart of Arabia." Logan was recruited in 1946 and worked as an assistant for two years before returning to ranching. Tom Barger brought him back as director in 1950 and Sam brought Mildred with him. "Madam Sam" reported on the adventure in a series of articles in the *Cattleman*.

> I feel compassion when I see the dirt, filth, and ignorance that exists among Bedouin people. Children wear gowns so lousy that their hair has to be shaved off, or it is so tangled it could never be combed. Women peer around corners like Halloween characters. Just two peepholes for their eyes in the stiff black masks—or veils, if you feel like being romantic.[23]

The Logans left again in 1952 and in the meantime the project had been taken over by Crown Prince Saud and his son Abdallah. But once Saud became king, he brought the Logans back in 1957 to try one more time to revive the farms. Mildred reported to her *Cattleman* readers that "I like being the Garden of Eden's first lady." There was increasing trouble in paradise, however. The Texans and the ill-fated al-Kharj venture were victims of a power struggle between the brothers Faisal and Saud. The winner, Faisal, gave the farms to a then-rising Saudi contractor named Mohammad Bin Laden to run. Bin Laden was apparently no more eager than ARAMCO had been to operate a loss-making venture, which led to protests by Logan and in June 1959 to the expulsion from Eden of all the American personnel.[24]

All on Board

The identical pattern is found in the other showcase modernization scheme that reporters were shuttled to in the 1940s and 1950s, the railroad that was built for the king from Dammam to Riyadh. He was apt to call it "the crowning glory of his reign" when trying to get Americans to pay for it. Once built, the oil company pointed to it as additional proof of the king's greatness and ARAMCO's own dedication to Saudi development, and it is commemorated in post–September 11 Saudi-American collective memory as ARAMCO's railroad. The reality is that it was government-owned, that the firm refused to build it—wisely—first because the we-are-here-to-modernize-the-kingdom line had left it vulnerable to appeals to loan the $50 million it would cost, and second because the feasibility studies made clear that the venture would run at a loss. The Truman administration had therefore turned the project down—twice. Prince Saud's 1947 trip, when he ran the little model railroad at dinner, was a failed effort to secure an Export Import Bank loan for the full-sized model. The problem was that the U.S. government had drawn the same conclusion as the firm and its consultants, the Western Pacific Railroad and Bechtel. Trains were a mistake. The country would be better served by a highway system, and time proved them correct. A highway system was adopted in the long run, but not before the Bechtel Brothers were hired to bring Riyadh into the railway age.

The king was, of course, sovereign, and he calculated costs and benefits differently from the firms. He came down squarely on the side of a railroad for two reasons, he told the Americans. One is that by the end of the war and in the face of increasing demand from a suddenly booming capital, the state had not yet found a way to master the logistics of truck transport, in which case an expanded highway system was a mistake. The royal fleet of over 150 trucks intended to run from Jidda to Riyadh were instead being run into the ground. A new British Transport Mission had tried to untangle the wreckage left by Palestinian profiteer Izzadin Shawa in the Bureau of Mines and Public Works in order to begin reliable deliveries of gasoline and other goods to Riyadh, but its members all quit after six months. In what was probably the first case of privatization in Saudi Arabia's history, the garage and its fleet were turned over to clients of the finance minister who owned the National Arabian Motor Car Company, the firm that monopolized pilgrim transport.[25] Little wonder, therefore, that as St. John Philby and others reported, Ibn Saud regarded trucks "as enemies, like the Germans and Russians."

The second reason the king gave for wanting a railroad is that highways provided advantages to enemies close by, leaving his capital vulnerable to invasion. "A railroad could be controlled—and if need be—blown up." He had flipped the same argument on its head in a meeting with Harold Hoskins, the World War II intelligence agent and chairman of board of the American University of Beirut, who was doing a risk assessment for Exxon in 1946 as negotiations to buy ARAMCO were under way (kept secret from the king). Abd al-Aziz told Hoskins that the railroad was a key means to weld the country together and "prevent the old regime of inter-tribal strife" from again destroying the Saudi state. Some guessed that symbolism may have also played a role in that the other great Arab king, Farouk, had a railroad, which provided Abd al-Aziz with his first train ride when he visited Egypt.[26]

Royal persistence paid off, at least with the Bechtel Brothers, the firm that was in line to earn millions in contracts for harbors, electric power stations, airports, and ever more palaces. After a conference with Abd al-Aziz in Riyadh, Bechtel's vice president, Paul English, one of the original naysayers, was won over, according to the U.S. embassy.

English seems to think that the "economic idea" is highly over-rated, stating that practically every railroad ever built was considered uneconomic in the beginning. He asked: "What great difference does ten million dollars make so long as the United States' best Arab friend gets what he is convinced is best for his country? After all, I think the king understands his people better than anybody does."

Bechtel's cost estimates suddenly dropped by $20 million. The Truman administration remained unconvinced, turning down repeated Bechtel-ARAMCO-Saudi requests for loans, and by the end of 1947 ARAMCO had agreed to pay for the king's railroad out of future royalties.[27] The oil company's hand was forced after it began to plan to build its own short line to service the Dam-mam-Dhahran-Abqaiq area. The king insisted on state ownership of any rail-way built in the kingdom while noting that ARAMCO's views on the economic viability of railroads must have changed.

As the war wound down, ARAMCO's owners were trying to move the firm off the path that the exigencies of the time had led it, and to withdraw as much as possible from work not directly related to oil production or outside the boundaries of the oil province. It was one thing to perform services for the palace when the royals claimed to be going hungry and the hundred men had time on their hands. The return to production in 1944 posed a dilemma. As the U.S. consulate put it, "Ohliger and MacPherson [the two managers of ARAMCO's Dhahran operations] were virtually resigned to becoming an 'East India Company' and to creating and staffing a new unit in ARAMCO tantamount to a 'Department of Government Management' (although such a name would probably have been carefully avoided)." San Francisco stopped them and turned instead to Steve Bechtel, who had been building pipelines for Chevron since the 1920s and by the 1930s had spun off one the world's first full services firm for the oil industry, Bechtel-McCone, and later International Bechtel (for the Middle East). They built Chevron refineries and laid pipelines from Richmond, California, to Caracas, and in 1944 were hired to build the re-finery in Ras Tanura. The entry of the Bechtel interests into the Saudi market resolved one dilemma while creating another one for those at the time who sought to cast ARAMCO in a kind of heavenly light and those who continue to look for its glow now.

THE ORIGIN OF A MYTH

Phil McConnell, a Texas oilmen and dedicated diarist, one of ARAMCO's hundred men of the war years that, he told the novelist Wallace Stegner, were the happiest of his life, took a moment in October 1944 to invent what I have been calling the idea of ARAMCO exceptionalism. The fortunes of the war were turning rapidly in favor of the Allies. ARAMCO oil would be critical to postwar European reconstruction. McConnell and the other senior staff in Dhahran were gearing up for the return of production, a mass of new hires, and a building frenzy, when the first attacks on the newly renamed Arabian American Oil Company appeared in the U.S. press, led by the *New York Times*. Texaco's president, Star Rodgers, told his contacts in the White House that the firm's enemies, Exxon and Mobil, had gotten to the *Times*.

Oil czar Harold Ickes' audacious project to have the American government enter the Middle East oil business was unraveling. He had to end negotiations to buy ARAMCO and his advisor James Terry Duce had to quit his position as director of foreign operations for the new Petroleum Administration for War in November 1943. Critics protested on the grounds that Duce was being paid a salary by Texaco while working for the government on the ARAMCO deal. These negotiations had ignited a firestorm of opposition inside and outside government. Every oil company in the country, most not friendly to the New Deal to begin with, joined forces with allies in Congress to defeat FDR's apparent socialist turn. The domestic producers feared that cheap Middle East oil would soon be flooding the U.S. market and driving them out of business. The giants, Exxon and Mobil, wanted a piece of ARAMCO for themselves rather than face a new, powerful government-backed competitor. The left (the "*New Republic* crowd") saw big oil once again pulling the strings in Washington.

Ickes though fought back. He and ARAMCO's owners concocted a new plan to build a trans-Arabian pipeline from the Gulf to the Mediterranean. The government would pay for it as a wartime emergency measure for the European theater. The only problem was that all the parties knew that the war in Europe was ending and the project would take years to build, hence the rush to sign the deal and the mobilization to try and thwart it. By the summer of 1944, the enemies of the pipeline deal had gained the high ground.

The *New York Times* hammered on two basic points. One was that the war had transformed America's role in world affairs, as could be seen on the streets of Washington, where "groups of strangely-clad individuals from far-away places" were now to be found, including those from the "little known Arabian kingdom . . . , the capital of Mohammedanism."[28]

It is evident that the Near East is no longer a remote region inhabited largely by backward Arabian nomads whose destiny is no concern of ours. Through a quirk of technology and economics the problems of the Near East become our problems, just as Mexican and Venezuelan oil problems become ours. The good neighbor policy must clearly be extended to a remote region. That policy will be put to the test when this country decides how and for whose benefit the Saudi Arabian oil fields are to be exploited.[29]

The *New York Times* argued that wartime involvement in the region was also leading some Americans who now knew better to revise the early and admiring portraits of "the Wahhabis as the Puritans of Arabia."

The thing [oil] may still be a novelty to the loyal subjects of the king of Saudi Arabia. Perhaps even the king himself and his chief counselors may be a little dazed by their new oil prosperity, but they are no doubt learning rapidly. Arabs are probably as quick to catching on to the enjoyment of wealth as the Cherokee royalty owners of Oklahoma were in acquiring a taste for Packards and Lincolns.[30]

The more important consequence, and the second point hammered on by the high priests of Atlanticism on the editorial board, was the prospect that the United States was moving backward to the "old imperialism," the era of "dollar diplomacy and economic imperialism."[31] If so, the editors argued, then military bases were likely to follow, although the government denied it.

The paper ran a front-page story on the oil scion and close ally of FDR, James A. Moffett, who was now calling for Ickes' impeachment for his role in the proposed government pipeline project, "the biggest scandal I've ever seen." Moffett had resigned as chairman of the board of ARAMCO's twin, the Bahrain Petroleum Company, over the New Dealer's plan to buy into the oil business. The *Times* took his side, calling instead for a new collectively developed oil regime under the auspices of "some form of international commission" or "world organization, with the United Nations as the initial nucleus." Ordinary Americans called for charting a more independent path, one that remains popular six decades later. "Why does it [the U.S. government] not use the $135,000,000 to $165,000,000 estimated for Saudi Arabia in financing the development of oil deposits on the north and south coasts of Alaska?"[32]

Back in Dhahran, McConnell began to craft the historic response to the charges that "Saudi Arabia is just another chapter in the ancient game of Yankee imperialism" and that dollar diplomacy ("the hated term") was back in vogue. He had discussed his plan with ARAMCO's president in San Francisco. They needed to counter "the welter of published misconceptions" about the pipeline project with "a reliable picture of what the Company actually is doing to assist the Saudi Arabian Government to develop its country." To that end, McConnell circulated a document titled "The Saudi Arabian Partnership," and invented the idea of ARAMCO as a dedicated agency of development in alliance with the visionary Ibn Saud.

The king has built his country within his lifetime, shaping it with the sword held in his good right hand. . . . He has cared for it and has held it by his own strength and wisdom. . . . He is no furtive sheikh, swiftly raiding and as swiftly withdrawing. He has been first a general, then a diplomat, and finally, an administrator who has exhibited a keen sense regarding the relation between his desert people and the march of the world around him. Where others have talked for generations of a united Arabia, he has set forth to accomplish it.[33]

The same 1944 campaign that led the owners to craft a specific identity for the firm also moved them to change its name—from California Arabian to Arabian American Oil Company or ARAMCO—and Casoc Town became American Camp.

D-Day

In June 1944, Colonel William Alfred Eddy of the Office of Strategic Services was named minister plenipotentiary to oversee the kingdom's transition from, he hoped, British protectorate to American partnership. Looking back on his life, he would recall the moment as the beginning of the "American Invasion of the Near East" in fulfillment of the commitment made to the Arabs by the missionaries of his parents' and grandparents' generations.

As the North Africa campaign began to wind down, Eddy moved to Jidda as special assistant to the American minister before taking over as head of the legation. In 1946 he left Jidda for Washington where he assisted in reassembling the pieces of the wartime OSS, first in the State Department and in 1947 in a newly created Central Intelligence Agency, where he remained for the rest of his life. Ever since, Eddy's cover story has been treated as fact by diplomatic historians—that he had resigned his government position in protest of Truman's capitulation to the Zionists and joined ARAMCO.

Eddy's wife, Mary Garvin Eddy, joined her husband in Jidda in late 1944. "It seemed as if the whole white population and many natives were there to meet the plane—it is always an event in Jidda." She wrote regular letters home from the front as Eddy did battle with his British rival Stanley Jordan and Jordan's replacement, Laurence Grafftey-Smith. The British minister held rooftop parties, and at the first dance Mary attended there were six women. "Everyone seems to enjoy themselves—all were in evening dress—I think they do it so much here just so as to cling to civilization."[34] The Eddys opened the rooftop of the U.S. legation one night each week to allow Saudis to mingle with the small foreign community and watch American movies. When FDR died in April 1945, they screened a documentary of the president's meeting just two months earlier with King Abd al-Aziz, when Eddy was translator.

Mary made her first trip to Dhahran the same month. By then ARAMCO was running a regular weekly flight across the kingdom. "The oil town at Dhaharan [sic] is just like a bit of U.S.A.—modern air-conditioned houses, swimming pool, movie theater etc. Six American wives have already arrived and more are on the way. About 1000 men (Americans) are employed and about 10,000 Arabs. This is only the beginning of a tremendous expansion."[35]

The invasion had indeed begun. There was a kind of seamless transition in 1944–45 from a focus on the oil resources in the Gulf in support of the war in the Pacific to the role of Saudi oil in the postwar world economy. The latter issue was a complicated one, the British Foreign Office had warned, when it still imagined playing a significant role in the kingdom jointly with the Americans. "It is important to remember how undeveloped the country is, and to prevent the impact of the sophisticated races of the West from spoiling an Arab race by rushing it too rapidly through a period in its history which in other countries has occupied about a thousand years."[36]

The Americans weren't looking back. They opened a new consulate in Dhahran. A showcase agricultural mission was growing food at al-Kharj for the king's palaces. Thousands of men and their families began migrating to the al-Hasa coast, where the U.S. government had assisted ARAMCO in building the kingdom's first major refinery as an emergency war measure. With Eddy's help, International Telephone and Telegraph had ended the British imperial communications monopoly. The British interests protested, but by 1946, Transworld Airlines (TWA) pilots would be flying Abd al-Aziz's planes under contract and starting a national airline. The *New York Times* had also guessed right back in 1944. A secret agreement with the king was signed and ground broken in August 1945 on a U.S. military base in Dhahran, another war measure, although the war was effectively over and the War Department no longer wanted it.

The course of negotiations and the rush to begin building the base are a reminder of the rapidly shifting balance of forces in the Pacific theater in early 1945. In March the Joint Chiefs of Staff had determined that a base in Dhahran was critical to the war effort, and the machinery of state began turning to get the White House's approval for negotiations with the Saudis. After Roosevelt died in April, the issue of aid to Saudi Arabia was one of the first to reach President Truman's desk in May, but the State Department and the Navy were frank about the dilemma they and their ARAMCO allies were confronting. "The problem is that as war winds down, the base seems less important, or only as important as the rapidity with which it can be constructed." The War Department was no longer going to pay for it, certainly, and so it pushed for the base issue to be included in the negotiations for a long-term aid agreement being sought by the king.

The evolving in-house definition of the national interest in the Dhahran base deserves highlighting. A key memorandum's original specification argued that the greatest immediate danger to the oil fields, "now in American hands," was the possibility that "one of the other powers involved might well decide to occupy the oil fields in order to prevent occupation by a rival power. The mere existence of an American military airfield at Dhahran would contribute to the preservation of the political integrity of Saudi Arabia and to the maintenance of our interest in the oil fields." This formulation was changed and the policy became one of protecting ARAMCO's concession.

> The continuance of that concession in American hands holds out the best prospect that the oil of Saudi Arabia will be developed commercially with the greatest rapidity and upon the largest scale, producing the revenues which will contribute to the betterment of the economic condition of Saudi Arabia and, in consequence, to its political stability. The manifestation of American interests in Saudi Arabia in addition to oil will tend to strengthen the political integrity of Saudi Arabia externally and, hence, to provide conditions under which an early expansion of the costly development of the oil concession can be proceeded with. The immediate construction by this country of an airfield at Dhahran, to be used for military purposes initially but destined for an ultimate civil utilization, would be a strong showing of American interest.[37]

Eddy received the green light to begin negotiations with the king on a comprehensive aid package in June, and Faisal arrived in Washington late in July to continue these talks. The Saudis learned that the large development loans they sought for the railroad and other works would require the invention of new institutional mechanisms, work on which could only start after the war. The principals would return to the question of aid in 1946. Meanwhile, Lend Lease aid would be extended for one more year. The trade-off, the Americans learned, was that a base agreement, for facilities that the U.S. government would pay for, the Saudis would own, and the Air Force could use, would be limited to three years following the war. The deal would then have to be revisited. Assured that the clause recognizing the right of base personnel to have "normal facilities for personal recreation and self improvements" did not include prostitutes, the king agreed to the presence of U.S. military forces in Dhahran on August 8. Eleven days later, the War Department determined that since the airfield would be "of doubtful military usefulness," State and the White House would have to get congressional approval for this extraordinary use of funds.[38] The dissembling to lawmakers may not have been the decisive factor in securing approval, but State acted as if it were.

The strategy would include obscuring as much as possible the reality that construction of the base was only beginning in the fall of 1945 by dating the process back to 1944 when the idea of the base was first conjured up ("preparations were started") by the Joint Chiefs. It was agreed to emphasize British efforts to hinder American military and civilian air transport, and the fact that some military personnel in China, Japan, and Iran were likely to remain in the Pacific theater for the foreseeable future. ARAMCO was assisting the U.S. military, not the other way around, and the royalties it paid the king were vital to securing the kingdom's stability. Inquiries by congressmen in response to constituents' complaints continued to be fended off with improvisations around the "general line," as, for example, when a soldier complained that he had been transferred from Cairo to Dhahran in October to build an airport for the benefit of the oil companies and TWA. "There was no connection between the construction of the airfield and the protection of the interests of the oil company," although, as we saw, the opposite was articulated as a policy statement by War and State in June. The line for the public was that the Americans working in the oil colony rather than the company that employed them were the true beneficiaries.[39]

There really was no choice but to lie outright when the *New York Times* reported on the policy of discriminating against Jews for work on the air base and the first of many rounds of protests began. So Gordon Merriam at State would tell protesters in 1945, "You will [be] pleased to know that there is no truth to the charge that nationality or religion are conditions of employment on the project." The dean of State Department anti-Semites, Loy Henderson, was even surer in 1946. "Insofar as the Department is aware, there is no agreement to which any agency of the Government is a party prohibiting the employment of workers in Saudi Arabia on the grounds of nationality or religion." Compare these typical claims to the warning Eddy sent to Washington when a Saudi official berated him for recommending a *National Geographic* reporter for a visa who turned out to be a Jew. He urged increased vigilance on this score and repeated the gist of a key October 1944 dispatch.

To prevent a rebuff to an applicant in the future as well as to avoid embarrassment to Jewish or Gentile Americans seeking entrance into Saudi Arabia, whose religion should not be subjected to official inquiry, it is recommended that all interested government agencies and private concerns planning to send personnel into Saudi Arabia be advised confidentially that the Saudi Arabian Government does not, at present welcome Jews into the country.[40]

The custom had already caused its overzealous guardians in ARAMCO some difficulty once shipping channels had been reopened and traffic through the Gulf was increasing. In December 1944 a member of Government Relations boarded a merchant marine vessel, the S.S. *Bellows*, in the harbor at Ras Tanura and warned the purser that under Saudi law Jews *and Negroes* were not permitted on land. The Congress of Industrial Organizations lodged a protest with the State Department, and officials at Dhahran who investigated the incident reported of course that no such law existed, that the Saudi government preferred not to be "put on the spot" by being forced formally to define its attitude with regard to Jews, and "as far as negroes are concerned, it has no attitude toward them whatsoever, for many of the Arabs have negro blood."[41]

The relative bluntness of the Saudis nonetheless created additional obstacles for their friends in the State Department and ARAMCO who had begun to draw the distinction between bad Zionists on the one hand and good Jews on the other. It took years before Riyadh came to appreciate the usefulness of this formulation. A reporter in Cairo seeking a visa to the kingdom in 1946 might be told simply that Shaykh Yusuf Yassin, the deputy foreign minister, "refuses to speak to you because you are a Jew." If he persisted by asking for a visa to "visit the American oilfields," he would be turned away summarily. "No Jews are allowed in Saudi Arabia."[42] The general manager of TWA's operation at the Dhahran Air Base where, by 1949, 92 percent of the work was devoted to ARAMCO and related civilian business, and only 8 percent to the U.S. Air Force, had underscored a particular difficulty in the way of running a commercial airline there.

As for Jews, a Jew is a Jew in the eyes of the Saudis, regardless of what passport he may be carrying. As Yusuf [Yassin] explained to us in no uncertain terms, Saudi Arabia is a holy Moslem country and no Jew is going to be allowed to pass through it and to contaminate it. They realize such an attitude is tough on international carriers but they just don't care—they don't want any Jews setting foot on Saudi Arabian soil, "period!"

Readers will appreciate the several levels of irony revealed in the past few pages. Most obvious, and not a little sad, the Allies had just fought a war to defeat a regime dedicated to the extermination of European Jewry. The full horrors of that murderous project could no longer be denied, yet here were oilmen and Arabists bending over backwards to accommodate a reactionary like Yassin with his "a Jew is a Jew" line. Yassin was, by the way, positively enlightened in comparison to some of the Wahhabi faithful in Najd. Ten years later, when he toured the pipeline complex in Lebanon and the oil fields in Saudi Arabia, Wallace Stegner would take special note of the prejudice against Jews among the ARMACO Americans, the ease with which they sprinkled their conversations with one Arabic word in particular, "yuhud."[43] We should not be surprised by the anti-Semitism inside American Camp nor imagine that it was due to the Saudi king having them over a barrel, so to speak. If you are unconvinced and think the Americans' attitudes are explicable most basically in terms of a growing conflict between Arab and Jew, there is still the fact of the ARAMCO official banning both Jews and Negroes from setting foot in Ras Tanura, and no one argues that white supremacy followed from Zionist colonization of Palestine. By the way, it is not clear when, if ever, the first African American joined the company and set foot inside American Camp.

Then there is the matter of ARAMCO's effort for the first but not the last time to set the record straight, as McConnell imagined it, after opponents began to mobilize against the firm's rent-seeking and the U.S. state's first steps to extend its power over the peninsula. The effort involved two distinct arguments: what ARAMCO was doing to assist the king and his country (and why the king, a slave-holding despot, warranted U.S. support), and what benefits accrued to the American people. Let's consider each of these in turn.

Much of the air is let out of ARAMCO's inflated claims about aiding the kingdom simply by reading the archival record carefully and reporting it accurately. The model farms were disasters, but ARAMCO had relatively little to do either with setting them up or running them. Perhaps future historians will tell it as a story of a firm's wise stewardship over scarce resources, even before the American government pulled the plug on aid for the king's private estates. The small railroad the company paid for was for its own use for transport from the port to the fields. It was a cost of production, similar to the refinery (that it was hoping the U.S. government would pay for), dock works, and so on. Mining firms built these regularly as part of the process of getting minerals to market without the pretense—at least until the 1910 revolution in Mexico—of claiming to be doing so on behalf of other people. The bigger railroad that the king bought against ARAMCO's recommendation was just one of many projects farmed out to Bechtel and others. Credit though is due to ARAMCO's public relations operation, then just getting off the ground. The company charged against future royalties for all services performed on behalf of the house of Saud in installing pumps at al-Kharj, in wiring the palaces, drawing boundaries, hiring chefs, and more, but then took to describing some of this work as a form of charity (and the rest it attempted to hide from view).

Most remarkable of all is the argument that all this modern-day missionary work was somehow "free for the American people." ARAMCO's owners managed to shift to the U.S. state responsibility for almost $20 million (some $190 million in today's terms) that the king otherwise would have borrowed from the company to be repaid once royalties began rolling in again. Had ARAMCO paid it, we would have called this additional cost risk, and the firms that we conjure in textbooks assume risk almost by definition. In the real world, however, firms like ARAMCO prefer rent to risk. The owners went for broke during the war. Before negotiations with Ickes to buy a piece of ARAMCO collapsed—and we still don't have a fully satisfactory account of this episode—the price that the owners had negotiated would have meant the recovery of the entire cost of their ten-year-old venture. Not bad. The owners failed though in their bid to have the New Deal state pay for the Ras Tanura refinery and Trans-Arabia Pipeline (or TAPLINE). The Dhahran airfield was a kind of consolation prize.

To repeat a point, in a democracy all such favors must come couched in an argument about the benefits to the "nation," the "people," or the "national interest." The owners and their allies advanced two in particular. One was the idea that in the absence of decisive action by Roosevelt, a British government fighting for its survival against the Germans and dependent on the United States for vital support might, nonetheless, try to take over the concession. The second was that the United States, the single largest oil producer, supplier of about half of the entire world's needs, was rapidly running out of oil, the same argument made at the end of World War I, when the largest and most powerful American firms began exploring overseas in earnest. America needed to conserve its limited domestic reserves in the event of another war and so the state should do all it could to bring foreign sources to market, in this case the vast reservoir in the Persian Gulf for which ARAMCO held the concession. Opponents, needless to say, rejected both these claims.

The United States did not build the refinery and pipeline, and the only consequence was that ARAMCO had to raise the capital to do so itself. We can't say for sure that the argument about British designs on "our" oil was wrong, although no one has found evidence yet to suggest that such plans existed.[44] Since the Roosevelt administration accepted the challenge of funding and arming the house of Saud, what one finds in the Foreign Office archives is a massive paper trail left by outraged British officials whose accounts of U.S. "economic imperialism" read much like the *New York Times* editorial from the same moment. Colonel Eddy's nemesis in Jidda, the British ambassador, Laurence Grafftey-Smith, at one point pleaded to Whitehall to resist the surrender to American power: "This is not Panama or San Salvador."[45] The point being that American domination of Saudi Arabia was moving rapidly in that direction. McConnell tried to dispute it, as we saw ("the hated term" dollar diplomacy), but his reasoning was uninspired. Back before World War I, the Taft administration pressed for orderly development of foreign nations through the mechanism of private capital, mining firms and other raw materials producers, banks, and public utilities, rather than colonial-style occupation armies, or at least that was the theory. The best evidence of continuity between the two eras is not in the actions of the state so much as in those of the firms, visible in the layouts of the camps and canal zones, and in the organization of work and life inside the oil fields.

The Frontier Closes

Wallace Stegner ended his history of the ARAMCO pioneers with a chapter on the late war years titled "The Frontier Closes." "American involvement in Middle Eastern economic, cultural, and political life," Stegner wrote, "would grow deeper, more complicated, and more sobering. . . . But that is another story. This one is purely and simply the story of a frontier, and the return of seven war-exiled wives to Dhahran . . . in February 1945, is as good a date as any to mark its passing."

Looking backward from Dhahran, where he had visited in 1955, and Palo Alto in 1956, where he banged the book out in thirteen exhausting weeks, the device—the turn to this set piece of much American Western myth—was useful and perhaps even fitting. ARAMCO did not want the politics of the postwar years discussed—we will soon see why—and Stegner had to write the book in a hurry. He was under contract. The wise and forward-looking warrior king Abd al-Aziz was still alive, if somewhat frail, in 1945. The Suez Canal was still in foreign hands. Nasserism in Egypt had not yet been born, and it would be another decade before Arab radicals in Cairo, Baghdad, Beirut, and Dhahran began to champion indigenization of the western oil firms, following the early lead of the Mexicans, Iranians, and Venezuelans. No doubt, Stegner had been sharing drafts of the thick brew of romance and nostalgia with the self-styled pioneers, the band of the hundred men, Phil McConnell chief among them. Stegner was handsomely paid as well. So, he ends *Discovery!* with a toast. "They were building something new in the history of the world: not an empire made for plundering by the intruding power, but a modern nation in which American and Arab could work out fair contracts, produce in partnership, and profit mutually by their association." Nonetheless he must have known, because it is recorded in McConnell's diary, which he had read, that just a few months later, in July 1945, the Arab partners rose up against the miserable terms that had been offered them.

4 THE WIZARDS OF DHAHRAN

> *But we must hasten on our journey. This that we pass as we leave Atlanta is the ancient land of the Cherokees—that brave Indian nation which strove so long for its fatherland, until Fate and the United States Government drove them beyond the Mississippi. If you wish to ride with me you must come into the "Jim Crow Car." There will be no objection,—already four other white men, and a little white girl with her nurse, are in there. Usually the races are mixed in there; but the white coach is all white. Of course, this car is not so good as the other, but it is fairly clean and comfortable. The discomfort lies chiefly in the hearts of those four black men yonder—and in mine.*
>
> **W. E. B. Du Bois, *Souls of Black Folk***

AN AMERICAN DILEMMA

Early in January 1944 ARAMCO's senior managers in Dhahran, including Bill Eltiste, William Burleigh, Roy Lebkicher, and Floyd Ohliger, met over a series of days with Vice President James Terry Duce, who had arrived from Washington, to try to formulate policy for dealing with the influx of labor and supplies for the new refinery in Ras Tanura. Tensions were rising inside American Camp. Housing was unavailable for all the new recruits. In addition, "the subject of wives is becoming one of the big personnel problems of the camp." The wartime American skeleton crew of a "hundred men" had rapidly grown to a thousand, they missed their families, and now were faced with the prospect of making room in their cramped homes for the construction crews. Recruiters had misled the new arrivals about when their own wives would be arriving in Dhahran—two years or more, rather than just two months, later. Outside the fences, meanwhile, the Arab workers were beginning to take matters in their own hands and settling their families alongside squatters living northeast of

Saudi Camp, in a place they named Nahadin.¹ Not surprisingly, therefore, the longest and most divisive debate among the planners concerned the inequities in the treatment of the Saudis and all other non-American employees.

Roy Lebkicher, who had come to Dhahran as Government Relations head, ought probably to be remembered as the first to question, however tentatively, the inequalities underpinning labor relations. In the January 5, 1944, meeting he argued against those who wanted Nahadin's settlers removed to someplace like Thugba, a desolate spot miles away from American Camp. Instead, he said, the company ought to plan for a permanent Arab settlement in Dhahran. "If the idea of an Arab town for Ras Tanura is considered sound, the idea is equally sound in the Dhahran area." The problem with Saudi Camp was that its existing palm-thatched huts or barastis could not accommodate all the new workers, Arab families were forbidden to live with the men inside the compound, and most of the workers were too poor to rent housing nearby. Lebkicher's point was sharp. "As long as we provide free housing we are going to be faced, one of these days, with the question of why we furnish the Americans with better free housing than we do the Arabs." Duce agreed that Arab workers had as much right as the Americans did to live with their families. So Lebkicher argued for a change in direction for housing policy. "The best way to avoid this problem is to discontinue free housing and to attempt to put housing on a more natural basis on which all have to pay for their housing and their type of housing will depend on income rather than nationality."

Lebkicher had gone too far, apparently, and his comrades dragged him back with reminders of the enormous obstacles to any such dramatic alteration in the existing housing regime. British Petroleum had built native-worker housing in Iran in place of more rudimentary camps, but costs had grown enormous. If local Arabs had the right to live with their families, then the firm would ultimately have to pay for and accommodate the families of its higher-skill Indian and Hijazi clerks ("one point for which the Indians have been agitating for some time").

Other pressing problems for which they had "been sort of marking time in the past two or three years" demanded attention, a decent Arab hospital for one. Another, the focus of this chapter, was a start on the education and training promised in the original concession in 1933 and reiterated in the 1942 Labor Law, the first major piece of legislation designed to secure some minimum standards in the oil industry, but ignored by the Americans. The depth of ARAMCO's commitment to hierarchy meant Lebkicher had no answer to the question raised about his proposal to settle Arab workers and their families in the vicinity of American Camp. "If we were to build a modern Arab city with all conveniences at the Saudi Camp we would be faced with the problem of employees living in Dammam and Al Khobar wanting to come to live in such a city."

We need to be clear, as well, therefore, about the distance even a relatively enlightened man such as Lebkicher still had to go. The only point about which there was no dispute, no dissent, and no doubt raised was that no Arab, whatever his skill level and seniority, should be permitted to reside inside American Camp itself. "The problem of importing Egyptians, Iranis [sic], Syrians, etc., as teachers brings up the question that if such a man is good enough to be a teacher he is probably high enough up the social scale so as to want to move into the staff camp. Such a thing would be undesirable and should be avoided at all costs."

At the end of the day, only one decision had been taken, or at least registered in the minutes. Duce requested that a plan be drawn up for building barastis for employees residing in Nahadin "in order that this 'sore eye' on our treatment of non-staff employees may be removed." Some months later, ARAMCO would begin to hire single American women for secretarial work, a solution that Duce and the others had considered for "the clerical problem," by which they meant "the higher classed Hejazi, Syrian, Iraqi, etc." worker, "the biggest troublemaker we have." By troublemaking Duce meant a propensity to agitate for equality with the Americans. Down the road, following strikes in 1945 and 1947, these new "American girls" would complicate the otherwise neat change ordered by headquarters: American Camp was to be referred to as Senior Staff Camp, Indian Camp as Intermediate Camp, and Saudi Camp as General Camp, as if all along who lived where and with what degree of luxury had been a matter of skill and not race. The intermediate-skilled white women, all of whom lived in senior staff camp at a dorm (building 1225) known as "Hallowed Square," were, well, an exception.

An additional strategy for reducing the troublemaking in the short term, one that had not come up for discussion in the January planning meeting, was in place by winter 1944, as ARAMCO imported approximately 1,700 Italians from Eritrea. Phil McConnell's journal supplies key details. Negotiations with Saudi authorities over terms for the Italians would be complicated, he recorded in November, but it was assumed that they would live more or less as they had been accustomed to in Eritrea while filling multiple jobs in addition to master masons. "They will bring doctors, dentists, male nurses, kitchen help, cobblers, barbers, etc. The problem of wives coming too is still in the air. We would like to have them to wash, wait on tables, stenos, etc. The groups will want musical instruments and movies, which will take considerable negotiating with the Saudi Arabian Government."[2] The tricky part would be fending off objections if the disparities between the two races—Italians and Arabs—proved too glaringly unfair, but it is clear that privileges of some kinds were being contemplated.

One month later, McConnell wrote that things had changed. "Their handling will be different. We don't dare show them any considerations than those given the Arabs, as the Arabs already have protested to Sheikh Abdullah Suleiman that we are hiring men who take jobs from the Arabs." The protestors were correct, of course. That *was* the point. McConnell continued with his account, for which we are indebted:

> The king allowed the Italians to enter against his better judgment and he might kick them out over night, if he learned that they were getting better food or living conditions than his people. Hence, food must be simple *and the housing poor*. We plan to place floors in the tents of skilled Arabs first, then in the tents of the Italians. We must discourage, perhaps prohibit Americans from visiting the Italians. The Americans will start feeling sorry for the Italians and will give them articles. We cannot allow it. One of the first requirements is a high fence separating the Italians from both Arabs and Americans. At first, Italians will work separate from Arabs. Later, we may be able to combine them, with caution, depending on how the two races get along. The Arabs haven't forgotten the atrocities of the Italians just across the Red Sea in Eritrea.[3]

In this passage, the formulation of the problem facing managers—anticipating a king's adverse reaction to learning of some kind of inequity—is quite different from what ARAMCO agents would soon begin to insist to workers and to prying outsiders. They would claim that their hands were tied by an explicit agreement with Ibn Saud barring improvements for any group of workers (except Americans!) while his own people were made to do without, say, running water, bare light bulbs, floors, decent food, water, and the like. Maybe some actually came to believe in the existence of such a strange compact. It is nonetheless extremely hard to believe that the Americans' option for treating all with equal injustice—"hence food must be simple and the housing poor"—is the one that any government official would accept as the best means for fulfilling a principle of just treatment for native Saudis. The reaction engendered generally by the workers' protests, as we will see, supports our skepticism. By 1947, the U.S. consul had come to the same conclusion.

The company began repeating another, to be perfectly blunt about it, lie around this time. Managers insisted that Saudi law made it illegal to organize unions. They did so with at least as much conviction as the relations man aboard the SS *Bellows* who had announced that the law barred Negroes and Jews. Various U.S. embassy officials would repeat the claim about unions being outlawed when reporting on the strikes in the 1940s.[4] A decade later, someone in Dhahran finally thought to dig up the statute in question, leading to a telegram to Washington reporting that there was in fact "no known basis for the belief."[5]

WE ARE ALL ONE

The first workers' strike in Saudi Arabia's history took place in Riyadh, not in Dhahran, and it did not involve ARAMCO. At the beginning of the month of Ramadan (September) 1942, some two thousand men engaged in hard labor for the state at a building site near the Shamsiya Palace had assembled in a mass demonstration to demand a shorter workday. They had been worked some ten to eleven hours a day, while fasting. For their insolence, twenty-five leaders were beaten on the spot and imprisoned for one year.[6] But ARAMCO's skilled Saudi drillers in Dhahran led the second strike.

Working conditions in the camps late in 1944 and early 1945 were tough, no doubt. The influx of labor to build the new refinery at Ras Tanura and to return the fields to production had outstripped the planning capacity of Barger and the others on site. Skilled Italian masons, as mentioned, had been brought in from Eritrea, Indians from the Gulf and beyond, thousands of other skilled and unskilled men from al-Hasa, Basra, and Bahrain, and roughnecks from the States. These men (and only in the case of the Americans, women and families) lived in strictly segregated compounds, divided by race, although we are likely to call these divisions "nationalities" or "ethnicities" now. The Italians lived apart from the South Asians. Later, Egyptians and Palestinians would be put into the camps with the South Asians. Saudis were kept separate from all others. I hope it has been clear throughout that conditions in each of the camps grew worse as you descended the ladder from whites, with their ranch-style homes, swimming pool, movie theater, and the rest, to Italians, kind of like whites but with many fewer amenities, then the Indians, in what some called Servant Camp, and finally the miserable surroundings of Saudi Camp.

Still, Americans hated it in Dhahran at first as much as Arabs, some better educated than the Oklahomans, who had tried working for the company before fleeing back to Beirut. Out of a total of 565 new U.S. employees who arrived between June 1, 1944, and June 1, 1945, 150 were fired and another 124 quit before completing their first thirty–month contract.[7]

The first, apparently spontaneous outbreak of labor unrest in ARAMCO occurred around June 11, 1945, when hundreds of workers at the Ras Tanura refinery rioted after delivery of their food rations. The company had promised wheat, rice, and flour, but the men only received the rice. A second reason for rioting was mistreatment by the company guards who had been searching refinery workers at quitting time. The men marched on the terminal, smashing headlights on cars and at one point stoning an American employee who happened upon them. McConnell wrote about the riot in his journal, and he concluded, as American personnel would after each strike to follow, that unspecified "outside agitators" had to have been behind the demonstration.[8]

One month later, on July 12, 1945, 137 Arab drillers in Dhahran began a strike against the unequal pay and benefits they received compared to Americans and other foreigners. They sought a meeting with supervisors about conditions, and when they were rebuffed, they picked up and went home. Ohliger sent members of the Government Relations Organization to warn the amir of the Dhahran district that a general strike was being planned and to urge him to intervene to end it. Ohliger was right. Seven hundred other workers had joined the drillers by July 16, assembling at a site east of American Camp while others stood at the gates to try and stop other workers from entering. The amir had ordered his "black slave soldier" to break up the demonstration and to bring all the workers to him. McConnell wrote: "The negro went wild," herding the mob out of the area with his rifle, knocking people out, and breaking the teeth of one old man. Barger rushed to the police station, where the amir sat waiting, and protested the beatings, according to McConnell. Barger would not ever do so again.

Facing the strikers inside the grounds of the police station, the amir demanded to know who were the leaders. One report claims that a man stepped forward and volunteered, "inasmuch as they were all equal, there were no leaders." The amir then ordered the police to "take him back and beat him." The eyewitness said a second man repeated the we-are-all-equal line and took a beating as well. McConnell, who was not there, says it was a crowd that started chanting "that they were all as one" and that the amir ordered the police to herd them back to the worksite. "If any turned . . . toward Saudi camp they were to be beaten."[9] The amir promised Ohliger that all the men would be back at work the next day and that they should not harass them when they returned. The strike continued the following day, and the numbers grew to over two thousand, before the men began trickling back. The strike was over by July 20. The amir then conveyed the list of demands to the company and urged it to do the right thing since the complaints of unequal treatment were true.

Walter Birge, the American vice consul at Dhahran who reported the demands to Washington, confirmed that Saudi drillers—"highly skilled and every bit as valuable as the average of the Indians and Iraqis who have been contracted abroad"—were being paid less than the Indians and Iraqis, let alone the Americans. The firm admitted the hospital needed improvement but that it was better than any other in the kingdom. Still, Birge was not "impressed with its cleanliness or with its facilities. Moreover, it is apparent that ARAMCO's medical director takes little interest in the health and care of Arabs." The vice consul also worked through the logic of what foreign element had spearheaded the strike. It couldn't have been the Iraqis because all the known "trouble makers" had already been sent back to Iraq, therefore the consensus was that the better-educated Hijazi clerks, the best paid of the Saudi workers, were chiefly responsible for "inciting the workers and organizing the strike."

> Several Company officials have explained this by saying that these clerical workers, largely owing to the Company's efforts, have developed a taste for Western ways and Western comforts, including better sanitation and modern houses; that they have now become dissatisfied with their present living conditions; that they have become jealous of the way the Americans live and of the large American salaries which are so much greater than their own.[10]

Ten days later, on July 30, all 1,700 of the Italians who were building the refinery put down their tools, protesting that they were "fed up" with being treated "just like the Arabs." Their demands included better food, clean water, a kitchen at their camp, decent medical care, and latrines. The Italians laughed off the entreaty by the American consul that in striking at a vital war project they were putting many innocent lives at stake. The copper companies had tried to make the same argument thirty years earlier. Later, Birge reported, "it was at once clear that most of the grievances of these men were just, though hardly grounds for a strike." The only one he scoffed at was the several claims of "ill-treatment" by Americans. The Italians agreed to arbitration of the issues and most went back to work. ARAMCO agreed to improvements, but the leaders themselves were fired and deported to Eritrea.[11]

Four days after the Italian strike, on August 4, the entire labor force of 9,000 Arabs employed in Dhahran, Ras Tanura, and the outlying worksites defied the amir and resumed the strike against ARAMCO. For the next three days drilling operations ceased, the building of the refinery was disrupted, and the Americans had to do their own driving, laundry, cafeteria duty, and the like. The demands remained the same. Workers resumed the strike because the company's offer of a raise did not deal with the underlying question of discrimination. ARAMCO was ultimately forced to begin negotiations with a workers' committee chosen by the amir in order to bring the strike to an end.

The firm scrambled to make sense of events. Signs of discontent had been growing for some time, but language barriers had kept management from understanding the depth of the grievances and individual complaints had been ignored. As Birge put it, "most of the Saudi workers apparently consider themselves ill-housed, ill-paid, and ill-used in general." More troubling still, the company believed the workers had support among some of the merchants in al-Khobar who were contributing to a strike fund.

Colonel Eddy sent a top-secret cable to Washington following a private audience with the king in Riyadh on August 9. Not even an interpreter had been present. The king asked that his remarks not be repeated, because he had been uncharacteristically blunt and direct. "I regret ARAMCO has not been keeping faith with us recently. I have discovered that the manager Ohliger is a liar. Some of my people have been spoken to as no man should speak to a dog." A week later, Floyd Ohliger received word that the home office in San Francisco wanted him back in the United States for consultations.

Love and Loyalty

The beginning of ARAMCO's research and intelligence-gathering group, the Arabian Affairs Division (AAD), is a direct outcome of the workers' strikes of 1945. The first head of AAD, George Rentz, had been flown to Dhahran to consider work as the company's first full-time Arabist. Rentz was an American working toward a Ph.D. in Near Eastern history at Berkeley when the United States entered World War II. He shipped out to Cairo to work in intelligence,

where he ran the research and translation section of the Office of War Infor-
mation. He turned down ARAMCO's job offer for a teaching position at the
American University in Cairo (AUC). After the strikes, the firm proposed to
bring him to Dhahran temporarily until the beginning of AUC's fall semester.
It was an emergency. ARAMCO needed at least one American with advanced
Arabic skills because the company did not want any of its Saudi translators involved in the exchanges with Dhahran's amir over treatment of Saudi workers.
No American working for ARAMCO, including Barger, knew enough Arabic
to handle the job.[12]

ARAMCO's basic response to the workers and their interlocutor, the amir,
had been to stall for time. The owners forced Ohliger to give up day-to-day running of American Camp. His personal relationship to Ibn Saud, built up over
many years of too slavish devotion, as some of his critics said, saved his job. He
was made vice president in charge of Government Relations ("and placed in a
sort of ivory tower"), and a Scotsman from Chevron, James MacPherson, was
brought in as a new vice president and resident administrator.[13] The firm also
began fingerprinting all its Arab employees to aid in telling them apart.[14]

In early August 1945, the king ordered ARAMCO to begin negotiations
to redress the workers' grievances, and the amir of Dhahran appointed five
workers and four government members to a negotiating committee. The
company at first refused to meet with the Saudis, arguing that the appointees, all clerks from the Hijaz, did not fairly represent the workers as a whole.
The amir deftly ignored the complaint. Once the meetings got under way,
two issues dominated discussion—the company's pay scale, which the Saudi
government's labor negotiator argued was lower than in Iraq and Iran, and
the discrimination faced by Saudis. Others were being treated better than the
men in ARAMCO, and the workers toiled under worse conditions than those
found at Abadan and Kirkuk.[15]

The firm denied the charge that it treated Saudis unfairly. Housing and
hospital care might stand some improvement but were still better than at
camps outside the kingdom and better than anything Saudis had enjoyed
prior to the discovery of oil. Their wages were also better than in the British
camps, management insisted. The problem is that the oil workers in Iraq were
indeed higher paid than their own men. The base pay for workers at the Iraq
Petroleum Company for seven days (they were paid for Fridays) was 18 percent lower than for six days' pay in Dhahran, but the cost of living allowance
added to the Iraqi workers' weekly wages translated into 13 percent higher net

pay. The firm would continue to insist that Saudi workers were better paid. The king responded with orders to adopt Iraq's more generous labor law and insure that statutes already in place were being adhered to, namely, the one passed in 1942 but ignored by ARAMCO that granted all Saudi workers seven days' pay for six days of work, reduced their working hours, extended health and injury benefits, and called for a training program for all workers.[16]

The visceral response both to the first serious Saudi labor regulations and the first signs of dissent among the workers themselves included dusting off the firm-as-family homilies that mining capital had accumulated down through the generations. At audiences with King Abd al-Aziz, Ohliger or others would proclaim that ARAMCO's purpose, like that of the Great White Father's approach to the Indians decades earlier, was to gain and maintain Saudis' "love and loyalty," while regulations would confuse workers about where their "gratitude rightfully belongs." Thus when a man named Hamad Abdallah Adhayf (employee no. 4733) trekked to Riyadh to deliver a petition to the king about the miserable conditions in Dhahran, Lebkicher could recommend leniency toward him and the other "disloyal ringleaders."

> Our first intentions were to terminate their services for poor citizenship and labor agitation, but on further consideration it was decided to retain them. All but one of the ringleaders were called into the Personnel Office, and before witnesses received a lecture on good citizenship, loyalty to the country, company, self, and fellow employees, and were informed that the Company would not tolerate labor agitators or a repetition of their recent conduct. . . . They were invited to present any grievances they may entertain to the Company for consideration and study.[17]

Negotiations with the amir's committee dragged on through September 1945, with ARAMCO's intelligence revealing that labor leaders were attempting to restart the strike. The company made concessions on two of four issues, the building of a hospital for Arabs and the start on permanent housing to replace the palm-frond huts. Back in the States, Chevron would soon crow that the new quarters, drab cinder block rooms with bare light bulbs, were "among the best in the Middle East."[18] On the third issue, the question of treating workers with similar skills and responsibilities the same, the State Department correspondence notes that there was agreement on the principle, but there is no record of any concrete measures being taken. On the fourth issue, however, the managers continued to insist that ARAMCO paid wages as "high or higher than those paid workers by other oil companies in Iraq and Iran, and with the food subsidy, [these were] adequate to enable [a] worker to live comfort[ably] even under present high cost of living conditions." The firm nonetheless conceded a pay increase apparently through a back door. "The [1942] Saudi Labor Law provides for full pay on Fridays. So far the Company has not complied with the law in this regard but a week ago . . . finally agreed to put Friday pay into effect on Jan 1, 1946."[19]

The American consul betrayed his skepticism in reporting on the firm's position. ARAMCO had insisted that the Italians were all higher skilled and so received higher pay, but the Saudis pointed out that many Italians were doing unskilled and semiskilled labor—driving trucks, running gas stations, and the like. However much the Italians might protest the lack of amenities, the consul was sure that the Saudis were treated worse than all other workers, and to illustrate the point he zeroed in on the so-called hospital, a fly-infested "disgrace to the company and indirectly to Americans in general," where the one decent doctor had just been fired for insubordination.

> A month ago, Dr. Mast, who was much loved by the Arabs here, was dismissed. . . . She had allegedly refused to travel to Dammam, a town nine miles distant, to treat the wife of the Governor of the Province [Ibn Jiluwi]. According to Dr. Mast, the Governor's wife was not in any real need of her visit . . . [while] a patient in the Arab hospital was critically ill and urgently in need of the doctor's attention. Company officials have maintained that Dr. Mast made remarks detrimental to the interests of the Company to Saudi government officials; that this, coupled with her insubordinate attitude, left the Company no choice.[20]

In Bahrain, meanwhile, thirty-two South African refinery operators working for the Chevron subsidiary went on strike in October 1945, demanding equal pay with the Americans. All of them were fired and deported.

The Color Line

By the time George Rentz reached Dhahran to take over the Arabic research and translation work in the summer of 1946, the strike wave had passed, but he was pressed nonetheless to accept a permanent rather than nine-month post. According to Bill Mulligan, his command of Arabic was "so superior that James MacPherson, ARAMCO's resident manager, was determined not to lose Rentz—no matter what the price," even after Rentz admitted that he could not actually translate documents from English into Arabic. MacPherson persisted, and Rentz lived in the kingdom off and on for the next seventeen years, first as administrative head of ARAMCO's Arabian Affairs Division, which he modeled on the Office of War Information in Cairo, and then as a senior advisor in Arabian affairs, which freed him to take on more scholarly-like tasks. In 1963, he retired and moved to Stanford to become a curator at the Hoover Institution.[21]

Rentz had made two demands before accepting the position. First, he sought assurances that he would have the resources to build and run a full-fledged research organization, including additional hires—three or four more Americans and fifteen to twenty Arabs in all. MacPherson agreed, but then went back on the promise, and for the next few years, Rentz would battle against competing claims for resources, space, and personnel as the company's operations grew rapidly. He struggled for a year or more with only three Arab assistants and waged an endless campaign for other divisions' old mimeo machines and typewriters. Nonetheless, his new organization handled all the company's confidential translation and correspondence work. In their spare moments, the AAD staff made desultory progress on the various early projects—standardizing place names used by the company's geologists, creating a list of all Saudi government officials and royal family members, preparing a handbook on Saudi Arabia for new employees, and starting a reference library. Rentz meanwhile earned a reputation as a hard-partying, hard-drinking Arabist with a penchant for what today we call sexual harassment.[22]

His second condition proved the more difficult one. In Mulligan's words, ARAMCO had to "swallow hard." In insisting on housing inside American Camp for him, his young daughter, and his wife, Sophie Basili, who had also worked for the OSS in Cairo, Rentz was challenging two powerful norms at once. Married housing, which was in extremely short supply, was considered a privilege earned only after fulfilling a first two-year contract in Arabia. At the time of his hire, there was a long list of already eligible senior staff still waiting two years or more to bring their families to Dhahran. Further, "not only was Sophie an Arabic speaking Egyptian, she was, though attractive, dark skinned." Rentz was in effect challenging the hierarchy inside American Camp at a time when his protégé Mulligan says a "Texas *herrenvolk* atmosphere pervaded the oil town."[23]

The color line in the American colony had been jealously guarded to 1947, and no other Arab would be permitted to cross it for at least five more years. As we saw, even the more liberal-minded senior staff members had insisted that the camp's interests were best served by keeping Arabs out. We know from an early internal and, unfortunately, redacted company history that ARAMCO fired Bishara Daud, the first university graduate hired to teach Arabic to the Americans and English to the Saudis, after only two months.

> Such cases do not workout at all satisfactorily due principally to the fact that a man born in the East and educated at an American university seems to be a complete misfit. . . . The only possible way of fitting a man under such circumstances is to set him off alone, in Daud's case providing a schoolhouse along with separate quarters for the man, give him his own servants, let him take care of his own food preparation, etc. This of course would be an attempt to set up a "middle" ground, which under the present conditions is hardly feasible. Thus, you can readily see that to try to fit a man in with such circumstances is not likely and any attempt to do so by importing such men will result only in unnecessary expense and difficulties.

Kemper Moore, the researcher who uncovered the early records of ARAMCO's refusal to hire qualified Arabs for senior positions, said they mark the "xenophobia" of the time. "This particular incident seems to have set a Company policy which was adhered to for the next 14 years. An educated Lebanese was not accepted on an equal basis by even the rough, uneducated Texas or Oklahoma oil drillers or roustabouts."[24] Not surprisingly, therefore, ARAMCO management also refused to hire any educated and qualified Saudis for senior staff positions in Government Relations. The test case was a Hasawi named Ibrahim Radwan, a government official who had "proven himself to be intelligent and honorable and was well thought of in Saudi Arabia," and yet was passed over by Ohliger. As the records show, the reason was the obligation to house him either in American Camp or in comfortable quarters outside the fence or accord him other benefits commensurate with the position.[25]

When Rentz returned from home leave in January 1948 he produced a scathing report to Barger on the costs that racism was exacting on the work of his research and translation division. He had finally hired three additional Americans, the office manager, Mulligan, who had minimal Arabic skills, and, after great difficulty, two analysts, James Knight and Charles Matthews, who had begun to supply the firm with materials on the kingdom's government, history, boundaries, and the like. Knight worked in Government Relations for the rest of his life. Matthews, an ex–military intelligence officer in Cairo with training in classical Arabic, was more of a scholar and loner—Mulligan called him "loony"—who left the firm by the early 1960s. The ballooning mass of translation work nonetheless required at least a dozen more employees working full time. Rentz had gradually assembled a staff of native Arabic language speakers from Yemen, Sudan, Egypt, Iraq, and Palestine. He warned this work was in imminent danger of "falling apart" because his men, all highly educated and definitely not "troublemakers," were "accustomed to standards of living not far inferior to those of the ordinary Americans, yet they are forced to live in surrounding[s] similar to those provided for the workers whom many of the American employees call 'coolies.'" Their salaries were too low. They worked under miserable conditions, thirty people in two small rooms, and were still living in tents and barastis without floors let alone electricity. The building of better housing, promised after the strikes, had not begun. "Arabs who had hopes of being quartered in such houses have seen the great number of portables erected for American employees and the attractive stone house built for Indian doctors besides the Arab Hospital."[26]

Rentz had emphasized his own men's loyalties because ARAMCO's remaining Italian workers, about 800 in all, had launched another strike in May 1947, which lasted a week, again over the inequities in wages and living conditions in comparison to the Americans. Some of the Italians working for Bechtel also joined the action because the two firms had colluded to set wage rates. ARAMCO immediately fired thirty-five men it identified as the ringleaders, refused an offer to negotiate—maintaining "that it was a Company policy not to negotiate with strikers"—and told the consulate that the strike was definitely communist inspired. It ended dramatically with the mass resignation of the remaining men and a demand for immediate repatriation. Many later reconsidered, but 40 percent of them ultimately returned to Eritrea.

ARAMCO's management ran down the full list of reasons why the Italians were misguided and wrong, but the arguments were beginning to ring hollow. When it was insisted that the Italians should have brought their grievances to management, the workers countered that they had been complaining for months without effect. The strike was a last resort. The American consul, Waldo Bailey, toured Italian camps at al-Aziziyah and Ras Tanura and called these places a "disgrace to American foreign enterprise." Two years after the first strike and despite the promises of improved conditions, the men still lived in tents or in barastis without electricity and thus without fans. The hospital was abysmal. They had cold water only, and Bailey said it was close to undrinkable. Only the Americans were given distilled drinking water. They were also "shockingly underpaid for their labor." The level of discrimination was galling, even to some of the Americans.

> A factor contributing to the dissatisfaction of the Italians before the strike was frequent statements by American employees that their fellow Italian workers were grossly underpaid for their skilled labor. Even Aramco officials admit that fact. I was told by several members of the Italian committee at al-Aziziyah that when they left their jobs to go on strike their American bosses expressed hope that the Italians would get what they were demanding. I have heard dozens of American employees here say they do not blame the Italians for striking.[27]

ARAMCO loyalists argued their hands were tied by the alleged agreement with the king, but the Italian workers at Jidda and those employed by the U.S. Air Force all lived in housing as good as the Americans. Those residing on the U.S. base, where the housing was air conditioned, lived better than did the American embassy personnel. Ultimately, the company said, the Italians had signed contracts, however unfair, to which they were committed to honor. Some improvements in the recreation hall and infirmary were carried out over the next six months, but not in housing. By then, the first Arabs had been moved from barastis into their new cinder-block dormitories. Nonetheless, conditions were such that in 1948 the embassy in Jidda warned, "a danger exists, through housing and through the provision of other amenities, of drawing a permanent caste line between Saudi Arabs and Americans."[28]

Officials of the State Department would rehearse the arguments again a year later, in 1949, as a barrage of cables reached Washington from Dhahran and Karachi about the sudden expulsion of fifty of ARAMCO's Pakistani employees who, when they reached home, began a campaign to publicize their mistreatment at the hands of the Americans. The men had organized a union to press for improvements in pay and living conditions, they told the newspaper, *New Orient.*

> They provided us deserted and thatched huts and worn out tents to live in, coarse food to eat and forced us to wade through ankle-deep sand to distant offices and yards . . . whereas they had for themselves air-conditioned houses, clubs, cinemas and deep-cushioned buses to carry them to offices and back.
>
> Besides, they inflicted relentlessly mental affliction where imagination itself staggers. Their color distinction and racial discrimination coupled with their haughtiness and high-handedness were at the root of every maneuver and machination devised to degrade and insult us.[29]

Another Pakistani newspaper, *Freedom*, drew the obvious comparison to conditions in the United States itself and, from there, to the lofty rhetoric of the U.S. government's claim that all the world's people have rights to the "four freedoms" (freedom of speech and worship, and freedom from want and fear) that President Roosevelt outlined in his famous State of the Union speech in January 1941. At the opening of the United Nations General Assembly in 1946, President Truman argued that the same four freedoms formed the basis of the UN charter. "But on reaching Saudi Arabia, the young Pakistanis find out that the American officers drunk with racial arrogance, are all too primed to subject young Muslims to an unscrupulous 'lynch-the-nigger' treatment. And when the 'niggers' muster enough courage to protest against the treatment meted out to them, they are first handed over to the Police and then dispatched to Pakistan without any regard for the terms of contract."[30]

For diplomats charged with promotion of U.S. interests, Jim Crow institutions were coming to be seen as the Achilles' heel in America's Cold War struggle with the Soviet Union. The concern with ARAMCO's discriminatory policies is an early manifestation of this sensitivity to the gulf opening up between the rhetoric of the four freedoms and the reality of racism. Officials at State convened a meeting to consider the problem, and Duce was invited in for consultation, while the embassy in Karachi warned that any promotion of the company, such as in the May 1949 issue of *Life* magazine, was a danger "in this section of the world. . . . Particularly dangerous are such phrases as 'a spectacular example of American enterprise abroad' and 'a prototype of the kind of thing President Truman had in mind in his "bold new program" of American guidance for "undeveloped areas."'"[31]

Parker Hart, who served in the consulate in Dhahran, briefed his colleagues on the conditions that had led to the "several instances recently in which ARAMCO's labor policies caused resentment with Indian and Pakistani workers" there and in Bahrain. "Hart emphasized that the Indians and Pakistanis made good servants and accountants and that it is impossible to get good servants in Saudi Arabia." The living conditions in the camps that housed the mixed group of semiskilled, white collar workers, clerks, and the like from Yemen, Egypt, Sudan, Pakistan, and India were miserable. "The housing is primitive, the food is poor and there are no recreation facilities, no movies, no swimming pool, etc." Another official reminded colleagues of the dilemma the firm allegedly faced, albeit with an interesting new twist. "ARAMCO has an agreement with Saudi Arabia to the effect that except for

Americans *and a small group of Italian skilled workers* all ARAMCO employees in that country will be treated on the same basis as the Saudi Arabs."[32]
Just a few months earlier the famous agreement applied to all the Italians.
The resident Saudi expert, Richard Sanger, held out the prospect that, over
time, neighboring Dammam would develop as a city rather than as a "company town" where non-Americans would be able to live "just as they would
anywhere else in the world, and the 'caste system' which has developed at
Dhahran will start to break down." Hart, however, argued that in the short
term ARAMCO still ought to be pressed to improve the lot of the Intermediate Camp denizens, and it was agreed to bring Terry Duce or his assistant in
Washington in for a meeting.

When Duce finally found time, two months later, he had Eddy in tow, a
stack of documents, and proof, he said, that communists were behind the
strike. "It had been found that an influential group of Pakistanis were concentrated in the Company's Government Relations Department." Moreover,
he alleged, there "was a connection between this group and Mr. Sweeney, an
American employee, who was alleged to have organized the Communist cell
in Dhahran and at two other oil installations." As if more proof were needed,
he brought a copy of the crude circular that the Pakistani workers had produced, "'Pakistanis Treated Like Dogs' which directly follows the Communist
line, particularly as regards evils of capitalism and racial discrimination."
When a labor expert began probing, Duce admitted that not all the men "followed the communist line" and that a "great deal" of the trouble was "based
on poor conditions and inequality of treatment and some of it undoubtedly
legitimate." But his company's hands were tied, the king didn't like the South
Asians, it wished it were possible to encourage employees' associations, the
government, however, insisted that only it would represent Arabs, Saudis and
non-Saudis alike, in labor affairs. Nonetheless, Duce promised that reforms
were already under way and that ARAMCO might start to use Sudanese as
servants in place of Pakistanis and Indians.[33]

Each of the strikes between 1945 and 1949 produced a mass of denials and
exculpatory materials and a minimum of reforms, implementation of which,
unlike the ratcheting up of production to record levels, moved at a snail's pace.
Duce, however, had not been forthcoming, nor did the State Department appear to know, about the new and arguably decisive source of pressure on him
and the other Chevron and Texaco veterans to improve conditions inside the
camps. The new partners, Exxon and Mobil, had launched a campaign, which

was ultimately successful, to replace the architect of ARAMCO's separate and unequal reforms, James MacPherson, who had taken over from Ohliger after the first strike, and Duce himself had come within an inch of losing his job.

THE PENNSYLVANIA COMPROMISE

Four years after the first planning meeting in Dhahran called to deal with the mounting labor problems, Duce reassembled the planners for a second round. The surroundings were better this time. Barger and the other heads of the Government Relations Organization met in the mountains of Pennsylvania, "sky high in the Poconos," at a resort called the Monomonock Inn. The issues were still the same: labor recruitment, housing, education, and the operation of the caste system in the camps. The circumstances facing ARAMCO, however, were far worse than anyone imagined back in 1944, according to the Saudi watchers. The workers' strikes had put the company on the defensive, complicating its relationship with the king and other government officials while revealing sources of potential opposition to the American presence beyond the so-called traditional or fanatical elements in Najd. That is, Saudis appeared to be "modernizing" more rapidly than some thought possible or wise. The first Saudi oil workers' strike actually predated those in the oil fields in Iran (1946) and Iraq (1947) that have long been seen as milestones in the course of radical nationalist and anticapitalist politics in the region.[34] The U.S. cultural attaché in Riyadh warned that anti-Americanism was on the rise among all classes of Arabs in the kingdom. The goodwill gained during the war years was being squandered, and the company's policies were mainly to blame.

> During those days labor troubles were afflicting the oil company in Dhahran by a series of strikes which affected the population even more bitterly than the political issues. Remarks by all classes of Arabs expressed the antagonistic trend against the American methods of handling the Arab workers in the oil fields. There was much talk of injustice, of discrimination in wages, in housing and quality of food. It was said that Arab workmen were treated in a more humane manner in the British oil fields.[35]

Ohliger reached the same conclusion according to a memo circulated at the Poconos meeting.[36]

ARAMCO wasn't doing so well at home, either. It was under attack by the right in 1948 led by Republican Senator Ralph Owen Brewster of Maine, an ex-governor, ex-klansman, and rabid anti–New Dealer who headed a special committee looking into big oil's ties to the wartime Roosevelt administration. On the left, ARAMCO was called the ally of the landed oligarchs, slave-owning kings, the Mufti, and fascist army officers of the Middle East in the war against Israel. Unfortunately for the company, the *Nation* editors had gotten their hands on a memo by Duce sent from Cairo in December 1947 to ARAMCO's president summarizing his contacts with Azzam Pasha and other Arab leaders. To the *Nation's* Freda Kirchway, the memo was "documentary evidence" of ARAMCO's efforts "to destroy the partition resolution" in "collusion with the State Department," and of how "American dollars . . . finance the war on the Jews."[37] Its authenticity was confirmed by the attack on Duce inside the company's board later in the year, when the new partners tried to fire him.[38]

The Poconos planning meeting was in essence a response to mounting criticism of and pressure on Duce's division by ARAMCO's new owners to improve conditions and reduce discontent in the camps—that is, after the company had ostensibly undertaken with great fanfare to do just that. The camp boss, James MacPherson, had made a hospital for Arabs a priority and symbol of this supposed new and more equitable order, but Winthrop Rockefeller, a major investor in Mobil and head of its visiting medical committee, attacked the project before the building was finished. While MacPherson had indeed sought to replace the old fly-infested shack, he had insisted on separate facilities—for Arabs only. After Rockefeller's report, Dhahran was forced to begin planning for a combined Arab-American hospital, although it is not clear when the change was carried out. MacPherson, however, quit ARAMCO soon after learning that he had been singled out in the report, while insisting that his own best efforts to improve Arab health and education were blocked by board members in San Francisco with eyes on the bottom line. Years later, Mulligan sought to set the record straight, describing "Mac" as another one of the imperial-minded early bosses who "acted like Britons in the Indian Raj. . . . They were nice to the king and to merchant princes, but they considered most other Arabs coolies."[39]

Harold Hoskins, Exxon's Middle East advisor and member of the board of American University of Beirut, arrived in Dhahran on the heels of Rockefeller, and his report reiterated and extended the criticisms.

> Local labor turnover also is high . . . due in considerable part to a lack of proper housing facilities. There are, at the present time, no married housing facilities for Arabs and far less than fully adequate hospital facilities. All of

these personnel problems should be supported by a larger and better training program both "on and off" their jobs, as well as a more complete apprentice training plan. A report, studying this whole education and training problem, was made for ARAMCO by an especially appointed group from the Near East College Association almost a year ago. As I understand it, no action has as yet been taken on their recommendations.

Hoskins's report compared ARAMCO's performance to Exxon's joint venture with British Petroleum (BP), the Iraq Petroleum Company (IPC), and BP's biggest investment, the Anglo-Iranian Oil Company (AIOC).[40] In Kirkuk, IPC had committed $20 million to a housing development for local labor, and in Iran, after the strike of 1946, a crash housing program was under way. Hoskins said the other services were even better than those in Iraq, and Iranians were being moved into senior positions. "AIOC has gone further with its education work, both on and off job training than IPC or ARAMCO." Hoskins stressed in particular the company's $500,000 Technical Institute completed in September 1939 for training Iranian technicians.[41] The Hoskins team urged ARAMCO to build the same kind of institute for Saudis, but the proposal was rejected out of hand.[42]

The ARAMCO case suggests that changes under way in the oil fields of Kirkuk, Dhahran, and Abadan were primarily due to the workers' demanding equitable treatment with the privileged minority of white drillers and managers. If Hoskins's ranking is correct, with British Petroleum in "the lead" and ARAMCO bringing up the rear, then the explanation may lie in the wider political institutions, the communist parties and the like, that distinguished Iran and Iraq from Saudi Arabia. I would look there, rather than believing, as a company historian might, that some national character trait distinguished more enlightened, forward-looking managers from less enlightened ones. At the same time, the Hoskins report suggests that the accuracy of Americans' beliefs about their British counterparts as more imperial-minded and patronizing than themselves was irrelevant. As Hoskins reported, British officials in Iran had "already moved competent Iranians into increasingly important posts, which now include Director of Marketing in Iran, Field Personnel Manger, who handles both Iranian and British personnel, and one of four Assistant General Managers in Abadan" at a time when the Americans in Dhahran would not let university-trained Arabs work for the firm, let alone live inside American Camp. Ultimately, it is the question of the scope and pace of equality that matters—when did the caste or color line give way in Saudi Arabia and why?

The importance of this question was confirmed on the second day of the planning meeting, when Barger reported on the back-to-back visits in May and June 1948 of Abd al-Aziz's two top officials, the finance minister, Abdallah Sulayman, and the foreign minister, Yusuf Yassin. Both had pressed the same point. "The living conditions of Arab employees must approach that provided for Americans. The discussions in this respect went much further than they ever have before. *They told us that we should plan our camp layouts so that eventually senior Arab family housing will merge into the American camp.*"[43] This explicit and unambiguous statement by the two most trusted allies of the king ought finally to dispel the myth that the Saudis preferred segregation. The opposite is true. "The Government is becoming more and more conscious of what is done in other countries. The precedents which have been established at Abadan and Kirkuk may set a pattern which we will be forced to follow."[44] In the case of Abadan, the precedent dated back to the 1930s.[45] The Poconos meeting report reveals a set of agents aware of just how profoundly policies in place in Dhahran—in wage scales, housing, and education and training—diverged from norms of justice and equality. The world was rushing ahead and this particular firm was lagging behind.

While Duce and the other managers admitted that pressures were intensifying to do something about education and housing, or what was coming to be referred to as "community development," nonetheless the basic decision that came out of the meeting was to stall. The Government Relations group concluded that further study was necessary in particular before adopting solutions that the new partners appeared to be pressing on them. Both inside the United States and at operating sites abroad, oil companies faced with the enormity of the costs entailed in extending the benefits of the old welfare work housing regime to all workers were turning to alternatives. The Exxon subsidiary in Venezuela, Creole Petroleum, for one had begun providing loans to employees to build homes of their own, and in 1952 ARAMCO adopted the same strategy. The Creole Company historians stretch in order to tell it as an enlightened story of turning away from "extreme paternalism," but at the Poconos meeting it was made clear that the bottom line drove the process and that Americans were to be insulated from any such changes, that is, from loss of privilege.[46]

One reason for temporizing is that the build-your-own-house-plan had been envisioned as a device in the effort to keep American Camp white.

Recall the earlier, wartime planning discussions about developing Arab town sites away from Dhahran. In the ensuing years, ARAMCO had in fact made the definite choice to keep Arab families out—"the possibility of enlarging Company Arab camps to house families was considered but was discarded as being both undesirable and very expensive"—and had begun to develop a model modern town site in the vicinity of Dammam. It was carried on the books as a service to the government and by 1948 about $100,000 had been expended in laying out a housing grid (99 blocks subdivided into building lots) and preparing for the supply of water and electricity. ARAMCO estimated that it would take five years and cost $2 million total to develop. The planners saw the Dammam project as a model for smaller settlements in Abqaiq, Ras Tanura, and in the pipeline districts, at a total cost of $5 million. Loans would make it possible over time for ARAMCO's Arab employees to build at these places. If the firm instead provided "free" housing that was comparable in quality to what was supplied the Americans, it would cost in the vicinity of $40 million.[47]

In case one is tempted to conclude that so-called markets alone and not racism guided these deliberations, note how, above all, the men sought to keep Arabs from settling permanently in Dhahran, where they and most other Americans and their families lived, and to maintain the native camps as labor reservations where other, less-privileged men would have to live separated from their own families. Note too that updated justifications for these preferences on the grounds that Americans were apt to be unjustly treated in lopsided "culture clashes" with Saudi nationals had not yet been invented. Such arguments only came into play in the 1960s once racism no longer worked as trouble free as before. If the Saudi officials were serious about wanting family housing built in Dhahran and the color line undermined in the camps, then ARAMCO would need to shift gears and back away from investment in development of Dammam, which it never did.

Only one decision by Duce and his colleagues did not result in a development question being turned over to a newly created strategic planning group under Barger for further study. ARAMCO had begun a school for the children of American employees in 1945, built the first dedicated school building in American Camp in 1946, graduated the first class in 1947, and by 1949 had schools in all American/senior staff camps throughout the oil province. The norm was also established that American teenage dependents would be sent abroad—the first cohort went to the American Community School in Beirut— to complete their education rather than let them remain in country.[48]

At the Poconos meeting the planners took up the question of education for Saudi dependents for the first time. Yet, even as the Government Relations men argued for a more aggressive public relations campaign in Saudi Arabia and the wider Arab world, which would "stress what hospitals . . . schools, and aid to local business have meant in terms of a better life" for the average Saudi, to Duce and colleagues the "answer seems quite clear that the Company should not engage in a general education program" for the children of Saudi workers. It did not seem to matter that the 1942 Labor Law required the company to do so. ARAMCO held the line until a new round of worker unrest in the early 1950s led a new king to demand that the law be obeyed. Doheny had done the same in Mexico decades earlier. The company finally built one elementary school in Dammam, while continuing to refuse to school the growing number of employees' children living "illegally," by which it meant against the "law" that the firm had invented, inside and nearby the main Saudi workers' compound in Dhahran.[49]

Cast Down Your Bucket Where You Are

Policy makers in Washington in 1945 viewed reform of the chaotic Saudi Arabian financial system as the single most important priority for development assistance in the form of an advisory mission. Such a mission might take charge of "a budget and controls over expenditures and receipts . . . , broad plans for capital development, planning for the institution of a modern currency system, and developing trade and exchange policies."[50] The model was the well-known one from decades of intervention in the Caribbean and Asia during the dollar diplomacy era and beyond. But educational reform came right behind finances.

> An educational program is of the greatest importance in developing a group of leaders who will be able to direct [the] modernization of Saudi Arabia. . . . The U.S. Minister in Jidda, his Economic counselor, and officials of the State Department in Washington must guide the Government of Saudi Arabia during this period of its emergence into the 20th Century world. . . . With the passing of the present King and the development of education, the way may be opened for a gradual change from the present absolute monarchy into a more constitutional form of Government. Every effort should be made to keep down the extravagance of the Royal family, to educate the Princes, to put them to work, and to develop an educated middle class in that country.[51]

As is often the case in theoretical statements of this type, all good things appear to go together. Not only would an educated ruling class turn to a more enlightened form of authoritarian rule, the U.S. Treasury would not have to pay the bill for all the trains, ports, planes, and still more palaces, although the king was hoping they would. Put the finances in order and get the princes into schools and "it would seem that the economic development of Saudi Arabia along the lines desired by the United States can be financed by the Government of Saudi Arabia."

Not a bad plan, although the authors knew it faced some obstacles in the short run. For one, King Abd al-Aziz was apparently not all that excited about it, as he told William Eddy, then the U.S. ambassador.

> I believe the King's statement to Colonel Eddy indicated some skepticism on the King's part of the value of an elaborate system of schools for his people. The King agreed that education in the manual arts and in trades would be most helpful in enabling his people to attain a higher standard of living, but that an educational program developed along the usual academic lines would probably result in developing a class of politicians and lawyers in Saudi Arabia, as happened in Egypt, which the King would view with great concern.[52]

For another, the American oilmen were at least as worried as the king was about the costs of creating a state-building political class. Experience in places like Mexico and Venezuela had taught firms that the idea of education for modern rule often ended up meaning an increased propensity for interference in and maybe even takeover of the industry. Even Abd al-Aziz's simpler-sounding plan, which could have been lifted directly from Booker T. Washington's speech at Atlanta for "education in the manual arts and in trades," was resisted more than embraced inside American Camp although this would be hard to tell by reading the retrospective accounts.[53]

The firm like all others on the world mining frontiers did have an abiding interest in making use of native labor where possible because it costs much less. This was as true about plantations as it was about mills, canals, and mines. ARAMCO's "foreign" labor costs (that is, primarily, the costs of employing Americans) were extraordinarily high and so was the rate of turnover in the early days, as we have seen. The phenomenon was well known in the industry and so there was definitely an incentive to develop cheaper and closer supplies of labor. This is true even though oil production is unlike the other industries

I named in that it uses relatively less labor and more capital. Drilling wells, unlike mining copper, relies on small, autonomous teams of men with relatively specialized skills. What Saudis mostly did around Dhahran in the beginning was the grunt work of building and maintaining the camps. Yet, unless there were schools to teach the specific skills used in oil drilling, learning the trade depended on apprenticing with more experienced workers.

"White" and "American" workers, as they described themselves in Arizona or California in the early 1900s and in Saudi Arabia a half century later, acted predictably in trying to preserve their privileged positions in the labor hierarchy. They fought with black, Mexican, Hispanic, et al., workers, at times they formed secret organizations like the Ku Klux Klan, and in the oil industry they often simply refused to train their replacements.[54] The last was the case in American Camp in the late 1940s and 1950s, which also helps explain the insistence there and elsewhere in Dhahran similarly to exclude Saudis and others from the housing, the swimming pools, the new elementary school, and so on.

Company historians and anniversary issues of *ARAMCO World* conventionally and quite correctly point out that the very first classes taught by the firm were for Saudi employees, to learn English, beginning in May 1940. Nineteen students showed up the first night at the home of an employee in al-Khobar who was paid extra as an instructor. The Khobar "school" English classes grew as more employees and some leading townspeople were admitted, and by March 1941 a barasti had been erected to house them. A second school, run by a Syrian employee, was opened in July 1940 in a barasti in Saudi Camp in Dhahran. Again, for a time classes were open to workers and nonworkers alike, although there were fewer of the latter in Saudi Camp than in al-Khobar. The Saudi Camp school held special night classes for houseboys, waiters, and telephone operators. The third and best-known "Jebel School" (evoking Jabal Dhahran, the Arabic name for the location of ARAMCO's headquarters) was opened inside American Camp in January 1941 and was run by a full-time teacher. Again, special classes were run for houseboys, and by 1944, many of the students were young Saudi workers age eight to eighteen. The schools began teaching written Arabic and basic arithmetic as well as English.[55]

The Jebel School is an irresistible piece of ARAMCO lore. Its early "graduates" include the Mecca-born scholar Ismail Nawwab. He appears in a wartime photo in *Life* playing baseball outside the school—an old bunkhouse. By the 1980s Nawwab had become head of Government Relations in Dhahran.

ARAMCO's first Saudi president, Ali Naimi, who was serving as minister of petroleum in 2005, was another Jebel School alumnus. Tom Barger and other managers pitched in at the night schools. By 1944 all the youngest employees-turned–school children in al-Khobar and Saudi Camp had been reclassified as Education Trainees who worked half-time at full-pay while attending the Jebel School. At the same time, the government appointed two shaykhs, Hamad al-Jasir, a founder of the Riyadh newspaper *al-Yamamah*, and Abdallah al-Malhuq, a future Saudi ambassador to Sudan (and hostage in a Palestinian operation there in 1973), to provide religious instruction to these coffee and office boys, and the two quite learned men took over Arabic instruction for the young Saudi workers.[56]

Predictably, however, the romance- and nostalgia-tinged portraits of the pioneering days are useful to those who believe that "in the annals of human manpower development, there has probably never been a story to equal" that of ARAMCO's commitment to educate and train the Saudi people in an "ever evolving mission" to ready Saudis for running the company on their own.[57] The same portraits look different in light of what actually happened in the 1940s and early 1950s, driven by the workers' strikes, on the one hand, and the resistance by the firm's agents to the linked demands for equal educational opportunity and Saudization of the fledgling industry. The most effective means to accomplish the latter were all known in 1945, discussed, and most significantly, rejected by ARAMCO. These means included schools for young Saudis, dedicated institutes for the training of craftsmen and technicians such as drillers and the like, and university study for what ARAMCO called senior staff positions. These means were ultimately forced on the company in the 1950s and 1960s. The steps taken in the early postwar years are best seen as ones away rather than toward them.

By 1945 two separate delegations of company managers had firsthand knowledge of the education and training programs in place in oil industry sites and in schools in Abadan, Basrah, Baghdad, and Beirut. Following the Saudi strike in 1945, the firm revisited AIOC operations in Abadan, where the most extensive programs were in place. As we have already noted, this British-U.S. oil company had paid for and staffed a technical training institute and built elementary schools for workers' children in the vicinity of the camps, which it turned over to the Iranian government to run. The pattern was the familiar one too in Latin America. Thus, let us first consider ARAMCO's retreat from the norm of the war years, when nonemployees and workers' dependents were accepted in the company-run classes, and its replacement by a policy *against* educating the dependents of its Saudi employees and others living in the vicinity of the oil installations. The new position was that the Saudi government should pay for and run schools for all those other than American dependents. This is the position that would be reaffirmed at the Poconos meeting in 1948 and then abandoned in 1953 after the next large workers' strike.

A new Personnel Department created in 1946 should probably be credited with the narrowing vision toward educating Saudis. Under its auspices, the so-called voluntary or opportunity schools in Khobar and Saudi Camp were closed down, and the Jebel School inside American Camp was renamed the Arab Trade Preparatory School. Young Saudis would be selected to study there before advancing to two new trade schools built, respectively, in Ras Tanura in 1947 and in Dhahran in 1948. A conflict with religious authorities in the Eastern Province, who were objecting to the boys' exposure to western teachers, led in 1948 to the withdrawal of any student under age fifteen and, in turn, petitions to the king by families who wanted their children readmitted to the Jebel School. A group of forty Saudis age seven to thirteen was admitted in fall 1948, but a new Personnel Department Planning Group would soon recommend ending the short-lived experiment in vocational education for Saudis and the schools were all closed in 1950.[58]

More remarkable still (and heretofore unremarked on) is the effect of these decisions in extending the racial hierarchy to a new domain. The first classes for American dependents in 1945 in Dhahran were apparently held in the same building being used for the Saudi students, presumably the Jebel School, and some of the classes themselves may have been mixed. This was the case according to U.S. Air Force colonel Harry Snyder, an intelligence official who headed the training program at the Dhahran Air Base and who briefed the embassy on ARAMCO's first American school.[59] Yet, the Government-Relations-man-turned-company-historian Bill Mulligan is also clear that the new school buildings that went up in Dhahran in 1946 and in all other senior staff camps in the next two years were for "American dependents" only. School-aged Saudi employees were instead assigned to the Jebel School together with stray numbers of boys who may have avoided for a time ARAMCO's increasingly formalized position that the kingdom and not the company should pay for and educate the sons of its workers. Segregation in the senior staff schools was adopted and preserved for the next decade or more.

One more piece of evidence speaks to the conscious creation of a separate and unequal education regime in the camps in the late 1940s as ARAMCO managers looked for ways to counter the idea that not just servants and office boys ought to be educated at company expense. In June 1946, ARAMCO vice president Terry Duce began arrangements with Snyder, who was an associate director of the Near East College Foundation, the consortium that operated the American colleges in Beirut, Cairo, and Istanbul, for him to lead a mission from the foundation to Dhahran to advise the firm on education policy. Duce wanted the foundation to take over the running of the senior staff schools. It took more than a year to arrange the mission, however, in large part because the foundation's position was that any educational survey of the kingdom had to involve the Saudi government.[60] Yet ARAMCO would not budge, insisting that the mission was a private one focused on "oil company activities." In 1947, the foundation's report was received and promptly shelved. When the Poconos group insisted in 1948 that ARAMCO should not be in the business of educating the Saudi population, they were responding to the foundation's idea, and in 1949 ARAMCO's Personnel Planning Committee produced a plan of its own, which, as we saw, led to the closing of the company's little Tuskegee-like Jebel School.

ARAMCO actually shifted from "an American-type school-oriented education and training policy" for Saudis in favor of "training by line personnel as an integrated part of the production process." The experiment lasted about six years, until ARAMCO's opponents in the labor movement and in the government forced it to pay for a system of schools, training institutes, and, ultimately, an engineering college.

> The model selected for adaptation was that of America in WW II, in which large numbers of persons without industrial experience were quickly absorbed into war industry, and brought production to the highest level that far attained by any world economy.
>
> The secret was minutely differentiated work organization, in which the entire production process was exhaustively studied, broken down into its smallest elements. These were regrouped into simple families of skills that could be quickly taught to inexperienced labor, followed by intensive instruction, largely by actual production supervisors, to give necessary single skills.[61]

The application of Taylorism or the kind of labor system we associate with assembly lines in factories, in this instance proved expensive because, as I have noted, oil production is capital- rather than labor-intensive and the jobs that workers at the refinery and the well sites must perform are highly varied. There were lots of different tasks and thus startup costs were high, but relatively few people were to be trained in each task, and thus the system was inefficient. Still, it worked tolerably well in two respects. First, as the postmortem on the new training system reports, it permitted ARAMCO to "utilize Saudis without previous experience and an astonishing percentage of work was quickly taken over." Saudis did indeed learn a number of "monotonous single-skilled jobs." Second, since the underlying approach was the one conventionally used in the oil industry for decades, it was hardly surprising to find that three, five, or eight years down the road Saudis were not being promoted. It is the same reason that U.S industry remained segregated for so long. Those already in higher-skilled and higher-paid technical and managerial positions would not willingly train themselves out of a job.

Reprise: Dusk at Dawn

The way to nurture a "talented tenth" capable of managing the company either directly or indirectly was of course known long before the 1940s, when arguments for taking the kingdom down that path had begun. It is also clear that ARAMCO's agents in Dhahran were as threatened by a class of advanced-educated commoners as was King Abd al-Aziz. We might consider in this light the evidence from the oldest records of the company refusing to employ educated Saudis and other Arabs or to accord them equal status with the Americans much less let them reside in American Camp. We might also consider again the story told in Chapter 2 of an ARAMCO public relations man in 1950 pressing the State Department's desk officer Richard Sanger to omit the description in his *Arabian Frontiers* of "a King who thinks like an oil company and an oil company that thinks like a King."

Finally, therefore, we might consider the report to Washington in 1949 from the young diplomat Parker Hart that "ARAMCO is awaiting with deep misgivings the imminent return to Dhahran of Shaikh Abdullah Tariqi," a twenty-four-year-old from al-Zilfi, in Najd, who received a scholarship from the U.S. government in 1945 to study petroleum geology. What the oilmen now feared was that he would begin to "pry into the technical phases of the operations of the company."

> Shaikh Tariqi received training at the University of Texas of a sufficient nature to make his presence embarrassing from ARAMCO's point of view. He is understood to have sufficient knowledge of petroleum engineering and geology

to be rated in Saudi Arabia as a Government expert, although ARAMCO's management considers his knowledge to be superficial. However, if his knowledge were more complete ARAMCO's problem with him might well be even greater, for he appears to have been given a mandate to probe into matters of a highly classified nature and, more recently, to recruit specialists in refinery, gauging and other matters.[62]

Tariki in fact came to haunt ARAMCO until his exile from the kingdom in 1962.

The steepest costs to educating a Saudi professional class couldn't be included in annual reports and other public relations materials but were instead discussed in private, acknowledged in meetings between company managers and State Department officials. Renewed labor unrest in the Eastern Province would combine with the rise of populist governments in places like Egypt and Iran to put ARAMCO on the defensive, leading it to make at least a minimal commitment to educating future technocrats. Yet it was also these first western-educated Saudis who became the leading critics of the Jim Crow system that ARAMCO built in Dhahran. A handful of them were sent to the United States in 1950 on work assignments. ARAMCO had opened a training center among the dunes of Westhampton, Long Island where, the *Saturday Evening Post* reported, new American recruits where "to learn the native habits of the Arabs and as much of the language as possible" for a few weeks before heading to the kingdom. Some of the Arabs brought to lend some color would return to compare the native habits of Americans in Washington and other cities and in Dhahran. Tariki himself would liken his time in Texas, where he was mistaken for a Mexican, to living in the vicinity of American Camp. In 1953, State Department officials interviewed ARAMCO executives in New York following the month-long strike against the Jim Crow settlements in Dhahran. The executives were found wringing their hands "over the fact that most of the original, known leaders in the strike had been trained in Beirut at company expense."[63] ARAMCO had awarded its first scholarships to twenty Saudi students to attend the American University of Beirut the year before.

In the preceding three chapters I have tried to develop an account of the institutions that the Americans built in Dhahran in the 1940s during the first decade of oil production in eastern Saudi Arabia. Doing so has meant stripping away the layers of myths that have accumulated about the time and place. Stripping those layers away has taken some work. The most difficult myth to deal with both because its layers are so numerous and its loyalists so tenacious in preserving whatever remains of its luster is the idea that ARAMCO was in some way special or unique—more generous, just, or progressive, less exploitative, bottom-line oriented, or self-interested than other firms. ARAMCO's record of accomplishments in Saudi Arabia, if plotted against the formidable physical and cultural barriers its agents faced, would place it far ahead of most other firm on the curve, apparently.

Many pages have thus been taken up with a detailed accounting of specific institutions, the layout of the camps, the kinds of houses that were built, the creation of ARAMCO's Government Relations Organization, the organization of the king's model truck farm in al-Kharj, the timing of the decision to build a U.S. air base in Dhahran for the company's benefit, and so on, using multiple archives and record sets, and comparing these for inconsistencies and incommensurable truth claims. This narrative method is the most useful way I know to respond to a mythology that depends on faulty memories, captive scholarship, poor reasoning, impoverished sources, sons and daughters who want us to believe the best about their parents, and the largesse of an old firm with a new set of owners.

By far the most critical institutions discussed have been, first, the system of privilege and inequality built in Dhahran and, second, the movement that emerged to challenge this hierarchy in the oil camps, beginning with the strike by Saudi workers in 1945, which has been recounted here for the first time. These institutions, which have appeared over and over again on the nineteenth- and twentieth-century world mining frontiers, are the worm at the core of the apple that a company's public relations division would hold out for us to admire. Together, they go far to undermine the idea that ARAMCO made miracles happen. Of the two, labor strikes have proved a problem for firms and, later, for the historians that firms buy across the long twentieth century. As we have seen, back in time, in places like Ludlow, Douglas, and Bisbee, the mining industry had perfected a protective belt of ideas to surround its preferred theory of business enterprise, in which strikers are not men toiling long and hard in mines and at refineries but agents of power-seeking politicians, alien ideas, and a country's enemies.

The men that governed American Camp were, by contrast, more nearly blindsided by attacks on the Jim Crow system that they had been busy building, at a moment when it was still not clear to all in 1945 that their theory of the firm (let alone the way they preferred to live their lives) was coming under a long siege. Movements for African and African American liberation were under way, and one can look back to find the books that would make the new world intelligible for white racial liberals, from Ruth Benedict's *Race, Science, and Politics* (1940) and Gunner Myrdal's *American Dilemma* (1944) to Wallace Stegner's *One Nation* (1945). Yet it would take another twenty years or more to undo the institutions of apartheid in America and in its outposts abroad, and as we have seen, the ARAMCO managers on the ground in Dhahran were hardly pioneers of equality in the oil fields—quite the opposite. They worked overtime it seems to keep Negroes out of Dhahran and Arabs out of American Camp. So the firm's legendary operative James Terry Duce would reach for the white South's peculiar variation on the Cold War's anticommunist creed when pressed by the State Department to explain the protest and deportation of the Pakistani workers in 1949. They were followers of "the Communist line, particularly as regards evils of capitalism *and racial discrimination.*"

Three not-very-rigorous kinds of arguments started to be produced by firms and their defenders in the 1950s intended to salvage the vision of ARAMCO as exceptional even in light of the investment in racism. One is the idea that where racism (the preferred term was "prejudice") existed it was because the members of its less-educated and cosmopolitan class brought their folkways with them to the Gulf. Another is the idea that American firms picked up some bad colonial-style habits from nearby British firms. The third is that the Americans built a segregated and inequitable order because Saudis required it of them. As I said, these claims should not delay let alone fool anyone for too long, which may explain why most writers until now either ignore the institution of racism or else try to imagine it as a short and fleeting moment in a longer and more noble course through time. Since these myths about the Texarkanan, Anglo-Indian, and Wahhabi origins of ARAMCO's Jim Crow camps exist, our account of the 1940s has presented what archival evidence also exists to undermine them. We saw the firm's principals, not just its lowliest agents, building racial hierarchy anew in American Camp and I have suggested why, even if British oilmen had never set foot in Abadan, Kirkuk, and Bahrain, American Camp would have looked the same. American Camp's racial geography was identical to that of every other oil installation Americans had built in three continents across one hundred years in places under widely different forms of rule. Saudis actually expected and then pressed for integration. We also saw that, prejudices aside, ARAMCO's own records show nearby British-identified firms moving faster and farther toward equality among native and foreign labor. We proposed the hypothesis that populist coalitions in Iran and Iraq account for the difference with Saudi Arabia, where a tribal-chief-turned-king reigned and where, unsurprisingly, ARAMCO's devout thanked God daily for His beneficence and mercy.

Finally, in these three chapters we began to chart the process of the making of the most popular myths about ARAMCO. The earliest and one of the most redoubtable is that the firm or its agents are better understood as an extraordinary "modern day development mission" for Saudi Arabian development, a "private Marshall or Point Four plan" and so on. We traced the invention of this myth to the first challenges to the firm's power and privilege *in Washington* by its various critics and competitors inside and outside the wartime and postwar New Deal state.

Myth production was just getting off the ground, however, and the firm's true capacity had hardly been tapped. Stegner had not yet been hired, movies not yet made, Middle East centers not yet built. But demand would explode in the 1950s and a letter from one of the old-timers written thirty years later recalls the moment when the analysts, operatives, and public relations men in ARAMCO's Government Relations Organization began to gear up for the challenges ahead.

Highly confidential undercover activities were going on during those days [September 1951]—just after Mossadegh nationalized AIOC, and the first thing he nationalized was their confidential files. Well, Relations Confidential Files were so confidential that no one man could destroy them—therefore Bill [Palmer] and I were keeping each other honest—taking turns turning the handle of the course [sic] wire cylinder till the last small scrap of burned ash had gone with the wind. As I remember it, we both had to sign a certificate to that effect—but only after we had been notified by New York that the microfilm had been developed, and yes, I was one of four who worked eight hour shifts taking the pictures, which meant that I was working at least 16 hours per day— not in the overtime bracket of course.[64]

Part 2

INTRODUCTION

Desire's Empty Quarter

One night in June 1955 ARAMCO's Government Relations men ushered guests to the Cairo Palace, the art deco theater in Egypt owned and operated by Twentieth Century-Fox. Oilmen mingled with the American ambassador, Henry Byroade, Saudi royalty, the cream of Cairo's new elite of lawyers and army officers, publicists, European businessmen, and at least one real Hollywood star, icy beauty Dawn Addams. The occasion was the world premier of ARAMCO's first and only feature-length film, *Jazirat al-Arab* or *Island of the Arabs*. It was the latest of Vice President Terry Duce's public relations projects, cooked up with the help of the professionals in "505," the brand-new Park Avenue, Manhattan, skyscraper where ARAMCO had relocated in 1949. Duce wanted a film that would compete with the campaigns of the Arab republican governments, above all, Egypt's, led by the army colonel Gamal Abd al-Nasir. Nasser, as the name was usually spelled in the western newspapers and magazines, had just returned from the landmark Afro-Asian conference in Bandung, Indonesia, two months earlier and since then had cranked up the volume on his country's Voice of the Arabs radio station. To the Nasserists, those who were attracted to Nasser's increasingly pan-Arab orientation and populist authoritarian politics, Saudi Arabia and the other, smaller Gulf states were not much more than "tribes with flags." Arabia was Arabism's weak flank in the confrontation with Zionism and imperialism—subservient to western oil firms, propped up by British power or, in Saudi Arabia's case, American aid and military personnel. In March 1954, Egypt's rising star Ihsan Abd al-Quddus published his exposé "Christ in the Land of Oil" in *Ruz al-Yusuf*, the popular Cairo weekly magazine he edited. He focused on the blatant inequities between the Americans

121

who, despite their privileges, were unhappy and homesick, and the expatriate Arab and Saudi workers, who were forbidden from entering paradise.

ARAMCO made *Island of the Arabs* as a kind of response that would celebrate Abd al-Aziz, descendent of the Wahhabi conquerors of the 1700s and the man destined to unite the peninsula. The Public Relations Department made some bizarre choices along the way. It hired a celebrated documentary filmmaker, Richard Lyford, to make *Island*. Lyford's film, *Titan*, an earnest biography of Michelangelo, had won an Academy Award in 1951 for best documentary feature. Then it economized by casting employees in the main roles. A Government Relations man, John Jones, played Ibn Saud. The movie began with a conceit worthy of Cecil B. DeMille, who would appear on screen to introduce the final film of his career, *Ten Commandments*, the very next year.

As the house lights dimmed and the projector began to whir, none other than Saud Ibn Abd al-Aziz, the late warrior's eldest son and new king of Saudi Arabia, emerged to introduce the story of the kingdom's founding, as if his own ascension to the throne was the real point of what was about to unfold. *Island of the Arabs* begins with the arrival of three bearded American geologists to Saudi Arabia, one of them the legendary Max Steineke. Steineke is recalling his first sight of Arabia and the Dhahran dome. At the campfire, their guide, Kahlid, begins to recount the history of Arabia, in a composite of all those long nights in the 1930s when Barger and others told their stories about the American Bedouin and when Abd al-Aziz himself regaled his visitors from San Francisco with the story of his conquest of Riyadh. The version for the moviegoers must have suffered by comparison. "Nothing more than a tasteless combination of travelogue, historical drama, and documentary about the Arab world. It does little more than re-create some of the more prominent moments in Arab history as it rambles across the desert. Frederic March serves as sort of commentator during part of the sojourn."[1] March, a two-time Oscar winner for best actor, had been the narrator for *Titan*, but his contribution was lost when the Arabic was added. The March-less sojourn over, *Island* ends back where it started, with Max Steineke looking back at all the changes he had witnessed in a land blessed with a wise and farsighted ruler. Then the poor guy who plays Steineke has to deliver a line so leaden that not even Frederic March could have carried it off. George Rentz tried to get it cut from the script. "Truly, the Arabian Peninsula is the island of the Arab. It is floating on oil."

New Yorkers had a chance to see basically the same film six months later when the English language version, *Island of Allah*, ran for a very short while. The revised title for an audience not so enthralled as Egyptians with Arabism deliberately echoed the old and wildly popular *Garden of Allah*, a phrase that reverberated in the U.S. mass culture industries for decades, but the change did little to enhance its box office appeal.[2] The U.S. version also omitted the opening scenes with King Saud. Instead the distributor insisted on adding a belly dance sequence at the end of the film before he would show it to paying customers. As a reviewer noted, "Princess Yasmina offers a torso-twisting dance which has nothing to do with the continuity."[3] *Jazirat al-Arab* was reprised when ARAMCO began Arabic language television broadcasting, the first inside the kingdom, at its Dhahran station in 1957. It was the first film shown on the air, but *Island of Allah* was never seen commercially in the West again.[4]

What the story of the film points out now, apart from the questionable media savvy of an oil-producing enterprise, is the moment when the firm began to market its story beyond the borders of the United States to Europe and, more significantly still, the Arab world. In doing so, ARAMCO mixed the century's older technologies of industrial public relations with newer ones, notably, sixteen-millimeter sound film.[5] The one it reached for first was the magazine. As we have seen, oil firms among many other kinds of enterprises had, in response to the rise in labor militancy, long ago turned to company publications to reinforce the identity of the firm as family. All of ARAMCO's owners produced magazines of their own. For instance, the *Texaco Star* dates to around 1913, as does Chevron's *Standard Oil of California Bulletin*, which eventually became *Chevron World* in the 1970s.

The first issue of *ARAMCO World* appeared in November 1949 although it began as a nameless four-page newsletter run out of the New York headquarters and targeted at first mainly at the Manhattan staff. The newsletter linked the men and women of 505 Park Avenue to the firm's operations in Dhahran. Rather quickly, however, it turned into a vehicle for production and dissemination of the idea of the firm-as-development-mission that was increasingly offered up to critics at home and abroad. The images should seem familiar. The color of the firm's paternalism is displayed in a series of ubiquitous uplift photographs that appear in virtually every issue of *ARAMCO World* for ten years. A white man teaches a dark-skinned employee safety practices or English or baseball. The future of Saudi Arabia is revealed in a remarkable account of the modernization of a Mohawk family in Brooklyn, accompanied by a map of the United States divided into tribes. The story of Chief Bright Canoe is the only time that America's own multicultural landscape is shown inside *ARAMCO World*'s pages unless one counts the photo from the ARAMCO Follies of 1951 of four New York employees in blackface, singing "Waitin' for the Robert E. Lee."[6] ARAMCO also began to publish an Arabic version of *ARAMCO World*—titled *Qafilat al-Zayt* or *Oil Caravan*—which included a mix of translated articles and original content.

The effort to provide Saudis and others with positive images of the firm really began with a campaign to mount a series of two-by-three-foot posters inside or near cafés where men gathered. ARAMCO called them "Foto Stories." The Public Relations Department produced two dozen different posters, each consisting of eight to ten smaller photographs with brief captions in Arabic and English. Development and uplift themes predominated. One featured the Saudis hired as language teachers in the Overseas Training School (about to be moved from Westhampton, Long Island, to Sidon, Lebanon). Another told of Palestinian refugees being recruited for work by ARAMCO. Another featured the railroad built for Ibn Saud. Others showed workers being "fingerprinted" and as the State Department put it, "indoctrinated" and given aptitude tests. Another promoted the new training regime. Still others developed the theme that ARAMCO would use for the next thirty years, highlighting the industrial training and contracting opportunities for Saudi entrepreneurs-in-the-making in running a laundry, supplying cement blocks, and so on. Again, according to the State Department, ARAMCO mounted these all over Saudi Arabia as well as in Rome, where the mistreatment of Italian workers was still an issue, Cairo, Khartoum, Aden, Jerusalem, Beirut, Aleppo, Bombay, Madras, Karachi, and Damascus.[7]

The campaign to sell ARAMCO's image in the Arab world expanded over the next few years, although we may never learn its full extent. ARAMCO had started down this road in the 1940s, as the undercover CIA agent Bill Eddy reported in a letter home.

> We are preparing a documentation of the idea which I have proposed briefly in the enclosed memorandum, to be translated into Arabic and "planted" by Arab friends in the hands of members of the parliaments of Arab countries, etc. It must be written as though an Arab were writing. The idea will be to describe the tremendous development in Arab lands which can take place in the next few years by private capital, British and American, if the Arabs will encourage such development instead of taking sanctions against companies: the hundreds

of millions of dollars which will be invested in construction, payrolls, goods and services, without a cent of cost to the Arabs since the companies take all the financial risk. The royalties and benefits to the Arabs, on the other hand, will arm them economically to withstand expanding Zionism, and give them a bargaining point with the Powers who MUST have oil from the Near East and will therefore have to cooperate with the Arabs.[8]

The head of the Federal Trade Commission, Stephen Spingarn, wrote to Secretary of State Dulles in February 1953 to request an investigation of the *Beirut Star*, the English language daily, which he believed ARAMCO was secretly financing. A Senate staffer had passed the report of an American reporter in Beirut to Spingarn, and he sent it on to Dulles, although I could find no evidence that the request was pursued. The State Department did, however, report that ARAMCO was paying for Arab writers to visit and report from Dhahran.[9]

The writer they paid the most money for, Wallace Stegner, the rising star of Western biography and fiction at Stanford who would go on to win the Pulitzer Prize in 1970, was hired in 1955 to do a history of the pioneer years in Saudi Arabia. Stegner's refusal to write precisely to company order, however, led ARAMCO's Government Relations people to bury the completed manuscript for the next decade. We thus have a measure of management's heightened anxieties as concerns with antitrust actions at home and nationalism in the host region led these men to make films that no one wanted to watch and to kill a book that had, in the eyes of Paul Hoye, the *ARAMCO World* editor who finally published it in 1971, answered the era's "blanket indictments of 'economic imperialism' that have obscured the enormous contributions that fair, enlightened, far-seeking companies like ARAMCO have made in the development area," and yet was still not good enough.[10]

5 AYYAM AL-KADALAK (DAYS OF THE CADILLAC)

*The Muhammedan conservatism is also rapidly being broken up. When
we lived here in 1944 . . . it was a primitive, closed city, without even one
Christian Arab, not much changed from the description given of it by Richard
Burton in his "Pilgrimage to Mecca and Medina" 1853. But the past six years
have lifted the city's face several times: the walls are down and new houses
with electricity and running water are growing like mushrooms on three sides
of the old town. A new pier to berth ocean-going ships has trebled the port's
capacity. The King's royalties have also resulted in paved roads, new schools,
pure drinking water piped from 35 miles away. You must get Carmen's English
thesis on the subject to read of all the changes the king's wisdom is working
with his new wealth. Today we saw the opening of a coca-cola factory,
permitted by the Quran as cokes are non-alcoholic. . . . Time marches ON.*

William Eddy to his children, April 7, 1950

*American imperialism occupied the heart of the Arabian Peninsula,
exploited our oil, set up the Dhahran Air Base where atomic bombs are
stored. . . . American imperialism has turned King Saud into an American
phonograph record publicizing anything that comes from the United States.
It advertises the luxurious American Cadillac automobile. King Saud even
gave a prize to a poet for his ode to a Cadillac. . . . Oh Amir of music sing us
a song, and drive us in the Cadillac slowly along.*

Voice of the Arabs, November 13, 1961

SECRET MISSION

In August 1950 Harold Hoskins, a cousin of Bill Eddy and another ex-OSS
officer turned business consultant, recently returned from the Middle East
on his annual trip for Exxon, sent a confidential extract of a report he had
prepared to Allen Dulles. Dulles, a lawyer who headed the Near East College

Association, the body that oversaw the American University of Beirut, where Hoskins was a trustee, would move to Washington six months later to take over as deputy director of plans—the covert operations arm—in the Central Intelligence Agency (CIA) that was Bill Eddy's secret employer. When Dwight Eisenhower became president in 1953, he would name Dulles director of the CIA.

The report, a kind of tour de horizon of the Arab world, touched on virtually every issue that would preoccupy policy makers in Washington for two decades or more: communism, the creation of Israel, the growth of nationalism across the region, and the threats that these developments posed to the oil that the western firms (and economies) needed. Things did not look too good to Hoskins, apparently. After the North Koreans invaded the South in June 1950, the fear was common among Middle Eastern elites that Korea was the opening campaign of a Third World War. Hoskins wrote that conditions were deteriorating everywhere he visited. Palestine and communism "must have taken up 90 percent of all conversations." Arab leaders he said found U.S. intervention in Korea and "inaction in the face of Israel's non-compliance with several United Nations' resolutions regarding Palestine" inexplicable save as a consequence of "continued Jewish domination of American foreign policy."

In the face of the postwar rise of new and often radical political currents, from the Baath Party to the Muslim Brotherhood, Hoskins upped his appreciation of the "value of monarchy in the Middle East as a stabilizing force in the area." Egypt, Iraq, and Jordan all seemed in much better shape than Syria, he said, a republic where a parliamentary system had given way to a series of military interventions. The one downside was that in the monarchies "the rulers themselves may not always be highly regarded by many of their own citizens."[1]

His reasoning was hardly out of step with those who steered U.S. foreign policy, as can be gauged by the subsequent course of events during the tumultuous 1950s, when America backed—at times successfully (Jordan, Saudi Arabia, Iran), at times not (Egypt, Iraq)—the regions' kings against the various military conspirators, revolutionaries, and communists who sought their overthrow. Take Egypt for example. In October 1951 the ruling party, the Wafd, cancelled a twenty-year-old treaty of alliance with the imperial power in eclipse, Great Britain. The outcome was a minor guerilla war in the Suez Canal Zone, where thousands of British troops were based, and the burning of foreign businesses in downtown Cairo. Dulles's secret agent, Kermit Roosevelt, entered the fray in the final months, but he backed a loser. Less than a year later King Farouk would be overthrown by a military junta and by 1954 Nasser would emerge as head of a military dictatorship that is still in place.

While Hoskins had left Iran out of his brief, the leaders of Egypt's Wafd Party had taken inspiration from the decision of the Iranian parliament in March 1951 to nationalize the massive Anglo-Iranian Oil Company's holdings, which were Britain's main source of foreign oil. Dulles's agents would perform better in Iran than in Egypt, as is well known, in a joint operation with the British called Operation Ajax in August 1953. There the shah was ultimately restored to power. Iran's prime minister, Mohammad Mossadegh, who had led the pro-nationalization forces, was imprisoned along with his allies in the Tudeh (communist) Party. Several of the Tudeh military officers were executed. Famously, U.S. firms gained entry into Iran's oil market for the first time.

Where populism was on the rise in Egypt and Iran, the region's two most powerful countries in 1950, and talk of nationalization was in the air seriously for the first time, in Saudi Arabia the king and his protector, the United States, tightened their embrace. Saudis wanted weapons and a formal treaty of alliance, but these were not in the cards with a Democratic administration in 1949 (wait until Kennedy and Johnson). Instead, the king got a military survey mission and America got a one-year renewal of the Dhahran airfield agreement. Colonel Harry Snyder returned to Saudi Arabia to head a modest military training program until ARAMCO hired him away for its own enlarged production training program. The Americans also upgraded their diplomatic equipment. A new plaque on the wall of the consulate in Dhahran was changed to read Consulate General, the one outside the legation in Jidda to read Embassy of the United States, and J. Rives Childs exited as minister and reentered as ambassador to the land of the Al Saud.

The outbreak of the Korean War, which increased the importance of the Dhahran airfield to the U.S. military, gained the Saudis some more resources in the form of a Mutual Defense Assistance Agreement, including the first real arms from America along with an upgraded training mission that continues under different rubrics until today. The Truman administration settled for a five-year renewal of the base agreement, after an opening gambit of twenty-five years minimum. But the king never got the treaty he really wanted for protection against his enemies in Jordan and Iraq. Instead, the United States gave him a big illuminated grandfather clock for the fiftieth anniversary of his conquest of Riyadh, although Ibn Saud was forced by the religious authorities to cancel all the planned public festivities. He also got the president's private physician, who was flown to Riyadh to try to keep him alive.[2] Most important, the war influenced the State Department to support revision of a key part of ARAMCO's original concession agreement, increasing the Saudi share of rents.

The conventional arrangement prior to the 1940s had been for a host country to accept a fixed royalty rate from the concession companies per ton of oil produced. The government of Venezuela had successfully challenged these terms in a series of moves in 1943 and 1947, resulting in an equal ("fifty–fifty") division of profits with the multinationals. The U.S. firms assented more easily than they otherwise might have because the additional rents to Venezuela were in the form of "taxes" that allowed them to deduct the payments from those owed to the United States Treasury. Across the Gulf, Anglo-Iranian Oil Company (AIOC) refused to agree to similar revisions when pressed by Teheran, where a Venezuelan delegation was visiting as negotiations began in 1950. One reason is that British law did not grant firms the same capacity to offset tax payments. The result was, first, the beginning of the drive in the Iranian parliament toward nationalization of AIOC's holdings and, second, a marked shift toward accommodation on the part of ARAMCO's owners, who signed a fifty–fifty agreement of their own in December 1950.[3] What remains a mystery is who among the Saudis ought to be credited with promoting the policy shift. The king was growing senile and the finance minister was in the advanced stages of alcoholism. As we will soon see, the obvious candidate is Abdallah Tariki, who was then supervisor of petroleum affairs in the Ministry of Finance.[4]

In places like Iran and Egypt, where the American star was rising, U.S. officials blamed the British for clinging to their outmoded imperialist ways and blocking the reforms that might check communism's advance. In Iran, the Truman administration had launched development (Point Four) and military training missions, and American consulting firms were assisting with seven-year-plan writing and the like. The most important Iranian adviser was Max Thornburg, the one-time Chevron executive who had run ARAMCO's sister company in Bahrain and designed the State Department's oil policy in World War II. He and others like Walter Levy, an up-and-coming consultant in the oil business, assisted the Iranians in their negotiations with British oil and statesmen.

The techniques the Truman administration used to support Ibn Saud were the same it used to support the monarchies in Egypt and Iran, including Point Four Aid, the Military Advisory Mission and, by 1952, a Financial Mission led by Arthur Young to establish the Saudi Arabian Monetary Authority. Yet the kingdom was by contrast a place where the Americans had displaced the British and where Harry Truman's "brother," the king, delivered remarkable value: key resources for the recovery of Europe and his "reasoned" course on the Palestine front. The Americans received a tremendous, positive return on their investment in fact if you consider that at the time developmentalist and populist coalitions in Latin America were challenging the privileges of foreign investors and, in Asia, wars and insurgencies threatened. Americans therefore tended to embrace the ARAMCO version of the wise warrior king and his kingdom, crowing about the crown prince's initiation of new reforms rather than criticizing the rampaging avarice and corruption in the frank way that others, the London *Economist* for one, did.

> Here are the facts. In 1949 the king received around $90 million, paid mainly in gold; it is estimated that by November 1950 he received another $70 million. Officially these funds go to "raise the living standards" of his people; in fact all but 10 percent, which he pays mainly to tribes to keep them quiet, goes into the bottomless pockets of the king, his immediate family and entourage. The embryonic administrative machine is financed almost entirely from the proceeds of taxation on pilgrim traffic resources. This in a good year yields over £4 million. Understandably enough, the Saudi family is only too ready to use its revenues in ways that preserve the medieval character of the country.[5]

A brutal account but true. Through mid-1951, the kingdom was borrowing heavily on future royalties from Chase in New York among other banks. It owed over $1 million to Bechtel for palaces and gardens, and the California firm was threatening to quit the country. Seven years later, in 1958, the same basic and unchanged facts about the political economy would become the main proof for Americans that King Saud needed to be replaced by his brother Faisal, who was, yes, a farsighted, modernizing reformer.

Modernization

While Bill Eddy advised and consulted across the wide expanse of Arabia, Mary Eddy would, as always, write letters home, and these invariably dwelled on the changes being wrought on places they first visited during the war. In Hofuf, the home of many of ARAMCO's workers, fifty miles inland from the oil enclave, where al-Hasa's amir Ibn Jiluwi ruled, the landscape might still appear like a "picture from the Arabian nights." She called it "the most Arab city I have ever seen and the most beautiful," "set in an oasis with beautiful

gardens and date palm orchards. . . . The old forts with high rounded towers stand out above the date palms." The amir's palace, where they lunched,
was "a real old Arab house but he had installed modern bathrooms, electric
fans, telephones, and we even saw a vacuum cleaner in the corner of the bedroom. The furniture was new heavy plush chairs arranged around the wall
and we saw a tag which said 'Sears Robuc [*sic*].' Even in Hofuf!!"[6] But as she
approached the Gulf, the Arabian Nights gave way to the future, California's
first gated community, City of Quartz, arising in Dhahran circa 1950.

> To look around this well-appointed comfortable house you would never know
> you were in Arabia. (Except that in America you would *not* have two good
> servants). Dhahran is so big now that it is easy to get lost. There are busses
> and taxis to take you anywhere you want to go but I usually prefer to walk.
> Sidewalks, trees, lawns, vegetable gardens, flowers, and even birds: All in a huge
> area surrounded by a high barbed-wire fence![7]

American Camp itself may have been fenced, and those inside the fence
may have thought themselves special, living in a liminal, not really Arabia,
not really America borderland, but the same was even more true about all
those drawn to Dhahran, who were on the outside looking in. The last scenes
in ARAMCO's movie *Island of Allah* trace the broad contours of the new landscape of power in al-Hasa, renamed the Eastern Province in 1953: from the pier
at Dammam to the railroad running from there to Riyadh, completed in 1951.
A Saudi Arabian Airlines DC-3 lands in Dhahran Airport, taxis ply new-laid
roads, and, of course, workers file into the refinery at Ras Tanura. The future
was still unfolding. "Throughout the year," the U.S. embassy reported, dutifully, every year, the kingdom "began to acquire other earmarks of a modern
nation": the first traffic lights in Riyadh in 1951, the first radio station in Mecca, the first stadium in Jidda for the new, wildly popular football matches, and
the Coca-Cola factory Eddy had mentioned in his own letter home, scheduled
to open in April 1951. The first hotel in Dammam, the Intercontinental, an $11
million investment by the Saudi Ministry of Finance, was going up on land
owned by the finance minister, Abdallah Sulayman.[8]

Business began to grow in Dammam and especially al-Khobar in response
to the influx of foreigners and concentration of population in these coastal
towns. A dozen merchant families handled 90 percent of trade, reported the
U.S. embassy, and contractors dominated the new Dammam Chamber of Commerce and Industry. The most prominent of these merchants and contractors

included the Fakhrus, the Qusaibis, and the Zahids, augmented by some who began new satellite ventures such as laundries for ARAMCO.[9] The company's new Arab Industrial Development Department provided support to Saudi workers-turned-entrepreneurs who started ventures to supply electric power to al-Khobar and Dammam, and, most famous of all, ARAMCO assisted its ex-employee Sulayman Olayan in founding the National Gas Company. Olayan went from selling bottles of propane produced by the Ras Tanura refinery to amassing one of the largest fortunes in the world. By 1952, the rudimentary Saudi provincial administration had relocated from Hofuf to Dammam.

Keep in mind that the government had expended vastly more in the west, the Hijaz, the richest and most populated part of the country, than it did in the east, where ARAMCO was to "carry out public works development." It spent much more too in Riyadh, "on the verge of a commercial and economic expansion undreamed of" just five years earlier.[10] The first great oil bonanza extended across the country, from the great merchant families of the Hijaz, the Alirezas and the Juffailis, to the new stars of Najd, like the Bin Ladens, who were building roads and putting up buildings everywhere it seemed.

Inside American Camp, George Rentz argued for more money and men for his Research Division to meet the challenge of the new world. His "division exists primarily for the purpose of providing guidance and research and translation services to other divisions of the Local Government Relations Department and all other branches of the company." Demand for its services was growing given the "substantially enlarged Saudi Arabian Government organization" in the Eastern Province. "The recent government reorganization will bring about closer relations between the Company and the government. The establishment of new Saudi Arabian Government departments and the appointment of new high-ranking officials require a considerable expansion of the research program."[11]

Ibn al-Jazira (Son of the Peninsula)
The government made its most important and fateful appointment by far in 1949, when a Najdi, Abdallah Tariki, moved to Dammam to become the Ministry of Finance's supervisor of petroleum affairs.[12] Tariki's rise in Saudi oil policy making is often attributed to Prince Faisal's influence.[13] A CIA-produced biographical sketch says that Tariki's promotion in 1954 to become director general of a new Office of Petroleum and Mineral Affairs inside the Ministry of Finance was Faisal's doing. Faisal created the office. Tariki filled it.[14]

Tariki was smart, dedicated, driven, principled, and apparently incorruptible, as even the CIA acknowledged. When he died in 1997, the exiled novelist Abdelrahman Munif paid him a glowing tribute. Tariki "tried to give the region and the era what they were worth, what they deserved."[15] Predictably, therefore, because the company would be made to pay, ARAMCO managers portrayed him as an emotional, hot-headed zealot. Barger once called Tariki a "superb demagogue." Perhaps, but as the CIA reported, "Tariqi's personality and attitudes have proven difficult for Americans who have to deal with him."[16] Little wonder.

Wanda Jablonski, who was emerging as the most influential oil journalist of the era, reported in *Petroleum Week* (February 22, 1957) on Tariki's long and unstinting "investigation into almost every phase of ARAMCO's activities." Tariki's first target was the company's evasion of the fifty–fifty deal. He had discovered that the posted price used to calculate the kingdom's share of profits, the price ARAMCO sold Saudi oil to the four owner companies, was being kept low, with the rents going to the multinationals. The Saudi share was effectively 22 not 50 percent of profits. The owners' net annual return after payment to the house of Saud averaged 57 percent through the 1950s. This was five times greater than average returns inside the United States.[17] Tariki's research led to growing pressure on the company in 1951 and 1952 and ultimately to revisions of the fifty–fifty deal, with compensatory payments to the government.

Tariki opened up a new line of investigation after returning from Venezuela in 1951, where he developed a clearer understanding of the unfair wages paid his countrymen, the lack of opportunities for Saudi workers, and the firm's substandard native housing. Tariki apparently learned a great deal by comparing ARAMCO to Creole Petroleum, a firm that had already been pushed to integrate, in both senses of the term. In industry parlance, it was an integrated company, combining downstream (crude oil production) and upstream (marketing of refined products) activities.[18] ARAMCO began to face demands for moving the company headquarters from New York to Dhahran, for holding board meetings there, and for appointing Saudis to the board of directors. In Venezuela, too, as in Iran, integration of local skilled labor, technicians, and managers inside the ranks of the company was much more advanced than in ARAMCO, and Tariki began to argue for creation of a talented tenth of the kingdom's own. ARAMCO of course resisted. It would take eight years before it invited him to join the board as the first Saudi director.

In an introduction to Tariki's collected works, the editor, Walid Khadduri, emphasizes in particular Tariki's interest in the education of young coun-trymen. His first published article, "Where Are We Heading?" called these Saudis "the true wealth of the nation." Tariki acknowledged the alienation among educated Saudis and the resistance they encountered upon returning home, but also exhorted them to uplift the nation, "raise the level of the poor, treat the sick, take the weak by the hand, teach the illiterate."[19] A Saudi lawyer, Muhammad al-Hushan, recalls that throughout the 1950s Tariki's home and office served as a salon for ambitious young engineers and others.[20] Khadduri recalls Tariki's efforts to build up the expertise of the nascent oil directorate and then ministry, recruiting talented Saudis and others, especially Palestin-ians, many of whom obtained citizenship in 1960 due to his efforts.

Tariki relentlessly pressed the issues of ARAMCO's evasion of its obliga-tions and responsibilities under the concession and its devotion to the hier-archy it had built in eastern Saudi Arabia. After his flight from Saudi Arabia, he expertly dissected the rationale behind ARAMCO's turn, around 1950, to supporting Saudi contractors nearby the camps, which it began to call "Arab Industrial Development" in the company's photo montages and the like.

> ARAMCO found that applying the Law of Work and Workers to all the workers they needed in their operations would cost them a lot, so they introduced the concept of contractors and vehicle owners. . . . And the mission of those new contractors was to collect willing workers, and the company would bring materials and engineers to train the contractor and his workers to do the work required of them. In this way the company was not responsible for arranging accommodation or health insurance or the care of the workers and their families.[21]

We have already seen the emergence of the arguments and strategies for avoiding the costs the firm would have to pay if it were to extend the benefits of its welfare work measures for whites to the entire Saudi labor force or fulfill its commitment to educate and train Saudi workers and their offspring. Thus when managers complained that it was the Saudi government that was delay-ing the beginning of ARAMCO's housing loan program, it means that men like Tariki were seeking to move the firm down a more equitable (costly) path, in housing, training, education, and the like. Doubtless, the firm's men saw things differently. "For a lot of people in the oil business back then . . . if you weren't pro-ARAMCO you were pro-communist. I remember being at parties in Dhahran and hearing general managers talking about 'that red bastard in

Riyadh' meaning Tariki—it certainly wasn't true."[22] Ultimately what matters is that following the strike wave of the 1950s and the rise of a kind of proto-populist tendency represented by Tariki and others, the firm would be forced slowly toward Saudization.

Tariki's own encounter with the racism inside American Camp is one of the few well-documented moments in this decade-long campaign to domesticate ARAMCO. Journalists in the 1950s and 1960s were, after all, witness to what Howard Winant would later describe as the "break" in the world racial order of the past century, and it became acceptable to suggest that perhaps Americans had been, on occasion, insensitive to their "partners" and "hosts." This was also a time when one or another company man resorted to shallow psychologizing in order to reconcile the beliefs about ARAMCO exceptionalism with Tariki's increasingly visible role in the campaign that by 1960 led to the creation of the Organization of Petroleum Exporting Countries.[23]

One clause of the fifty–fifty agreement specified that ARAMCO would pay $700,000 annually "towards expenses, support and maintenance of representatives of the Government concerned with the administration of ARAMCO's operations." Tariki used the clause to demand the right of him and his American-born wife to housing inside the senior staff camp in Dhahran. ARAMCO refused, offering instead to build him quarters in one of the native camps, but ultimately the Tarikis were admitted inside the fence.

> "I was the first Arab to penetrate into the tight ARAMCO compound" he said last week, "and I never saw such narrow people." American matrons took his wife aside and reproved her for marrying an Arab. Says Tariki bitterly: "It was a perfect case of an Arab being a stranger in his own country." For "purely personal reasons having nothing to do with nationality," Tariki's marriage broke up, and his wife and son Sakhr now live in Poughkeepsie, N.Y.[24]

Over a decade later, American officials would still complain of the breach in the wall between them and the Saudis who sought special access to ARAMCO facilities after Tariki took up residence in Camp and thought that gave him the right to swim in the pool, attend the free movies, and so on. It is Tariki's living with a white woman that obsessed his neighbors most, however. The failure of his first marriage in the 1950s was absorbed into the evolving lore about Tariki and the psychology behind his Saudization campaign, which was repeated for decades. Imagine Barger sometime in the 1960s or 1970s glancing down at his notes to offer a gathering of petroleum engineers in the United States the inside story on Tariki and the campaign against the firm. "A. Tariki—complex guy, strong ideals, *married a tramp*—worst thing happened—he did love her and felt inspired to do this," meaning bring her to Arabia with him, but he "couldn't do—young, just out of school, an ardent Nasserist." Now imagine someone in the Nixon White House reading the CIA's biographical profile of OPEC's founder.

> Tariqi was discriminated against in Texas bars because they thought he was Mexican. Married an American, Mrs. Eleanor Nichols, a Texan, and was sometimes hassled when trying to use ARAMCO facilities for American employees when entering with his American wife. Before their divorce in 1954 he had to tolerate her indiscreet behavior . . . with American ARAMCO employees and some members of the royal family.

Her name was Eleanor Ramnitz, not Nichols, and she was from Poughkeepsie, New York, not Texas. But no one ever forgot what led to the divorce.[25]

COVERT MISSIONARY

William Eddy unveiled another piece of the ARAMCO mythology in an address in 1952. We don't know the audience, but we know the context. The Federal Trade Commission had charged ARAMCO's owners along with the owners of a handful of other companies with conspiring to control worldwide oil production. Around that time another of the oil cartel members, the U.K. government-owned British Petroleum, which in turn owned the Anglo-Iranian Oil Company, was accused of economic exploitation and political intervention in Iran, hence the push to nationalize its holdings there. So Eddy set out to correct the record, because, he said, "We do not teach political democracy, nor try to replace prime ministers, nor monarchs, like the Anglo Iranian Oil Company." Eddy's dislike for the British in the Middle East had never abated. "We do not claim extra-territoriality. We admit the King can kick us out any minute. Nor would we invoke the World Court, nor ask for warships. We do not teach social revolution nor tear off veils. We do not aspire to be an East India Company. We prefer banks and Bechtel."[26]

When Wallace Stegner added a late introduction to the company history that he first drafted in 1955–56, he made a slightly different claim. He wrote that the company gradually learned to "give up political influence" in Saudi Arabia, but by 1970 the mythology was so deeply rooted that a Government Relations man, Malcolm Quint, had Stegner's version struck out, insisting

"ARAMCO did not give up direct political influence—it never sought such influence."[27] Yet just a few months before Eddy's lecture where he professed that ARAMCO was not in the business of replacing prime ministers, two top ARAMCO officials—Floyd Ohliger and Garry Owen—met with U.S. embassy officials in Dhahran to discuss whether "pressure might be applied to force out the Financial Minister," Abdallah Sulayman.[28]

The men at State tended toward the company's increasingly apprehensive view of the situation in Saudi Arabia—neither for the first time nor the last—as events in Egypt and especially Iran played themselves out. Scholars have generally missed this moment or misremembered it under the sway of the official ARAMCO story about the past. That version holds up the fifty–fifty agreement negotiated in December 1950 as proof of the firm's difference from British Petroleum in Iran, just as Eddy said, and its sagacity in avoiding a similar fate. But reading the moment as one in which the relatively enlightened firm resolves the most pressing problems with the government, stays on the course it had set, or even makes some slight adjustment to it, makes it difficult to explain the new, more explosive round of labor unrest in 1953, as if a month-long general strike of the entire labor force just happened out of the blue.

The reality is that ARAMCO was subject to steadily increasing pressure from above and from below after concluding the fifty–fifty agreement. The government demanded new loans and advances and accused the firm of violating the agreement, as we have seen. ARAMCO blamed its troubles on the finance minister, and U.S. embassy officials basically agreed with the too-simple story. Eddy wrote his wife about "the trouble hitting the company" that would keep him away from home.

> Trying to destroy the influence of the Crown Prince and to get finances and power wholly in his hands, Abdullah Sulayman (who is again drinking heavily) told the King a pack of lies and the King told Ray Hare [the U.S. ambassador] that ARAMCO has been insulting him, deceiving him, cheating the Saudi Government, and that "all those with whom he has been dealing recently in the Company must leave Saudi Arabia." . . .
>
> The Iran business is going to affect us very shortly in some very serious ways. Iran is demanding and getting Iranians on the Board of Directors, as is IPC in Iraq, and Saudi Arabia is sure to demand that right away. Also, there are questions about the price of oil when ultimately sold, compared to the lower

price on which royalties are figured (the price for oil when it leaves ARAMCO) which are being raised bitterly by the Saudis.[29]

So Sulayman represented the main source of corruption (conveniently forgetting the royal family for a moment), a chief obstacle to "reform" of the archaic and hemorrhaging financial regime, and a front of resistance to the crown prince, whom ARAMCO was supporting. ARAMCO may have been "burdened before" with bailing out the house of Saud due to "financial irresponsibility," but the situation was now dire, the state so decrepit, the finance minister so erratic because of his alcoholism, that it was an open question "whether any real government exists."[30] This nonexistent government dragged the firm into negotiations over TAPLINE (spring 1951), guarantees for loans to its creditors (August 1951), where, as we saw, the king attacked the company for its disloyalty, and, as the Iranian nationalization went forward, over Saudi participation in governance of ARAMCO (February 1952).

The firm's analysts argued that the king's senility had made him vulnerable to venal Sulayman's machinations against the firm and the man it was banking on to protect the concession, Saud Ibn Abd al-Aziz. ARAMCO Government Relations had in fact been shooting footage of Saud systematically since 1951, when the king also appeared about to die. The White House chief of protocol had a telegram of condolences prepared for delivery in March 1952. "The American people were proud to count him and his nation among their most trusted and valued friends." While the king hung on for another year, ARAMCO readied more movies "for immediate and worldwide release upon the death of His Majesty Abdel Aziz as a psychological contribution to insuring the general acceptance of the accession to power of the Crown Prince as a worthy successor to his great father."[31]

Colonel Eddy, the firm's most knowledgeable adviser on Middle Eastern affairs, was sure "that the crown prince would make a very capable leader and would be sympathetic to the oil company's cause in Saudi Arabia, but aside from the crown prince, the other people in power, notably the minister of finance, were entirely incapable and selfish in desires."[32] The firm thus proposed more than just a public relations program designed to sell the story of Saud the modernizing reformer. As the new Saudi specialist at State put it following consultations with Eddy and James Terry Duce, ARAMCO's vice president, "Shaikh Abdallah Sulayman must necessarily leave the Ministry of Finance which might pave the way for a more orderly system of collecting and expending Saudi Arabian revenue to be inaugurated."[33] Duce later had a series of meetings with Saud and reported that the prince planned to replace Sulayman with an ARAMCO favorite, Abdallah bin Adwan, head of the Ministry of Finance office at Dammam.[34]

The government was only half the problem, however. During a week-long visit in May 1951, the State Department's petroleum attaché reported that the company faced increasingly discontented workers among all segments of its labor force, including the Americans and other foreign staff. He guessed that better employment conditions in their home countries made the terms offered by ARAMCO increasingly unattractive, but all we can say for sure is that turnover costs were high, and dramatically high for women. As we saw, these costs explain the kind of training program ARAMCO turned to, teaching Saudis non-transferable single skills quickly, which it would soon tout as its contribution to the country's development.

In a section of the report labeled confidential, that is, representing information from ARAMCO sources or an account that the embassy ought not repeat back to the firm, the attaché singled out the company's continued failure to build any married housing for Arabs even after the Saudi government had pressed them on this point because other countries had successfully extracted this concession from oil companies. ARAMCO blamed the government in turn for rejecting its proposed alternative building and loan plan. The attaché pointed out, however, that married housing would still be necessary in those parts of ARAMCO's expanding operations where no town sites yet existed, and he urged the company to go to Egypt and see the married housing being built by firms operating on the Red Sea coast.[35]

The firm faced several "localized work slow downs and stoppages" in the early 1950s. One was a strike in 1951 by dockworkers in Ras Tanura, in which the king intervened directly. This fact may explain his deepening anger, requiring emergency tending to by Ohliger, who knew Abd al-Aziz longest, but also the firm's inability to acknowledge any source for this anger other than manipulation and deceit by its imagined enemies.[36]

The U.S. labor attaché flew from Abadan to Dhahran a few months later, in August 1951, as the Iranian nationalization was unfolding, and filed a remarkable report comparing conditions and prospects in the two oil states. In Khuzistan, at a "last supper" before their flight out of the country, executives of Anglo-Iranian were contending that the "social time table of nationalization" in Saudi Arabia was about one year. The attaché wisely rejected this conclusion, which he thought was based on a superficial reading of political dynamics.

> However backward the Persians might appear and however politically unsophisticated outsiders might think them, they are immensely more alert and mature in their economic and political policies than are the Saudi Arabians. . . .
>
> The government of Saudi Arabia is of course non-apologetically autarchic [an absolute despotism] and the number of those controlling the destiny of the country and the purse strings of the oil revenues could be counted on two hands. Iran is notable to westerners for the limited number in the ruling class, but that class, whatever its numerical limitations . . . is broader based than that of Saudi Arabia. There exists no rebellious force, even of negligible consequence in Saudi Arabia. . . . As a result, the ARAMCO officials can ignore public opinion, which is neither a force nor a voice, and concentrate their attention on the production of oil and purely economic considerations. . . . There appear, in the immediate future, no barriers to continued domination of the few . . . , no evidence of any force or individuals even working clandestinely, capable of and interested in leading a labor movement.[37]

Consistent with the analysis advanced in the past few chapters, the attaché turned on its head the loyalists' account of the firm's own more generous and progressive policies as key to its fortunes. The one dimension in which ARAMCO stood out in comparison to British Petroleum in Iran was its investment in a public relations program. ARAMCO made a display of consulting with Saudi officials and publishing colorful reports "in green and white," but in other key respects the firm lagged far behind, and the firm knew it. Thus "Hamlin [George Hamlin, ARAMCO's new assistant director of personnel] is reportedly most interested and concerned by the backwardness of ARAMCO industrial relations policy, at least as compared with Venezuela, where he had been serving earlier."[38] Hamlin was banking on the new single-skill-training program and the Venezuela-type reform of the housing regime, although he and his colleagues were still debating whether Saudi workers were to be directed to build their new homes "within established Arab communities or . . . in a new area remote from other older villages."

A few months later the chief of the Petroleum Division at the State Department, Robert Eakens, found himself on a plane with Abdallah Tariki en route to the States from Caracas after attending the Venezuelan National

Petroleum Convention. Eakens drew Tariki out on Saudi Arabia's relations with ARAMCO, and his assessment echoed those found in the analysis of the U.S. labor attaché. Asked to name the most pressing problems with the firm, he replied the low wages and lack of opportunities for Saudi workers and the failure to provide decent housing for them and their families. Tariki was simply more explicit about the inequity undergirding the oil industry in Saudi Arabia and the attitudes familiar to him from his time in Texas. "He said a director of ARAMCO was asked why the company had not developed plans for family housing and the director's reply was Haven't you read your bible? It says that Saudi Arabians are supposed to live in tents all their lives."[39]

The U.S. labor attaché in 1951 understood, as did virtually anyone else who stopped to analyze the situation, that ARAMCO's industrial relations policies remained "backward" because no labor movement had so far emerged that might push the costs to the firm of maintaining the hierarchy of the camps above those of building "Arab housing" on par with American housing and educating its workers' children, let alone paying what it paid white people. These were all, as we have seen, already understood and discussed alternatives to the "classic" American Camp regime. They were known not least because they were already in play in places like Iran, Iraq, and Venezuela where there indeed were, as the attaché put it, "forces . . . capable of and interested in leading a labor movement." In Saudi Arabia, ARAMCO instead could temporize "and concentrate [its] attention on the production of oil and purely economic considerations."[40] It would in fact try to get away with a lower-cost alternative to training workers until forced to adopt the British Petroleum model after strikes in 1953 and 1956.

Some Americans, arguably including even some company officials, might have believed that by doing more at the margins ARAMCO could perhaps preempt the rise of nationalist and anticapitalist forces like those found in other parts of the world oil frontier. The trouble is, it was a gamble at best, and others inside and outside the company believed that a consequence of educating and "westernizing" Saudis would be to accelerate rather than contain the growth of radicalism. Some accounts of the Iranian nationalization tended toward this conclusion.

> The experience of the Anglo-Iranian Oil Company in Iran proved to the satisfaction of the oil executives that the element of altruism which they introduced in their labor training program could "boomerang" to the

dissatisfaction of all concerned. For example, the AIOC deliberately trained more personnel, especially at the intermediate levels, then they could absorb within their own organization (unless they jeopardized the positions of their own nationals). It was intended that the surplus be absorbed by the Iranian government to steady and make more effective the Iranian bureaucracy and that other trained Iranians would be sources of more trained personnel for private Iranian industry. For reasons largely beyond the company's influence, neither the Government nor private industry had absorbed sufficient numbers of trained people to eliminate what then became a sore core of frustrated, educated Iranian workers.[41]

As the pressures on multinationals from populists increased elsewhere, both ARAMCO and the U.S. government increased their investment in the security of the Al Saud. As we saw, ARAMCO was promoting Crown Prince Saud Ibn Abd al-Aziz as the able successor, and was even ready to bargain upward on the share of oil rents, banking on the Saudis' fears of their enemies, Jordan and Iraq, to check any tendency to strong-arm the firm too much. The Truman and later the Eisenhower administrations did their parts as well. A now long-forgotten U.S. foreign aid mission (the Technical Cooperation Administration, or TCA) paid for the organization of the new Saudi Arabian Monetary Authority, which the Americans hoped might finally begin the process of fiscal reform—data gathering, budget making, and stabilizing the price of the riyal—talked about since World War II. Otherwise these earnest New Dealers looked in vain for "grassroots" organizations and initiatives to support.[42]

The Americans also built up the air base, guided the creation of a Ministry of Defense (the minister, Prince Mishaal was about twenty-five when appointed), and beefed up the military training mission. Arms were another matter. The Saudis wanted new tanks and planes, but were staggered a bit when told the cost: well over $150 million to outfit and train one 18,000-man division and one air force squadron. The United States provided $15 million in military aid in 1951 (with additional funds for upgrading the Dhahran base).[43] It was going to be a long haul.

Americans Who Know the Arabs

Other means existed to protect "a very large, rich and powerful American industrial corporation operating at full blast" in a country that seemed surrounded on all sides by treacherous national and populist political currents.[44] ARAMCO and in particular its Government Relations Organization became a primary cover for a growing number of Central Intelligence Agency assets in the 1950s in addition to Eddy, who had suffered a heart attack in 1952 and would not return to the field until 1955. One was Richard Kerin, who was involved in special operations and served as the CIA's station chief in Baghdad in the early 1950s. He moved to the ARAMCO settlement at Abqaiq in 1954 and was working side by side with Duce at the United Nations in 1958 during the Iraq and Lebanon crises. Bill Mulligan identified a handful of his own friends and colleagues in Government Relations as agents. Homer Mueller, one of the company's "special employee-trainees" at the Middle East Institute—the track that David Dodge had declined to take—headed the Riyadh office through most of the 1950s. Robert Headley, Jr., arrived in Dhahran in 1952 and eventually became chief of station in Oman in the 1970s. Ron and Helen Metz were recruited together. Doubtless, there were others. Another Relations man, Harry Alter, the son of missionaries and a new recruit in the 1950s, had a brother who was exposed as having been a CIA agent in this same period. We know much less about Arabs inside the kingdom who were recruited. I have only been able to identify one man who was turned. He was Hassan Yassin, the son of Deputy Foreign Minister Yusuf Yassin, a trusted advisor of Abd al-Aziz.[45]

Thus when Eddy told his audience in 1952 that unlike British Petroleum the oil firm he worked for did not intervene in Saudi political life, the disingenuousness was finely layered, as we might expect when a man made a career of deception on behalf of his country. Eddy sometimes lied outright. For instance, in 1947 his friend the king asked him in private, "Do you have any connection with your government any more, and are you a channel through which I may reach officials of your government?" Eddy said he was categorical. "No. I told him I had resigned from the service of my government completely."[46] The efforts by ARAMCO to back the crown prince and later king against Abdallah Sulayman would continue, and Eddy's pulse might not quicken. Perhaps he inflated the idea of Britons making and unmaking prime ministers, in which case his participation in discussions about the fate of Sulayman hardly seemed significant. Nonetheless, the close ties forged through Duce to the Central Intelligence Agency blurred the lines once again between ARAMCO and a state that would seek actively to make and unmake prime ministers in Iran and elsewhere around the world.

ARAMCO also began to increase its investments in those institutions at home and elsewhere that provided it and the CIA with specialized knowledge. As Eddy wrote to his son,

> I am glad to hear of the students in the Near East Department who are preparing for careers in this area. I don't know whether you know, but over the past eight years we recruited one by one, a number of Near East specialists from Princeton who are doing very well here. . . . As a matter of fact, as you probably know, ARAMCO contributes to institutions like Princeton, the Middle East Institute, at Washington, and the American University of Beirut not only because these centers prepare future employees, but because they also equip men to come out to the Near East in the Foreign Service, or in teaching or other capacities, which strengthens the small band of Americans who know the Arabs and understand them.[47]

The firm had come to believe in senior Arabist George Rentz's view that engineering or technical know-how alone was not sufficient to secure the company's position in Saudi Arabia. "The problem of establishing a common meeting ground for two vastly different civilizations must be faced and overcome," Tom Barger wrote in 1951, through an organization devoted to "securing the full truth about the life, character and background of the Arab people."[48] This faith in the Arabian Affairs Division's increasingly "intimate knowledge" of the Bedouin was put to the test very soon, however.

The U.S. labor attaché reported that only the year before, as the Iran crisis was unfolding, ARAMCO's management assumed that its investment was generally "safe for the next five years and hoped that the Company could remain in profitable operation for a further twenty years." It therefore chose to rely on the "Crown for the control and discipline of workers. They felt they could afford largely to ignore worker demands, having passed the obligation to the Crown, and in effect to resolve their own industrial relations difficulties by being completely solicitous of the Crown and Crown intrigues."[49] It didn't work.

SLAVES OF ARAMCO

On January 23, 1953, Crown Prince Saud addressed a letter to the president of ARAMCO on behalf of his father, the king. One of Abd al-Aziz's loyal subjects, Abd al-Aziz Abu Sunayd, an ARAMCO employee (no. 49128), had sought the king's help to save his job. Abu Sunayd had started with the company in 1949 as a teacher's assistant, was promoted in 1951, and sent in March of that year to the United States to teach at the new employees' training camp in Long Island. The same summer ARAMCO sent him to the American University of Beirut, one of the first Saudis approved to attend special summer sessions there. Since then he had worked as a teacher at the Dhahran Training Center, but claimed he had been suspended when his supervisor learned that Abu Sunayd was reporting to Crown Prince Saud about ARAMCO's education system. If Abu Sunayd was a source, his version couldn't have been much worse than the one by the Point Four administrator to Washington who visited the site a few months later ("facilities were inadequate in many instances and were housed in improvised shelters. There was the usual weakness in available instructional materials"). When the U.S. embassy officer reported the story of Abu Sunayd to the State Department ten months later, he also said the account wasn't true. That is, either Abu Sunayd hadn't been suspended or had been but not for the reasons he claimed. We don't know more than this about the incident, unfortunately.[50]

We know that in March the Saudi papers *Bilad al-Saudiya* and *Umm al-Qura* reported that the crown prince had made great progress in many reforms designed to secure the welfare of the Eastern Province, which included amending the 1947 labor code and gaining ARAMCO's recognition, finally, of a responsibility for educating the sons of its workers. The company had agreed in principle, the U.S. embassy emphasized—the details to be worked out in future negotiations—to fulfill its obligation under the 1947 law to build schools for Saudi children and contribute to their operating costs.[51] The crown prince was also apparently linking prosperity to security because he ordered the creation of a new Public Security Agency for the Eastern Province. An Egyptian mission was visiting the kingdom to advise on improved policing methods, and the American aid administrators were expanding their definition of grassroots initiatives to accommodate leadership grants for Saudi police officers to travel to the States.[52]

Abd al-Aziz Abu Sunayd may not have gotten the news of the impending reforms and prosperity or may have hoped to speed things along because on May 23, 1953, he and 154 other "intermediate skilled" Saudis, Palestinians, and various third nationals signed a petition to ARAMCO management calling on the company to pay a cost of living allowance and to provide better working conditions and privileges. Management took more than a month to clear its calendar for a meeting with the employees who had agreed to the condition that they send a small group to represent them rather than attend en masse.

The point is important because the company would then claim that these men had no right to speak for the others. When they finally held the meeting on June 30, Abu Sunayd and six other Saudis, Saley al-Zaid, Abdullah Ali Al-Ganim, Ibrahim al-Faraj, Abd al-Rahman al-Bahijan, Umar Wazna, and Abdul Aziz As-Sufayyan, presented themselves as spokesmen.[53] All seven were men whom the company had sent to Beirut for study in the past year or two.

At the meeting Abu Sunayd and his comrades returned to the issues that had been the focus of every action since the strikes of the 1940s, namely the failure to promote Saudis to senior-level positions and the treatment of them as inferiors inside the camps. The men went further, however, insisting that they be recognized as speaking for all of ARAMCO's Saudi workers. The company refused and called an immediate halt to the meeting. It directed them to pursue any grievance through the Dammam office of the Ministry of Finance, the workers' only legitimate representative, a procedure in place since the 1945 strike. They did, and the director of the Labor Office, Abd al-Munim Majdhub, in turn, directed the company to return to the negotiating table with the delegation. On the morning of September 5, the date of the next scheduled meeting, Majdhub forwarded an official letter endorsing the group's agenda, surprising ARAMCO's managers because he seemed to be opening the door to formation of a union. As an embassy officer later put it, "this is the first time the Government has allowed even the semblance of a labor organization to become established."[54] It was, in fact, the first and the last time.

The workers' collective action may have come at a propitious time and so was welcomed by various officials in Jidda and Riyadh. This argument runs counter to the one ARAMCO's Government Relations men were making in September and October. They said the movement was designed to take advantage of a weakened and distracted central administration, preoccupied with the king's health, at a time when the king's most trusted ally in Dammam, Ibn Jiluwi, had gone to Paris for emergency medical treatment.

Perhaps, yet it was also a time when the Ministry of Finance and the company were battling over many issues, including new "discriminatory" railroad rate charges, new TAPLINE duties, and an order that ARAMCO start to provide the eleven thousand additional Saudis working for subcontractors the same wages, medical treatment, housing, and vacation allowances that it provided its own workforce. The firm was asked to surrender the files on all its non-Saudi workers to a government committee charged with increasing the ranks of Saudis on the company payroll. Finally, the government ordered the firm to use Arabic in all its own records. All these measures, the U.S. embassy reported, were "plainly designed to force ARAMCO's acquiescence to the Government's oil pricing proposals."[55] Allowing the workers' representatives to press on was tantamount to the opening of another front in this campaign. It might be the case, too, that some Saudi officials actually thought the workers had a point.

The company took a hard line. It called off all further meetings with Abu Sunayd and the others, while offering to discuss solutions to the original demands with the government. Then, on September 22, it informed Shaykh Abdallah bin Adwan, the highest-ranking official in the Ministry of Finance with a direct line to the king and crown prince, that it rejected outright the main demand. It would not pay a cost of living allowance. The company would later be forced to reverse this decision. The workers responded by recruiting others to their cause. Broadsides against "our masters, the Americans" appeared on the walls in camp, and men began to hold meetings at the mosques. The would-be representatives filed more complaints with the government and sent around more petitions for signatures. Almost half of all Saudi workers signed a petition in September calling upon Crown Prince Saud to recognize the committee as a collective bargaining agent. Later that month, sixteen of the original petition signers sent a telegram to Saud protesting ARAMCO's suspension of their Ras Tanura delegate, Abd al-Rahman al-Bahijan, and reiterating the request to allow the workers to form a union. By early October, the Bahrainian weekly *al-Kamilah* was reporting on the mounting disturbances in the Saudi camps.

ARAMCO labor relations and intelligence units were by then in full swing, their informants inside the camps reporting that the threat of a strike was real, and that organizers were exhorting others to rise up. "One irresponsible agitator is reported to have said that the aim of the group was to overthrow, first, ARAMCO, second the Saudi Government, and third, Islam. An argument frequently used was oil belongs to the people." Others "have used, most regrettably, the religious issue of 'Christian dogs' in appealing to the newest ARAMCO employees, often Bedouins just arrived from the desert."[56] The firm in turn began to exhort the government to act to head off the threat. The acting amir, Ibn Jiluwi's brother Abd al-Muhsin, finally agreed to leave the comfort of Hofuf for Dammam and ARAMCO's representative began supplying him and the chief of police copies of the bulletins, tracts, and newspaper stories circulating inside the camps. He was reportedly unconvinced ("and clung to the 'it can't happen here' attitude") despite the police chief's warnings to take stronger measures. Bin Adwan too, allegedly ignored the entreaties of an ARAMCO envoy but gained the crown prince's approval to form an investigative commission to consider the workers' complaints. Bin Adwan, two other notables, and two assistants reached Dhahran during the first week in October.[57]

As the workers began to testify, ARAMCO provided the commission members with the personnel files of the leadership, and if the American embassy's melodramatic account, which is basically the firm's own, is even semi-accurate, then Abu Sunayd and his brothers overplayed their hand by demanding formal recognition as a condition for participating in the commission's proceedings.

> Their conduct became more defiant and their statements more acrimonious as deliberations proceeded. They stated that what happened to them was unimportant compared with [the] movement they represented or the welfare of their fellow Saudis. After accusing members of the Royal Commission of having no regard for the welfare of workers, of having sold out to the Americans, of being "sons of ARAMCO," they stated that unless they were recognized as the representative of ARAMCO Saudi employees and unless their demands were met, the workers would be called on strike, that oil operations would cease, and that enough American blood would flow to show the Royal Commission that workers were the real masters of the situation.[58]

The chairman of the commission instead had the leaders arrested and imprisoned in Hofuf on October 15. At the same time, the Saudi colonel in charge of the Dhahran Air Base was given control of all military forces in al-Hasa, reinforcements were sent from al-Kharj, and 1,000 troops were moved into the vicinity of the camps. The next day, on October 16, as the workers' committee had promised, demonstrations broke out in Saudi Camp and in nearby villages, and after protesting at the police station where the leaders were imprisoned, a crowd staged a march to the American consulate and the air base. American citizens hid in their compounds. Private cars and a U.S. Air Force bus were stoned, although no one was injured, before Saudi troops were ordered to crush the protest. The soldiers were brutal.

The Brightest Boys Are Always Unhappy

A strike was called on October 17 to protest the arrest of the movement's leaders. At 2:30 PM, a crowd, some said 500 and others said 2,000 strong, gathered in Saudi Camp to listen to speeches. As a participant insisted decades later,

"We did not respond" when the workers' committee first tried to mobilize around "more cash and better food. . . . But when they told us to ask for political rights, we all responded and joined the strikes in 1953."[59] Less than 10 percent of the labor force showed up for work, and the strikers staged massed meetings in various camps and towns. What shocked the company was the effectiveness of the boycott and the size of the crowds that assembled in public. The causes were obvious to ARAMCO and to the consulate: the police were ineffective in rooting out the "intimidators" and rabble-rousers. The local army commander refused at first to be drawn into the search for "labor agitators," and the senior police official, Abdallah bin Issa, deputy director of public security, was at one point himself "beating workers in Saudi camp with a stick in an attempt to disperse the crowd." The movement succeeded, the firm insisted, only through "threats and beatings" of the other workers. Nonetheless, by October 19, the strike had spread to the U.S. air base.[60] Reuters reported that the basic issue was a demand to be treated the same as the Americans.

The workers stayed out for ten days, many having returned to their villages. In order to continue pumping oil, ARAMCO reassigned men, relying on the Americans and third nationals working overtime and doing without the services that Saudi general labor supplied as drivers, gardeners, garbage haulers, kitchen help, and the like. The senior managers held two contradictory views of the strike at once. One was that it hardly mattered, the strikers had barely disrupted operations, and that the local authorities above all should not force them to return to their jobs. The Americans feared the prospects of sit-down strikes or sabotage if the men were beaten and dragged back to the workplace. The second view was that the government was rapidly losing control of the situation. Loyal workers were being threatened, they said, and the strikers had even roughed up the amir's secretary in Hofuf after he read to them the crown prince's reply to their telegram calling for the release of those in jail. By October 20, around one thousand Saudis had been arrested. The U.S. embassy reported that the amir had in effect been replaced by the crown prince's confidant, Ibn Adwan, head of the royal investigative commission, and the army commander Colonel Muhammad Awartani, as the real power in the Eastern Province. ARAMCO negotiated directly with them in working out a plan for the army to send troops into the company's camps to break the strike, they hoped, without recourse to undue violence.[61]

As the army prepared to end the strike, the company's public relations arm swung into action, reporting that ARAMCO had wanted to continue negotiations with the original workers committee back in May but, tragically, could not "since labor organizations are illegal in Saudi Arabia." The opposite was true, as we have seen, and in reporting this remarkable piece of revisionism, the U.S. embassy finally dug out the labor law and corrected the long-taken-for-granted belief about unions being outlawed.[62] ARAMCO assisted the army in its raid on the workers' camps. Suspects were interrogated on the spot, leading to twelve more arrests, and "communist literature" was confiscated, but the camps were also mostly deserted, the men having gone home. Saudi officials planned next to go into the villages and root out the troublemakers. The U.S embassy reported the beginning of a plan to deport the strike leaders to Najd. Ibn Jiluwi was troubled by the fact that his deadline for the men to return to work had come and gone, and bristled when he heard about radio broadcasts quoting the oil company's line about the strike being in essence a "revolt" against the government rather than a strike against ARAMCO.[63] No one seemed to notice the difficulty in drawing such sharp distinctions between these two theoretical hierarchies.

The Saudi officials continued interrogating camp residents while most workers continued to ignore the government order to return. More men ("unwilling workers and suspected agitators") were hauled off to jail, and ARAMCO's Garry Owen warned that the Saudi public security system had failed to do what was necessary to track down the leaders—a complaint Americans have never stopped making, apparently. The first deportations of fifty-six "undesirables"—forty-five workers and eleven others—to Riyadh took place under armed guard on Friday, October 23. A large crowd of Saudis stood and "booed and jeered the troops." Shades of Bisbee and Jerome in the U.S. copper belt. Other workers finally began returning to the camps. Thirty percent, those mostly in the lowest classifications and, it was guessed, those most strapped for cash, were back by midweek. By then, Barger and other ARAMCO officials had met with the royal commission to reopen discussions over the original demands for a cost of living allowance and transportation subsidy. The commission stressed that the agenda would eventually expand to take up the issues raised in subsequent petitions, including "better housing, more Saudis in jobs now held by foreigners, more hospitals, more schools, more drinking fountains, Saudis trained at universities abroad at company expense, and many others."[64]

While the U.S. embassy records may not be the last word on the course of the strike itself, they are remarkable for what they reveal about the resiliency of discursive traditions dating back at least half a century and found across the wide expense of the turn-of-the-century mineral frontiers. The firm's agents basically argued as did the embassy officers either that the men who challenged the hierarchy of the camps were power-hungry self-seekers using any grievances they could drum up to promote their political agenda or that the Saudis were being led by outsiders and were under the influence of alien ideas and ideologies. The king had suffered a heart attack on October 9 and was believed to be dying. It was thus hardly a coincidence in the Americans' minds that a strike was taking place at that moment, ignoring the fact that it was a process months in the making. "There is no question but that it is motivated almost entirely by political considerations." "Although ARAMCO executives believed that some outside force was guiding much of this activity, there was no concrete evidence at the time that it was Communist inspired and it is possible that the committee was genuinely interested *only* in forming a labor union which they would manipulate for their own personal power."[65]

The U.S. embassy reported that local authorities publicly dismissed any idea that outsiders had been involved in the strikes while privately expressing fears that it was communist led. After all, there were tracts in some of the rooms, the terms and phrases used by the most articulate had Moscow (via Beirut) written all over them. The embassy offered no examples. Instead the argument was theoretical. How else could coordination across the camps have been achieved? Why else would men be willing to defy government orders and become "martyrs"? "Communist involvement . . . in a manner and degree not now ascertainable, is a distinct possibility," the consul general reported, although he did not "consider a majority of the strikers sympathetic or even aware of the communist line." Rather, he believed the most zealous leaders were driven by a "personal craving for power" and, only last on the list, motivated by "general dissatisfaction" with working and living conditions.[66] Once the crisis began to subside in November, American officials would write longer and more detailed analyses without ever altering the basic analytical framework.

The Americans in ARAMCO would also insist that the strike was ultimately directed at the kingdom rather than the company and at what they called "political" ends. The initial report of CIA agent Bill Eddy was a rare exception in acknowledging that the grievances might in fact have been real. "The labor unrest was seen by Col Eddy to be the result of the dissatisfaction of the workers who compared this primitive land of low pay, slaves, eunuchs, and harems to the comfortable conditions of U.S. residents in Dhahran, plus probable Red stimulation."[67] Thus most would also ignore the letter that Abu Sunayd, whom the firm's agents judged the smartest of the committee members, had written to ARAMCO's president in June, complaining

> that he was denied admission to the Senior Staff theater to see Charlie Chaplin in *Limelight*, and went on to criticize severely what he termed "American persecution of Charlie Chaplin," and to describe an incident in Washington in which he and a group of fellow Saudi students were refused admission to a theater because of a color ban. This letter was widely circulated throughout the area.[68]

Not every protesting Saudi worker spoke in terms of color bars or had the chance to compare life in Beirut to life in Dhahran but few would have thought that the hierarchy brought by the Nasranis (as Saudis referred to Christians) was just, and in the crucible of the camps, men were forging new identities. ARAMCO's Government Relations men knew this too.

> But to get back to the cause of the strike and the unhappiness. No matter what we would do, the Saudis in the front of the development will be unhappy. Although they are leading their fellows in progress, they are accordingly all the more keenly aware that they are not receiving the benefits, and are not treated by most Americans, the same as an American employee. They cannot see why they shouldn't be treated the same and receive these same benefits. . . . The apparent leaders of this strike (and I'm sure there are others under cover) are the bright boys who we have been able to bring along fastest. They are the intelligentsia. Tom Barger tells me that Toynby [*sic*] says that the intelligentsia always are unhappy and discontented.[69]

The Cause Will Go On

The best one can say about the subsequent analyses is that over time the role of ARAMCO in events would be given some attention. This is particularly true once the American labor attaché in Abadan, C. C. Finch, visited ARAMCO headquarters in New York and returned to the kingdom to assess the post-strike situation. Finch stressed that the strike followed the arrests and probably would not have occurred had the company continued negotiating and, of course, had the commission not jailed the workers. The New York executives though seemed no better informed than the State Department about the situation "in the field," not least because Rentz's Arabian Affairs Division, or "intelligence" as Finch described it, had been completely in the dark about the strike. It had no real contact with Saudi security officials and little awareness of the kinds of "propaganda" circulating among the workers.

Enhanced intelligence gathering and the rooting out of workers with suspect loyalties would not be enough. The workers had real grievances. The reality is that, regardless of what the company claimed in public, it was clearly understood that Saudis would learn single, non-transferable skills, leading down a dead-end road rather than up to Intermediate and Senior camps. ARAMCO executives had been frank about their reluctance to reduce the disparities between Saudis and the most privileged class of workers. First, ARAMCO managers "approached the provision of more amenities and further cash increases to Saudi Arabians and to other non-American personnel largely as easily postponable incentives to further improvements in production." Executives also indulged their prejudices, arguing that the Saudis had embraced a "let ARAMCO do it" philosophy—for hospitals, public health, housing, and so on—and evaded "even partial responsibility" for the heath and welfare of their own next generation.

The labor attaché concluded that ultimately ARAMCO's policy was designed to "postpone westernization" rather than promote it, since the firm would be paying the costs. Executives had thought the investment was safe, even in the wake of Mossadegh's nationalization in Iran, and their basic strategy for insuring the future was the cheaper one of continuing to pay the king off. Finch concluded that the firm probably ought to try something else. "The reporting officer believes that there can be no alternative to more direct participation in the deliberate westernization of its Saudi Arabian employees. . . . Since the westernization will eventually come it might just as well be direct and controlled as indirect and uncontrolled."[70] But it was a gamble, at best. It would be expensive, and ARAMCO would have to start anteing up almost immediately.

It looked at first that the old rules would prevail. On November 1, the Saudi government deported more foreign-born workers, including three Palestinians, a Bahraini, and a naturalized Saudi citizen from Aden, Abd al-Rahman Farah Salim, who was stripped of his citizenship before being exiled. The same happened to Abu Sunayd before he was dumped at the Iraqi border. All the arrested labor leaders were released from prison in Hofuf and were either deported or banished from the Eastern Province, as Finch had advised. The newly vigilant Saudi security service pressed ahead with its investigation, ordering ARAMCO to fire workers that the government suspected of disloyalty. One of these targets was Nasir al-Said, a recent hire from Najd and an original signer of the petition, who would soon reemerge as the most important figure in the nascent Saudi labor movement.[71]

The rules began to change, however, upon the long-awaited death of the old king Abd al-Aziz on November 9, 1953. The government's Labor Commission investigating the strikes began pressing the firm to make concessions on virtually every point raised in the workers' petitions. ARAMCO learned that all the workers fired at Saudi orders at the beginning of the month were to be rehired, including Nasir al-Said. By the end of November ARAMCO had agreed to a wage hike for all categories of workers and various subsidies: paying the costs of workers' breakfasts, buying them bicycles, paying for their uniforms, and so on. The new king forced the company at the last minute to increase the size of the raise before solemnly announcing the results before an audience of loyal subjects of the Eastern Province on January 14. His aide brought a tape recorder and ARAMCO rebroadcast the speech to listeners of its radio station in Dhahran. ARAMCO was to begin building houses and schools for the workers and the royal commission was to be made permanent, demonstrating the new king's commitment to the welfare of the workers and also to their proper guidance.

> Let each and every one of you make it his primary goal to cherish truth and to work for the progress of the country. Whosoever of you shall leave the ranks or pose obstacles before or spread dissention among the people shall be dealt with by us in a way that will set him once more on the right path and will protect the nation against his evil influence.

ARAMCO vice president Garry Owen told the U.S. embassy that the company "on the whole was quite pleased with the results."[72] Not once did an American go on record, at least in the records I have found, saying that Saudi workers needed to organize in defense of their rights or that a labor union was a reasonable goal in and of itself.

In a strange echo of the 1940s when the Italian workers piggybacked on the Saudi movement, the government in Rome moved to protect the interests of the 1,300 or so Eritrean Italians who still worked in the kingdom. The pay raise won by the Saudi workers lifted their wages above that of the Italians in some classes, notwithstanding the firm's earnest professions that skill level alone determined what it paid the men, where they lived, and so on. To pressure ARAMCO, the Italian Bureau of Emigration began to require migrants to sign a disclaimer before validating their passports. They had to acknowledge the following: "that salaries are inadequate for one's needs," "that living quarters are not always provided with air conditioning," "that workers do not receive social security benefits granted under Italian law (family allowance, sickness insurance, etc.)," and "that Italian workers must live, as far as work, housing, mess, and sanitation facilities are concerned, in promiscuity with Arabs and other colored men and not in 'senior camps' which are reserved for higher categories of white technicians."

Although ARAMCO protested, the U.S. consulate in Dhahran informed the embassy in Rome that the claims were more or less correct.

> No American is in intermediate camp. Although a few Americans and some Dutch nurses are intermediate grade they live in senior staff camp. One Italian doctor employed by ARAMCO lives in senior staff camp. The Italian Minister to the Saudi Arabian Government and his wife visited Dhahran last spring and are reported to have protested placing Italians in Camps with Arabs, Pakistanis, and Sudanese, and maintained that Italians and Europeans should not have to live with colored men but should have the same privileges as Americans and other white people. When Italians moved from exclusively Italian camp to new intermediary camp early in 1953 many protested living and eating in camp with Arabs.[73]

The American officials concluded, accurately it turns out, that ARAMCO would do nothing because it planned to replace the Italians over the coming years.

In February 1954, three of the original strike leaders had returned to the Dhahran area where they were hunted down by ARAMCO and Saudi police and rearrested. Abu Sunayd had flown into Dhahran airfield but was kept on the plane bound for Kuwait. One of the others, Abd al-Rahman al-Bahijan, actually applied for work at the U.S. consulate general! He admitted he was one of the leaders, and when the police were summoned to arrest him, "he was so abusive in his attitude toward local authorities that they threatened to beat him." The "cause will go on," he shouted.[74]

In August, the U.S. embassy reported that copies of a one-page leaflet printed in Arabic and headed with a hammer and sickle were found in al-

Khobar, scattered on the streets. It was not clear who was producing or distributing them. It contained the harshest criticism of the royal family ever seen, the embassy said.

> Certainly the King and the royal Family have died after the people killed them because of their joint actions with the foreign imperialists. Certainly the people killed them because they were reactionaries and corrupt and exploited the workers terribly. The days of the Cadillac [*Ayyam al-Kadalak*] and the palaces are finished to give way to popular democracy for the workers.

> O workers!

> Get rid of the American pigs and seize the profitable [exploiting] oil company.

> O people

> You have nothing to do but follow your sincere leaders who seek your welfare and who will reveal their faces to the public.

> O Arabs

> Unite because the Arab Peninsula is for the Arabs.[75]

6 EYE OF THE DESERT

September 28, 1955
Secretary of State, Washington D.C.

Honorable Secretary Dulles:
I am a seventh grade student of Hartman-Main. We read an article in the
Kansas City Star of September 14, about selling slaves in Saudi Arabia and
Yemen. The class and I would like to know what could be done about it.
Sincerely yours,
Sandra Grousnick,
7935 Kenwood Avenue,
Kansas City, Missouri

Reference is made to your Excellency's communications regarding the
attitude of the Government of Saudi Arabia toward the Economic and
Social Council's resolution 575 A/XIX/. On behalf of my Government, I am
instructed to inform you that the Saudi Arabian Government considers
that the matter referred to in the allegations which led to passage of the
above resolution are solely within the domestic jurisdiction of Saudi Arabia.
Inasmuch as Saudi Arabia is not a member of the International Labor
Organization nor a party to the convention on Freedom of Association,
it does not believe it would be appropriate for a subsidiary body of
ILO to make inquiries. Under the circumstances, therefore, the Saudi
Arabian Government is unable to give its consent to the forwarding of the
communications containing the allegations to the governing body of the
ILO. This view also applies to the allegations contained in document E/2951,
dated 12 February 1957.
At the same time my Government is pleased to inform the Council that
the condition of labor is steadily improving in Saudi Arabia. My government

*is doing what it can to assist in these improvements. Council members will
recall that Saudi Arabia has within only one generation been confronted
with many new problems. The Saudi Arabian Government believes that the
conditions of labor will now easily stand favorable comparison with those in
many other countries.*

**Text supplied by J. T. Duce of ARAMCO to Saudi Arabian
UN Delegation, 1957**

WHEN SAUD WAS KING

On June 14, 1956, a battle broke out in Ras Tanura as Saudi workers stormed the
gates of ARAMCO's Intermediate Camp. They were demanding to be let into
the company movie theater, a symbol of the privileges that foreigners, in this
instance Pakistanis and Palestinians, enjoyed but were denied Saudi workers. For years, ARAMCO's managers insisted that they had been forbidden to
grant to foreigners other than Americans those things Saudis themselves were
denied. Now that the market and other factors worked in favor of the better-
educated Arab and Pakistani employees, the story changed to one about being
forbidden to show movies in Saudi Camp—and presumably to provide its
residents decent buses, recreation facilities, schools, and other benefits.

For two weeks there had been "symptoms of discontent," as the U.S. consulate's report put it, at the entryway to the makeshift theater. Groups of Saudis
attempted repeatedly to enter the building but had been kept out. On June 13,
some hundred or more massed at the theater and began to chant, "Down with
Pakistanis; they are Jews and friends of Jews." The next morning, Americans
learned that the theater was going to be stormed and possibly burned to the
ground. The amir of Ras Tanura, Salih Ibn Utayshan, ordered the local police
and his private guards to preserve order in the camp, but about a hundred
workers broke through the main gate at 10:30 PM and overwhelmed them.

Reinforcements arrived and they beat the workers back, but the demonstrators returned in larger numbers an hour later, lobbing stones at the guards
and at the amir himself when his car pulled up to the camp. The riot was broken up finally around 2 AM after additional police and guards arrived, sent by
Ibn Jiluwi, the governor in Dammam. ARAMCO officials assisted Ibn Utayshan's men in rounding up the "agitators," and later that day, a Friday, following noon prayers, ten workers were brought out, whipped 100 times apiece,
and dragged off to jail in Qatif. One was only thirteen years old. The consulate

general reported the rumor that two prisoners died from the beatings. Later, the U.S. Air Force's Office of Special Investigations reported that the jailed men had been tortured and that "several" had died.[1]

The brutality used against the Ras Tanura demonstrators marks a turning point in the course of the revived and momentarily emboldened Saudi workers' movement, following months of petitions, protests, work stoppages, and boycotts against ARAMCO beginning in May 1955. The workers were demanding that the company fulfill the commitments made in the aftermath of the 1953 strike and dismantle the system of privileges Americans enjoyed. They never ceased claiming the right to elect their own representatives as well. But the latest petition in May 1956 had expanded the agenda to include the demand that the king close the American Air Force base at Dhahran.

On June 9, just days before the attempt to storm the theater, workers had gathered at the entrance gate to ARAMCO headquarters in Dhahran to intercept King Saud and his entourage en route to a banquet that the company was holding in his honor. As the royal Cadillacs approached, his subjects cheered Saud but also appealed to him for redress of these long-standing grievances. Some had hoisted a banner, "Aramco Saudi Workers Welcome the King. Death to Imperialism and Traitors." Later, when the cavalcade once more passed the chanting crowd, which had grown even bigger and more boisterous, two of Saud's guards leaped out and ripped the banner from the demonstrators' hands before speeding off.

The effrontery of these men, mere laborers who would embarrass the House of Saud in front of westerners who had captured it all with their cameras, was too much. Two days later, a new Royal decree (no. 17/2/23/2639) was readied. Dated 13 Du al-Quid 1375 or June 11, 1956, it banned all strikes and demonstrations by workers and made incitement to demonstrate a crime punishable by two years in prison. Unfortunately, it took almost two weeks before the decree was read to the ARAMCO workers. The appearance of the demonstrators at the banquet is doubtless what led Ibn Jiluwi to punish the men who later rallied in Ras Tanura. His security forces hunted down the movement's leaders. Nasir al-Said, an organizer of the 1953 strike and ARAMCO's prime suspect in instigating the new disturbances, says he was tipped off by a contact in the king's bodyguard, went into hiding, but then fled the country.[2] Those of his comrades still at large in Dhahran on June 17 began to mobilize workers in case any of the new arrestees were brutalized the way the Ras Tanura workers had been. Thus began the last large labor strike in the kingdom's history.

By the time of the 1956 strike, all the diverse populist currents and tendencies that were remaking the political order in Egypt and the Arab East were taking root inside the kingdom, and the oil workers were only one of a number of forces beginning to crowd the political field. The one that may have worried the Al Saud most was the military, where the first conspiracy among a "reform group" of Saudi officers was discovered and broken up in Taif in May 1955. Lt. Colonel Ganim Madhya Hadi and his comrades had planned to overthrow the royal family. Most of those arrested had been trained in Egypt.[3] The intelligence officer who served as the U.S. embassy's military attaché was soon reporting on the major new hurdle in front of the American military advisors. "Both Prince Faisal and King Saud did not want the actual fighting capability of the Saudi army increased at this time because of its questionable loyalty to the royal family in event of an emergency." The U.S. Military Training Mission became the instrument "necessary to justify keeping an American air base in Saudi Arabia and would assure American military protection against any outside threat."[4] Any irony in reporting on the emerging institutional configuration of the Saudi defense forces was lost on the Americans. Imagine: princes that did not trust their own troops and so purposely kept them divided. It was exactly the same strategy that owners of mining firms had used for decades against the threat of a unified labor movement. Its traces were still visible in the layout of Dhahran and the other settlements, and Saudis were still wrestling with its effects.

The same month that the plot in the army was being uncovered and workers in Abqaiq were beginning a bus boycott inside the camps, dozens of Palestinians inside ARAMCO were rounded up under suspicion of belonging to the Parti Populaire Syrien (PPS), also known as the Syrian Social Nationalist Party, and to a rival emerging tendency in the 1950s, the Arab Baath (or Rebirth) Socialist Party. These were populist and pan-Arabist in orientation, close in some respects to European national socialism or fascism.[5] The prime example outside the officer corps of the cancer spreading across Najd and al-Hasa, however, was the arrest in May of Abd al-Aziz Ibn Muammar, a Najdi and the son of a close confidant of the first king. He had graduated from the American University of Beirut and was a rising intellectual in the kingdom.

Ibn Muammar was serving as director general of the Ministry of Finance's Labor Office in Dammam in 1955 when he was seized, taken to Riyadh, and thrown in a dungeon. No one inside ARAMCO or the U.S. embassy appears to have reconciled the competing accounts of the reasons for his arrest. He was called variously a member of the PPS, a Baathist, and a communist.[6] Rumors had him playing a behind-the-scenes role in mobilizing the workers against ARAMCO (the bus boycott in Abqaiq had begun the day before his arrest), or of being jailed for his "nationalistic ideas," or for circulating tracts denouncing the Royal family and calling "on people to throw out the Americans and preserve Arabia for the Arabs," or for inciting revolution.[7]

A visiting Texaco executive called him "an amiable man, rather well disposed to the company, but weak and perfectly capable of being used by stronger personalities." Another American admired his "inquiring mind" but said that after moving to the Labor Office, Ibn Muammar was beginning "to talk about the downtrodden Saudis and complain about extravagances of the Royal Family and provincial governors while average citizens received no benefit from country's vast oil reserves." Plus "he had a copy of Marx's Capital in his bookshelf."[8] Another Saudi, Faisal al-Umbra, a member of the Committee to Promote Virtue and Prevent Vice, more commonly known as the "religious police," who was arrested with him, was executed—this according to Muammar's friend, the Saudi oil expert Abdallah Tariki.[9] He had died in prison, the embassy corrected. Muammar was luckier, released in a general amnesty in March 1956. It would not be his last time in the Saudis' dungeons unfortunately. These two views of Ibn Muammar are significant because in the next year he would be routinely described by ARAMCO's local Government Relations men as an irresponsible radical and extremist in comparison to whom even the firm's public enemy number one, Tariki, was someone who could be reasoned with.

Ibn Muammar's criticism of the profligacy of the princes and their failure to make good use of the oil rents that ARAMCO produced hardly distinguished him from virtually anyone inside the kingdom brave or foolhardy enough to describe what was in front of one's face. Old King Abd al-Aziz's loyal adviser, Abdallah (ex–Harry St. John Bridger) Philby was thrown out of Saudi Arabia that same fateful month of May 1955 for about a year after criticizing the court in print outside the kingdom, in the New York journal *Foreign Affairs*, and, worse, inside the kingdom in lectures he gave in Dhahran, Abqaiq, and Ras Tanura.

The new Republican Eisenhower administration was the outlier in an era when foreign critics of the reactionary oil monarchy were growing more rapidly than the kingdom's oil revenues. Dwight D. Eisenhower and his secretary of state, John Foster Dulles, viewed the Middle East as a region increasingly open to Soviet expansion unless the United States stepped in first, and they get the credit for the first attempt to extend American preponderance or hegemony beyond the kingdom into Syria, Lebanon, Jordan, Iraq, and Egypt—although the effort failed, spectacularly. Historians of foreign relations associate that moment above all with the Eisenhower Doctrine, the promise of assistance, including the use of military force, to any Middle Eastern state threatened by "armed aggression from any nation controlled by International Communism."[10] The one aspect of the Eisenhower Doctrine that all those most closely associated with it tried hardest later to forget was the attempt to promote the new Saudi king as a rival to the region's charismatic, increasingly popular and powerful Egyptian leader, Gamal Abdel Nasser.

King Saud signed on to the Eisenhower Doctrine—although he would later try to deny it—and made the cover of *Time*. More important, he extended the agreement that permitted the U.S. Air Force to use the base America had built in Dhahran, and he was rewarded with tanks, planes, upgraded facilities for the airfield, and an upgraded, if, as I have already suggested, complicated training mission. But Saud may also have been America's most inept or at least unlucky ally. He would be forced by his family to turn the government over to his brother, Crown Prince Faisal, and go on leave in March 1958, and Faisal would strive to repair relations with Nasser. Six months later, U.S. combat troops would be sent to the region for the first time, to Lebanon, to back the Right against the Left in a civil war.

America's tightening embrace of one of the most reactionary dynasties in the world launched a debate in the United States that continues until today. Catholics and Jews condemned religious discrimination by the puritanical Wahhabi rulers. The New York Council of the CIO protested Eisenhower's "inviting to our country the feudal despot who rules Saudi Arabia . . . with medieval cruelty and who permits the enslavement of many, many thousands of men, women and children in his feudal domain [and thus] is the very antithesis of everything that American democracy stands for." The Executive Council of the national AFL-CIO followed with a motion of censure put forward by A. Philip Randolph, the president of the Brotherhood of Sleeping Car Porters, for hosting the "slave owner and plotter against the free world."[11] Meanwhile, in New York City, the curtain was raised on a piece of political theater that has been revived more times than *West Side Story*. Mayor Robert Wagner refused to meet the Saudi ruler when his ship docked in Manhattan (Governor Averell Harriman refused as well), which outraged Secretary of State Dulles.[12] Future Democrats would play this role with equal verve.[13]

The Saudi human rights record was, needless to say, abominable, although it did not stop groups like the American Friends Service Committee from applauding the king's "pursuit of peace," which was bizarre, because there was no pursuit of peace anywhere in those years. Saud was, though, reliably anti-Zionist, and ARAMCO helped those, such as the Quakers, who helped the refugees in Palestine. What is striking is how, while we would expect silence from various of the kingdom's friends and allies, even the critics who protested Saudi slavery and religious discrimination had nothing to say about ARAMCO's treatment of Saudi workers, the failure to uphold international norms about labor's freedom of association, and the still stark hierarchy inside the camps. By way of comparison, when the distinguished historian A. J. Toynbee was flown to Dhahran in 1958 for a series of lectures, he also wrote an eyes-only report for the firm's managers. In it, he compared the social order in Dhahran to the Gold Coast on the eve of independence and advised the firm to dismantle the institutions that kept the Saudis separate and unequal.[14]

Such were the terms (and limits) of dissent from the Republicans' embrace of King Saud, the notorious consumer of Cointreau who was named honorary president of the International Commission for the Prevention of Alcoholism in 1957. The United States guarded its "special friendship" with the Al Saud because of ARAMCO, "*the most important single American interest on the face of the earth outside the U.S.*," as George Wadsworth, II, the U.S. ambassador to Saudi Arabia, put it.[15] The phrase was destined to become a mantra for policy makers over the next decade. For those who thought like Ambassador Wadsworth, there were basically two kinds of threats to the Saudi-American relationship: Americans who seemed intent on harming it, say, by fixating on human rights, by refusing to sell Saudis weapons, or by letting ethnic lobbies steer the ship of state straight toward the trap set by the Soviets. More threatening still were Saudis like Tariki and Ibn Muammar who seemed bent on taking ARAMCO away from "us." Tariki's influence was growing, and he and his comrades would take the kingdom over the edge of the Free World if they weren't stopped.

Playing Populist

In February 1956, the Central Intelligence Agency circulated a new analysis of the kingdom, Intelligence Report No. 714, *Saudi Arabia: A Disruptive Force in Western Arab Relations*. From the title alone we might surmise that the high hopes the Americans had for King Saud back in 1953–54 when he was being championed as a modernizer dedicated to a crash program of developing

human resources, building up the army, and ending corruption were not pan-
ning out. The sharper reading would be that the king was not doing all that a
good client should, leading to the downward revision of his performance as an
overall wise leader who would take the kingdom "out of the twelfth century
and into the twentieth," as Americans often put it. There are good grounds for
thinking that this more cynical reading is the right one, apart from its proven
value in other cases where patrons suddenly discover (and publicize) a client's
grave flaws.[16] A year later, when Saud signed on to the Eisenhower Doctrine,
there were those who of course knew better and yet were hoisting Saud Ibn
Abd al-Aziz up on their shoulders, briefly, including ARAMCO's chairman.

> I was going to tell you what the King said when he first became King. He said, "my
> father's reign may be famous for its conquest and cohesion of my country. My
> reign will be remembered for what I did for my people in the way of their welfare,
> their education and their health." And what he has been doing in the past two
> or three years bears that statement out very much. He is building schools. He is
> building houses. He is building mobile clinics. He is building clinics that are not
> mobile all over the country. He is building roads, drilling water wells, developing
> agriculture, he is repairing the mosques. Too much is said here in the press and the
> magazines about the gold-plated Cadillacs. I have not seen gold-plated Cadillacs.
> I have seen Cadillacs but I have seen a lot of Cadillacs in Texas.[17]

What is "development" one day is "profligacy" the next, depending on the
client's value, loyalty, and his not turning into a liability, as Saud did when an
attempt to bring down Nasser, which Eisenhower encouraged, backfired. But
we have gotten ahead of ourselves.

The 1956 CIA report followed two years of a foreign policy designed to
emphasize the autonomy of the kingdom from its protector, the United States,
which, as the Cold War intensified, tended to display less patience with Saudi
sensitivity to the charge that it was, in fact, a client. To appreciate the Saudis'
dilemma one need only juxtapose Prince Faisal's profession to Eddy that in
the region only Egypt and Saudi Arabia were truly free and the American
military attaché's reminder of the kingdom's utter dependence on the United
States for its survival.

Two related policy strands had made the kingdom a disruptive force,
in the CIA's phrase. The Saudis had resisted the Anglo-American effort to
bring the states of the region into a Western defense alliance. In particular,
they refused to join the British-led Baghdad Pact, which brought together

their enemy, Iraq, and a group of non-Arab states, including Turkey (Saudis in Jidda had taken to calling Turks the "second Jews"), Pakistan (ditto), and Iran.[18] Nasser of Egypt had turned away from a nascent client relationship of his own with the United States to lead the opposition to the Cold War alliance strategy under the banner of "neutralism." He raised the stakes considerably when, as the cease-fire with Israel began to fray, he signed a $200 million arms deal with Czechoslovakia in the summer of 1955, and soon after orchestrated a counter-alliance of his own with Saudi Arabia and Syria.

The Saudis also sought ways to reduce their vulnerability to the charge that their country had become an American colony. In 1954 the king actually closed down the Point Four mission.[19] One reason is that the Eisenhower administration had concluded an arms deal with Saudi Arabia's rival, Iraq. Predictions at the time were that the Dhahran Air Base would be closed down too once the current agreement expired in 1956. Saud escalated tensions, however, when he approved a plan to launch a fleet of Saudi oil tankers that were going to built by the Greek shipowner Aristotle Onassis. He issued a decree that henceforth all Saudi Arabian oil would have to be transported in Saudi tankers. In response, the U.S. National Security Council met on July 22, 1954. Eisenhower himself presided. He and his advisors agreed to take, as the new Middle East policy directive, NSC 5428, put it, "all appropriate measures to bring about the cancellation of the agreement between the Saudi Arabian Government and Onassis."[20] ARAMCO was determined to fight the deal for infringing on its concession rights and the four owner firms were lobbying as a rare united front.

With intrigues by the CIA and Onassis's rivals crisscrossing Geneva, Nice, Athens, Beirut, and Jidda, it has been easy to lose sight of what actually happened, but basically ARAMCO cut a deal. The company signaled its readiness to settle the question of retroactive royalties (back to 1951, worth $70 million, as discussed in Chapter 5) and the king smoothed the way, delivering on the promise to ARAMCO to dismiss the finance minister, Sulayman, who was chief architect of the Onassis deal.[21] The government accepted ARAMCO's proposal to submit the issue to international arbitration, as provided for in the concession. This allowed the king to "finish with Onassis." The middleman in the deal, the merchant Muhammad Alireza, who was King Saud's minister of commerce, saw fit to quit the cabinet over this stunning defeat.

Abdullah Tariki's efforts to monitor ARAMCO's accounting stratagems had paid dividends, obviously, but otherwise the Saudi government's moves to identify itself with the emerging axis of Arab nationalist states by throwing out the U.S. aid mission and so on were largely symbolic. They annoyed Dulles and of course did nothing to alter the dependence on the United States for security. The prescription for such errant Cold War clients was routine. Leave basic issues of regional and world order to America. Attend to your mounting internal problems and the development of dissidence in all levels of Saudi society. Otherwise you run the risk of being overthrown.

"Neon Lights . . . Are Like Atomic Bombs to The King"

The scripts that were being written to make the kingdom a better place were all but useless, however. The king might be advised that his government ought to concentrate more resources on development, which he did after a fashion, although problems of absorptive capacity were a serious issue. The harbors were clogged with ships waiting days, and owners charging a premium, to unload goods. Others might counsel the Saudis instead to reduce spending, pay off their debts, and get their financial house in order. The king welcomed an International Monetary Fund (IMF) mission to Saudi Arabia in November 1957 and agreed to an austerity plan. Most people forget this fact now. Yet he also agreed to subsidize the rule of the new, young twenty-one-year-old king of Jordan. The Americans sold King Saud weapons. New York bankers loaned him all the dollars that ARAMCO allowed.[22]

The full measure of the incoherence in American prescriptions for getting the kingdom off its binges and on a twelve-step program becomes evident when we add the prime directive of protecting the ARAMCO concession and the basic means to insure it. Companies would pay more to governments of oil-producing states, in essence placating the nationalists, as an alternative "to defending the postwar petroleum order through military or covert intervention."[23] Exxon's fifty–fifty arrangement with Venezuela was the exemplar, and by the 1950s the logic was thought unassailable. Thus, the firm's response to Abdallah Tariki's Saudization proposals circa 1957–58 would be to offer some millions of dollars more a year in additional income from TAPLINE and a share of the profits in the owner companies' proposed new oil trading business. When these offers were turned down, the Americans chided the Saudis for not understanding their own best interests. Tariki's answer was that more revenues would simply be squandered in the same old way. The argument, not surprisingly, usually left the Americans speechless.

Imagine, therefore, the reaction as Tariki and like-minded intellectuals rolled out blueprints for the institutions that might make a difference: writing a constitution, electing representatives to a national parliament, building industry, investing seriously in human capital formation, expanding the ranks of Saudis with advanced degrees, and, above all, taking over ARAMCO in stages. The embassy always talked about these intellectuals as a group. There was Umar al-Saqqaf, Ahmad Abd al-Jabbar, Abd al-Rahman al-Huaisi, and Abd al-Aziz Ibn Muammar, the last invariably singled out as "fiery" or "idealistic" or, as Exxon's hired hand, Professor George Lenczowski put it, a "radical and emotionally unbalanced Arab nationalist."[24]

Here is how al-Saqqaf, rising in the ranks of the Foreign Ministry and considered by the American embassy the most cautious of the group, described the Saudi development dilemma:

> The King and his crooked advisors believe that spending billons of dollars constructing the ministries in Riyadh meant progress for the nation. In point of fact, it was a senseless waste. All of the ministries combined would not fit fully into one floor of any of the mutli-floor ministries constructed in Riyadh. Instead of the slow, solid progress of schools and hospitals and industry, the King threw money around recklessly and was unable to move to a form of government that would establish a trend toward modernization. Since Saudi Arabia depends on oil, there must be an industrialization of the country, Saudi Arabians must be prepared to work their own major industry and the country must have better communications, primarily through roads. The paradox is if Prince Faisal and the King go ahead with the modernizing program, they will bring closer the day when the monarchy must be severely curtailed or eliminated.[25]

Many others rejected outright the priorities of Tariki, al-Saqqaf, and other Saudi converts to developmentalism. Issa Sabbagh was one: a Palestinian-American who worked in ARAMCO's Government Relations Organization (he was a translator who played Khalid the Bedouin guide in *Island of Allah*) before joining the United States Information Service as public affairs officer in Jidda. A year after Saudi Arabia's first university ("the so-called King Saud University," he sneered) opened in Riyadh in 1957, officials unveiled plans to add a college of science by 1959 to the college of arts.

> The whole existence of this "university" is more akin to candy than to bread as far as the actual educational needs of Saudi Arabia are concerned. Every intelligent Saudi the PAO [public affairs officer] has talked to volunteers, or

agrees, that vocational schools are more essential to the *uplifting* of Saudi life. Such vocational schools would certainly be the quicker means to lighten the everyday spirit of the country.[26]

Booker T. Washington could not have put it any plainer.

As Umar al-Saqqaf's account makes clear, it was the same massive three-year-long building boom in Riyadh that fueled the critics' charges of extravagance, bribery, waste, and corruption and yet also inspired poets' (and ARAMCO's Public Relations Department's) paeans to the king's great works on behalf of his people.[27] By the mid-1950s the construction industry had grown into the third largest employer after the government and ARAMCO, although the boundaries between any two were at times as easy to lose sight of as the roads stretching from Riyadh across the desert to Dammam after a sandstorm. Consider the two building types that generated most of the expenditures, jobs, and goods orders in the mid-1950s: palaces and ministries. The sons of Abd al-Aziz were tearing down complexes built only a few years earlier and replacing them with ever more grand-scaled homes and gardens, and the king might donate last year's model for conversion into a school or hospital. The biggest of these palace works projects was the king's compound at al-Nasiriyah outside Riyadh. ARAMCO had been pressed to sell equipment to the man known as "the king's builder," Muhammad Bin Laden, in order to finish the massive rebuilding project on time. ARAMCO's estimate of the final cost of the project ($50–$100 million) is the equivalent of $300–$600 million in current dollars. The U.S. embassy reported dozens of these compounds going up around the country at the start of Saud's reign.

King Saud was determined to remake the town into a proper capital of a modern nation-state. A new four-lane highway was built from the airport to the race track, planted with palm trees, and designated as a future embassy row. Seven new ministries would be sited along a second road from the airport. By 1956, ARAMCO estimated that $100 million (more than $600 million today) had been spent in Riyadh on buildings and infrastructure. The embassy reported that some 700 new houses were being built for government officials, along with several hotels and a new airport to replace the not very old one.[28] Housing turned out in fact to be a bottleneck that delayed the move of the government from the Hijaz to Riyadh, and the King began gently coercing men and their families out of Jidda and Mecca into Riyadh. By the end of 1956, most ministries had moved parts of their staffs to temporary quarters in Riyadh as most of the ministry buildings themselves were not much more than shells. The Ministry of Interior moved but then moved back to the Hijaz, since it main policing work was there. The complete transfer of embassies, government agencies such as the Saudi Arabian Monetary Agency (SAMA) and Saudi Airlines, and Faisal's foreign affairs operations would ultimately take years more, but Riyadh now took its place among the modern, sprawling steel and concrete capitals of the Middle East.[29]

It was surely with an eye to places such as Cairo and Baghdad that Saud's regime announced its follow-up to the building of Riyadh, a five-year project to expand the Grand Mosque in Mecca—razing parts of the old quarter, widening streets and the like. No doubt it was important to answer the grumbling of the Hijazis about the costs of all the buildings going up in Riyadh by giving them a development project of their own, but Saud's ambition to be a leader in the region turned on two things: spreading his wealth around to a widening set of clients beyond the tribes and borders of Saudi Arabia itself and promoting Islam as the "third way" between communism and capitalism. Nasser could hardly compete in either arena. The work in Mecca would cost close to $1 billion dollars today. Muhammad Bin Laden, one of a score of manual laborers-turned-contractors in Riyadh, but the one with the good fortune of having been named Saud's minister of state and director of government construction, would direct the project.[30] His son, Osama, was born there the next year.

Any minimally objective analyst could point out the pervasive problems in Saud's approach to state building, so imagine what his critics said. Leave aside the obligatory although true complaints of rampant bribery and corruption. Virtually no economically productive (that is, self-amortizing) investments were being made, either in agriculture or industry, and since Abd al-Aziz's disastrous decision to forego highways for a railroad, commerce had been held hostage to a wholly inadequate system for transporting goods across the country. The new king's mounting debt followed the old patterns of borrowing from merchants and rolling over the loans. Government employees began to complain once again about salaries going unpaid.

In the last years of the Abdallah Sulayman regime, the Saudi Arabian Monetary Agency, first under Arthur Young and then (1951–54) under Young's hand-picked successor, George Blowers, managed to restore some order to expenditures, stabilizing the currency but also managing the internal debt and arranging to pay off the creditors by installment. Still, bidding for the role of regional hegemon did not come cheap. On top of the usual "burden of conspicuous royal consumption at home and abroad," there was the new look to foreign policy, "divided between limiting the possibility of increased Iraqi influence in Syria and Jordan, and bolstering Egypt as a counter-weight against Israel and Hashemite influence." In 1956 Saud upped his loan to the Syrian government by $10 million on top of $16 million due for repayment but defrayed, which was considered the price of "Syria's abstention from joining the Baghdad Pact and for full political cooperation with Egypt and Saudi Arabia." The United States suspected but could not confirm that additional sums went to Egypt and to Yemen early in the year.[31] Premiums would definitely be paid in the fall, however, once the Suez Canal war began, even as the king's oil revenues began to decline.

One of Saud's advisors and troubleshooters, Abd al-Rahman Azzam Pasha, rehearsed the then common criticism of the king in a meeting with the State Department's David Newsom at the Plaza in New York late in 1957.

> [He] imagines himself as acting in the patriarchal manner of his father when, in fact, the problems of the country have grown. . . . King Saud has a mistaken idea of what constitutes progress. In his bedouin outlook, the palaces he builds are symbols of national progress. "Neon lights," he said, "are like atomic bombs to the King." His tribal followers come to admire them, also, proud that their king has built something as fine as any King in the modern world. The educated peoples of the cities and the Saudis who have studied abroad see things differently.[32]

This view was a far cry obviously from the fantasy that ARAMCO's president spun in his testimony before Congress a few months earlier about the king's commitment to development, but Azzam Pasha was also far shrewder than all those who seemed to believe that Saudi Arabia teetered on the brink of revolution as a consequence of this wastefulness. Azzam said "that the wealth of the country, while not distributed wisely, had, through legitimate expenditure as well as corruption, been distributed widely and he felt that the country was basically sound and the King's position stable." This had always been the safer bet.

YEAR OF DECISION

On December 1, 1955, the police arrested Mrs. Rosa Parks, an African American seamstress, in Montgomery, Alabama, for not moving to the back of the bus and letting a white bus rider take her seat. The Montgomery bus boycott—black people quit riding the buses until second class treatment of them ended—is celebrated today as a galvanizing moment in the U.S. civil rights movement and the beginning of the end of Jim Crow in the South.

Months earlier, in May 1955, a much less well-known bus boycott began in Abqaiq by Saudi workers opposing the same thing—the Jim Crow order that ARAMCO had built in the Eastern Province. Transportation to and from the labor camps was one of the concessions made in the aftermath of the 1953 strike. The company handpicked a group of workers in each of the main camps to serve on something they called Employee Communications Groups (ECG), and it was through these groups that the workers were informed of management's plans for a new bus service. The buses in Abqaiq, though, were more like trailers, they ran infrequently, made few stops, and the company would not drive workers back to camp for a midday meal. On May 2, ARAMCO received a petition signed by 3,231 Intermediate and General Camp workers demanding the same quality buses and service provided American senior staff as well as the right to elect representatives to the Employee Groups. "These so-called committees . . . do not represent us, nor speak in our names. . . . The committees represent the interests of those who formed them." The same morning a boycott of the new buses was begun and by May 4 no one was riding them.[33]

ARAMCO sent a representative to Abqaiq's amir on May 2. The amir first requested that the visit be kept secret but soon after, in a panic, demanded an official report about all the workers involved in the boycott, the names of those who served on the ECG as well as the specific government authorization for it, and, finally, the whereabouts of the strike leaders from 1953. The panic was triggered by news of the impending arrest of Ibn Muammar, the director general of the Labor Office, on May 3, and the fear that the boycott was part of some wider conspiracy.

ARAMCO intelligence dutifully supplied the amirate and the embassy with news on the whereabouts of the 1953 leadership. It was hard to believe. Abu Sunayd had made his way back from Iraq, with a side trip to Beirut where, the firm said, he worked for a communist newspaper. Now he was in Dammam again, working for Ibn Muammar in the Labor Office. A second member of the original committee of seven that led the 1953 strike, Salih al-Zaid ("known to have leftist connections"), worked in the government's Labor Office as well. Juman Bin Farraj worked for ARAMCO in Abqaiq. Abd al-Rahman Bahijan had gone to England on a government scholarship, through the intervention of his cousin Sulayman Olayan, ARAMCO's main success story among the first generation of Saudi contractors turned moguls. The source said Bahijan's major in England was "communism." The whereabouts of the others among the seven were unknown. The other usual suspect, Nasir al-Said, was definitely in Abqaiq. An unnamed American intelligence agent working at Dhahran airfield said al-Said was behind the bus boycott.[34]

ARAMCO's management was mystified, according to a visiting Texaco industrial relations expert, by its Saudi workers.

> Much of our top management is frankly puzzled as to why we are having labor dissatisfaction when the Company has done so much for the workers and provided them with first rate hospitals and many other services of which they never dreamed before ARAMCO came on the scene. The answer I suppose is that we Americans place tremendous emphasis on material benefits whereas Saudis are much less impressed than we are by that sort of improvement.

The arguments never changed, it seems, from the first strike in 1945 to the day the men refused to climb on the flatbed trailer that ARAMCO called a bus. These workers were not interested in bettering their lives, Brooks said, but in controlling "the oil company's labor force" which "was the surest and possibly the only feasible path to political power in this country," although he had also just finished telling the consul, Grant McClanahan, that "we don't know enough about [our workers'] real interests, needs, and motives." I cannot be sure, but my guess is that even the embassy officer began to doubt ARAMCO's account of a project aimed ultimately at seizing state power that, nonetheless, had the government's and the palace's protection.

As the government continued to arrest and deport Arab nationalists and communists inside the ARAMCO camps, the bus boycott crawled to a halt. The firm's experts believed it was the presence of the one thousand or more *mujahidin* armed with Enfield rifles and bandoliers and encamped on the outskirts of Dhahran that had momentarily "cooled the ardor" of the "labor agitators." ARAMCO was paid to supply the Wahhabi guard with water and to haul them back and forth during their stay.[35] As the fears of insurrection receded and the mujahidin withdrew to Najd, however, workers in Dhahran began a petition drive of their own. The new appeal was addressed not to ARAMCO but to King Saud. It began with a criticism of the amir Ibn Jiluwi for failing to secure the company's compliance with the demands in the original petition and sought the king's help in achieving the workers' right to pay and benefits comparable to those of the Americans—including the right to choose their own representatives to the ARAMCO Communications Groups—and in putting an end to the aptitude tests that Americans administered to Arabs. The firm said the petition originated with two of the highest-paid Saudi intermediate employees, working in the "Heavy Duty Section of the Dhahran Garages," and that most of the signatories were Saudis as well.[36] The palace turned to a favorite in its repertoire, reviving something called the Permanent Committee for Labor Affairs to investigate, taking over from the Labor Office in Dammam, whose head, Ibn Muammar, had in the meantime been put in prison.

ARAMCO was troubled enough by the events of the past spring to send its senior spy, Bill Eddy, who had spent the past few years in Beirut while recovering from a heart attack, back to Dhahran for eighteen months. He told his family,

> several of you have complained that I did not tell you what I am going to do in Arabia, a flattering interest in my work. I have been a Middle East Consultant to Tapline and ARAMCO, with an area job of political reporting and contacts. Now I am going to Dhahran to be a Consultant on local government and industrial relations in Arabia, where, unlike Lebanon, Arabic-speaking Americans are few although Arabic alone is understood by the nationals.[37]

We don't know what Eddy advised when ARAMCO received word of a new twenty-nine-point petition delivered to the king in December 1955. It was a busy time, close to Christmas, which may explain the firm's decision to scoff at its demands ("unreasonable in the extreme") and declare the petition a problem not for ARAMCO but "for the Saudi Arabian government, and specifically as one which the Royal Committee and Labor Department will have to deal." But then swing-shift workers in Abqaiq began two nights of "protests, work stoppages, and disturbances" on December 18 and 19, which ended only when officials of the Labor Office agreed to take up their complaints.

The same day Mary Eddy wrote a letter home to her children from little America in Dhahran.

> Though the temperature is not Christmasy here, the houses are very gaily decorated and many Arabs come just to drive through the town at night and see the lights. Tomorrow evening we are going to see the Christmas pageant, out of doors and aided by real camels and a shepherd with real sheep. It is a long tradition in Dhahran and we are glad to be able to see it. Santa Claus

usually arrives by helicopter, but the General at the Air Base was afraid that a new regulation made this impossible this year.[38]

Back in Abqaiq, masons followed the example of the transport workers and put down their tools on December 22, demanding an audience with company officials to present their grievances before they would agree to return to their jobs. Having just proposed that its workers' grievances were problems for the Labor Office to handle, the firm panicked when a Saudi Labor Office representative stepped in before the workers tried settling disputes "within the company," and because both it and the masons had ignored the norm against "mass action outside the established grievance procedure." The firm took a step or two back from the hard line, agreeing to prepare a response to the petition for the use of the new Royal Labor Commission.[39]

Don't Fence Us In

The ARAMCO managers would have another chance to profess surprise about their workers "real interests, needs, and motives" as news leaked out that the company was preparing to build fences around the Saudi workers camps in Dhahran, Abqaiq, and Ras Tanura. On January 4, 1956, some two hundred Saudis in Abqaiq appealed to the amir to stop what the company said was a program to improve the quality of life inside the camps. As long as they went unfenced, officials had told the amir, the food that some of the men grew in small garden plots would be at the mercy of the goats from the surrounding settlements. The workers instead saw it as an effort to pen them in and further isolate them from their families and friends. Work had begun with the sinking of postholes and the erection of some posts and crossbars. The goats around there were apparently very hungry and resourceful. The "posts were topped with Y's to carry barbed wire and thus make them somewhat more difficult to climb over."[40]

After his confrontation with the workers, the amir left to confer with officials in Dammam. The amir's brother, Turki Ibn Utayshan, head of the Permanent Committee, confronted him. "Why do you always have so many troubles in Abqaiq?" he wanted to know, and the amir shot back, "we didn't have so much trouble before the labor office was established there." Turki took off for Abqaiq on January 7, 1956, armed with a list of seventeen names of workers his office had identified as leaders of the movement there and hoping to head off trouble. He was too late. As he reached the outskirts of al-Mansur, the name Saudis used for the camp in Abqaiq, workers were pulling up the posts and pulling down the Y bars.

When Turki reached the ARAMCO trailer that served as the Labor Office he was met by a crowd of 200–300 workers who shouted him down and, some said, roughed him up, before he retreated, at which point the crowd resumed tearing down the fence.[41] ARAMCO officials had to summon the police and later would complain that they were "not forceful nor particularly effective, but kept the crowd in bounds." In Barger's record "the police chief arrived on scene with pickup with which he charged the groups engaged in tearing down fence." Some men shattered a few windshields later that night, but the only serious damage was to the fence. One ARAMCO Government Relations man said "it was sort of like Halloween."[42]

The next day workers held mass meetings in front of the Labor Office both at Ras Tanura and at Dhahran in protest against the fences. Barger's memo reports "Dhahran meeting organized by one or two men at each bus stop in Saudi Camp who informed [others] of mass meeting at 5 PM." Barger, like the other pioneers, still used the old name for General Camp and what Saudis called al-Salamah. But, then, Barger also thought it important to specify how "after 5 PM a group of some 30–40 men attempted to destroy section of fencing north of Abqaiq Boulevard which was to extend around the *domestics' quarters*." Barger said he and another Government Relations official, Harry McDonald, had convinced the amir to drop his plan to divert all the noon buses to the football field where they would be told that the king had ordered ARAMCO to take down the fence. "In our opinion it was a mistake to make a public announcement to employees of a victory over the Company *and the Government*."[43] At the rally in Dhahran a representative of the Labor Office did present it as a victory over ARAMCO. Surely no one outside the company believed that the government was the target of this or any other action.

Inside ARAMCO, managers read the authorities' indifference to the destruction of its property as proof that the norms "of good order" that had been established over the last few years to settle disputes were giving way. "An organized group of employees, possibly a minority, have demonstrated that to some extent they can make the Company and the Government do as they desire." The Americans began to portray Turki Ibn Utayshan as weak or, worse, in league with the agitators. He had reversed himself during the latest troubles in Abqaiq and told the company that it could not enforce its "no work no pay" rule against the recent strikers. The work stoppages would continue. On January 8, a group of thirty-two heavy-shop workers staged a sit-down strike. Wire line operators followed the shop workers on January 11.[44]

Americans in Abqaiq were clearly growing jittery by the end of January. New petitions and broadsides were circulating, and the company had gotten hold of one. It portrayed ARAMCO as draining the blood of the Arabs and it attacked the Americans for "living in luxury while Arabs are living in huts in Thugbah."[45] When Ohliger finally met with King Saud about the trouble in Abqaiq, he was hardly reassured. "The King was reportedly not attentive and appeared to have a somewhat distorted idea of what occurred. Most important from the company's standpoint, he seemed to have the erroneous impression that Amir Turki . . . had brought the situation under control."[46]

The Americans' reading of events grew increasingly rigid, judging from what ARAMCO sources reported to the consulate and what officials there parroted in their reports to Washington. ARAMCO dismissed the workers' complaints of discrimination as ridiculous and absurd. The firm had submitted its response to the Abqaiq petition in January at the request of to the Permanent Committee. It was then learned that one of the men, Sayid Abdallah al-Hashim, had actually read the petition to King Saud in Dhahran the previous November. When two ARAMCO managers arrived at the Permanent Committee's meeting in March to discuss the firm's response, they were "surprised and disturbed" to see the same employee, Sayid Abdallah, waiting for them, armed with a long rebuttal to the company's document.[47] The firm was shaken by this turn and tried hard to have him removed.

To his credit, the last section of Barger's long memorandum for Ohliger back in January, titled "Action to Be Taken and Lessons Learned," expressed some reservations about ARAMCO's traditional paternalism in the matter of labor representation. We should "guard against our tendency not to accept the Government's conception of its role as a representative of the employees. . . . Lacking any representative employees' groups to deal with the Company, the employees tend to turn toward the Government as protector and Government officials tend to assume [the] role of bargaining agents on behalf of employees rather than as impartial arbitrators."[48] At the hearings, though, it was impossible to tell if his views had any effect for all the smoke from ARAMCO's barrage of inconsistent arguments against al-Hashim's presence.

The first argument turned on the fact that Turki had recently informed ARAMCO that the palace was ordering it to disband the worker-management Communications Groups. ARAMCO guessed this was intended to head off the demands for elections, similar to those taking place in Bahrain. Nonetheless, the consulate's report noted, the "move is a hard blow to ARAMCO officials who have nurtured the Communications Committees since their inception more than two years ago."[49]

ARAMCO's Harry McDonald first tried trapping Turki. If al-Hashim was not speaking on behalf of the Permanent Committee but because the petition had widespread worker support, then "this would seem inconsistent" with the order to close down the Communications Groups. The argument seems spurious to me, but the key point is that Turki ignored it. Barger himself, the consul's main source, then complained about the Permanent Committee failing to take "an active role" but instead acting "as an impartial third party while the issues are discussed between the Company and Abdallah"—turning on its head ARAMCO's standard complaint that the state was all too ready to side with the workers rather than acting as an honest broker.

Turki told the company to take the matter up with the governor if it desired, and ARAMCO did, in a letter that portrayed the hearing as a clear violation of the king's directive since a worker had been allowed to attend what his highness obviously meant to have been a meeting between the government's representative and the company alone.[50] The consulate's gloss on the accusation that Turki was in direct violation of a government directive was that ARAMCO had no choice "because judging by proceedings at the last two Committee hearings, it now faced the prospect of debating each of the twenty-nine demands, nearly all of which it considers unreasonable if not ridiculous, with this self-styled representative of workers while the members of the Royal Committee sit by as bored spectators."

We have seen this last point argued before. When ARAMCO appointed a worker to its Communications Group it is a remarkable instance of a forward-looking firm doing the right thing for its employees, but when someone other than the firm appointed a spokesman it was a rigged process or worse, without a shred of legitimacy ("this self-styled representative of workers").

The company was, in fact, made to endure the indignity of debating the issues and addressing "the unreasonable demands made in the petition" with Abdallah al-Hashim in a series of meetings that McDonald and others portrayed as hopelessly stacked against ARAMCO. Worse, al-Hashim was only twenty-three years old, a teacher-trainee with good English and a suspect travel record during his summer vacations. "A year ago he took a two month vacation trip to Lebanon and perhaps to Egypt. . . . Some Americans in Abqaiq consider him more self-seeking in his labor activities than other agitators or embryo labor leaders there."[51]

The "Demands of the Saudi Workmen in ARAMCO," as the petition was known, rehearsed all the issues that firms had been fighting for decades, across the oil frontiers: The failure to promote local labor and to provide adequate housing and other benefits comparable to those of the Americans. Thus, after a quarter century the "Saudi workmen do not occupy principal technical or clerical jobs in the company." It highlighted "discrimination between races: An American, for example, is given everything such as furniture and other facilities and assistance while a Saudi does not get any of these." For over a hundred years the racial hierarchies in mines and refineries had been reinforced through the payroll, with different races paid in different currencies (the old "gold and silver rolls"). The Abqaiq workers' petition "demanded that wages be paid in Saudi currency to every employee of the company" and "that houses be built for them at company expense similar to those built for Americans in quality, location, and observance of modern engineering principles," along with the right to rent these, as the Americans did, rather than having to buy them.[52]

ARAMCO endured the encounters with Abdallah al-Hashim and then fought back, using one of the heaviest guns in its arsenal. The head of government relations, Ohliger, went to Riyadh to see the king. We don't know what transpired, but a follow-up meeting with Ohliger, Barger, and McDonald was arranged during which the king issued an "informal policy statement" emphasizing the need for "full cooperation and mutual assistance between the government and company at all times." The consulate reported ARAMCO's satisfaction with the progress toward devising an alternative means of management-employee communications, an objective specified in the Barger memo. Most important, Ibn Jiluwi received the order to rein in the Labor Office.[53]

The real payoff came one week later when ARAMCO fired the labor leader Nasir al-Said for violating the terms of his probation, it being something ARAMCO invented when it had been forced to rehire him a few months after the November 1953 strike. The grounds were "disseminating false information and causing dissension among the workers." Al-Said worked hard and had been promoted to intermediate status in Abqaiq, but the firm suspected him of leading the bus boycott and so transferred him to Dhahran. On March 31 the company announced that a new private Saudi bus company was taking over the route from Dhahran to al-Khobar, and the next day workers began a new boycott. ARAMCO said it had specific evidence "that on April 7 or 8 he mimeographed and distributed a notice concerning the refusal of employees to ride the same contractor's buses between Dhahran and Jubail." The amir of the Eastern Province approved in advance the sacking of al-Said. The same day, al-Said was picked up, questioned, and released. An ARAMCO representative told the U.S. consul that "the police say they have 'secret orders' to keep him under surveillance." The overworked Government Relations Organization was also tracking al-Said's movements among the workers in the three districts as he tried to rally support for his reinstatement.[54]

If by firing him the company hoped to end the boycott, it was disappointed. A month later most Saudi workers were still refusing to ride the contractor's buses, and both the Labor Office and the Permanent Committee ordered ARAMCO to restart the company-run buses. The firm refused, arguing it had the right to subcontract services according to the terms of the 1933 concession, that the buses had been improved to address workers' specific complaints, and that the new service represented support for local enterprise. One week later the chief of police in Dhahran ordered ARAMCO to stop the new service, known as the Blue Line buses, and resume running the line itself. ARAMCO ignored this order as well.[55]

Labor struggles had gone on continuously for a year, and the argument most frequently heard outside ARAMCO was that these actions were due to the failure of the company to deliver what had been promised by the settlement of the 1953 strike. The question of the rights of the ARAMCO workers to organize was now out in the open as well, in advance, in fact, of elections that would take place in the kingdom for local administrative councils a few years later and of the calls for the creation of an elected national consultative council. Many factors may have played a role in elevating the question of elections. In Bahrain, labor unions participated in the Committee of National Union and the general strike that it had launched in March. A member of the Egyptian Revolutionary Command Council, Anwar al-Sadat, had traveled to Manama to lend support to the strikers but had been blocked by British authorities.[56] The same month, in Damascus, a draft constitution of the Federation of Arab Labor Unions referred specifically to the would-be Saudi union.[57] The workers themselves matter most, however, and some day it may be possible to say more about these matters.

ARAMCO obviously preferred being left alone to determine what was in its workers' best interests, but inside the palace the company's Communications Groups were viewed as a danger to the established order.[58] In April the Saudi government had made clear once again that it "was opposed to worker representation, either elected or self-appointed, and would itself represent the interests of workers." ARAMCO preferred the palace's position though over that of an emerging pro-union view among some Saudi officials and intellectuals, Tariki most prominent among them, let alone over the position of the workers themselves.[59] The company warmed to the palace's position even more after it disbanded the ironically named Permanent Committee and created yet another new royal commission to report on the causes of the labor disturbances, composed this time of men more closely aligned with the reigning opinion and headed by Prince Musaid Ibn Abd al-Rahman, the king's uncle (Philby said that he "is probably one of the best members of the Royal Family").

Not all was rosy, however, according to the U.S. consul. ARAMCO workers were circulating three new petitions to coincide with the arrival of the commission. One expressed support for the bus boycott, the second called for elections for a workers' committee, and the third appealed to the king to throw the Americans out of the Dhahran Air Base. "Word concerning the Musaid Commission's impending arrival was spread to ARAMCO workers on May 19 by Nasir al-Said, a worker recently discharged for agitation, and by his cohorts. They spoke to groups of workers gathered at bus loading points in Dhahran's General Camp, urging workers to stand firm and remain united . . . and see that their rights were protected."[60]

ARAMCO's agents delivered copies of the petitions to the police, who had begun arresting workers.

Hirelings Who Have Gone Astray

There is too much we don't yet know and may never know about the course taken by the workers in June 1956, where we first began this chapter, when King Saud ended his vacation cruise in the Gulf and arrived in Dhahran. Doubtless the point in unfurling a banner "ARAMCO Saudi Workers Welcome the King. Death to Imperialists and Traitors" was not to have their movement beheaded. There had been crowds to greet the king earlier in the week when he had come ashore, although presumably they were not made up of men shouting, "We are oppressed, We are not satisfied, We want company buses, not Blue Line buses." Earlier in the royal visit, a smaller group of workers had gathered at Dammam where the king was to inspect an ARAMCO housing project. The men tried to deliver a petition. The "King drove past them and then sent a member of his party back to inquire what they wanted."[61] Workers had also reportedly met with Prince Musaid earlier in the week, and he promised them he was "looking into their rights and their rights would be protected."[62] They may not have believed him. The workers began another demonstration two days after the Musaid Commission announced its findings.

Word spread on the afternoon of June 16 that all ARAMCO workers should gather at the main gates of the three districts at 5 PM. The local Labor Office representative read a "Proclamation to the Workmen of the Arabian American Oil Company." It asserted the government's undiminished commitment to their welfare. It ordered ARAMCO to expand benefits and meet workers demands for better health care, education, and work conditions, although the bus issue was finessed a bit. The declaration ended with a warning:

> It has come to the Government's knowledge that there is among you a small group who have gone astray and who are hirelings who have taken destruction and destructive methods as a part of their ways and who have taken disruption as their goal and method. This group comes to you with honeyed words to lead you astray from the path of God and the good of their homeland. They come either as hirelings of your enemies who covet your country, or as those embittered because of their own failings, or they are seeking personal interest, or they are ignorant. So beware of them.[63]

That evening, the police began searching for thirty-nine suspected organizers, and eventually found thirty-seven of them. They arrested Sayid Abdallah al-Hashim and, the rumors said, up to sixty or even ninety others, but Nasir al-Said had disappeared. A month later, U.S. Air Force Intelligence had confirmed that about ninety-five were in jail, with many banished to the interior or held in Riyadh. Justice was rough. Al-Hashim was later reported killed. The U.S. consulate wired news the next day. The bus boycott was over.[64]

ARAMCO senior staff had started preparing for a strike on the day after the demonstration at the banquet. So had the local authorities, led by the governor, Saud Ibn Jiluwi. Government Relations was convinced that one of the factors that had contributed to the rise in labor militancy had been Ibn Jiluwi's absence for the past six months. His protégé had failed to rule with the same iron fist. Ibn Jiluwi was back and, praise be to God, vigorous, as was evident in the handling of the rioters at Ras Tanura when the men there tried to storm the movie theater on June 14.

About 1 AM on June 18 an ARAMCO official learned from a loyal Saudi worker that a strike would begin in Dhahran that morning "and that it would be accompanied by mischief and trouble." Organizers were rallying men around the issues of the harsh punishment meted to those in Ras Tanura and those who were now being jailed. Reports were coming in from all three districts of men fleeing the camps in taxis or on foot with their belongings.

Those who have described the strike of June 18–19, 1956, as the beginning of a mounting crisis have gotten it wrong. The morning shifts in Dhahran and Abqaiq began with about 25–30 percent of the workforce absent, and in Ras Tanura, where the demonstration had been crushed so brutally, numbers were down by only 10 percent. A normal absentee rate was from 2 to 8 percent. On June 18, Ibn Jiluwi's security and an ARAMCO representative went room to room in Saudi Camp in Dhahran and those who had stayed away from their normal shift were told to report to work or else go home. The police dragged away any outsiders found inside the camps. The next day, June 19, the men began to return and Dhahran reported turnout to be "almost normal."

The royal decree outlawing strikes was read to workers on June 21 as they picked up their pay and headed home for the weekend. Stray rumors came in that the men would not be returning from their towns and villages when work began again after the weekend, but the promise of a year in jail for striking and demonstrating (two years for inciters) and two years for using force or destroying property (three years for inciters) made all the difference. The Saudis were back on the job on June 23 and ARAMCO called the strike a complete failure. The U.S. consul reported that it was "abundantly clear that the firm hand of Amir Saud Ibn Jiluwi was a key factor in maintenance of order," and anyone who might have been troubled by the amount of force used was reassured by the amir that the government knew "best how to deal with its own people." Barger believed that the morale of the company's longtime Saudi employees had risen to new heights as a result of the government's firm action.[65]

Two years later, Harry McDonald, a career Government Relations man and member of ARAMCO's new policy planning staff, assured a member of the embassy that its labor force had showed no sign of dissent since the crushing of the movement by Ibn Jiluwi. Workers still circulated petitions, but these now were delivered directly to management rather than to the government. ARAMCO employees were, he said, "more content than usual." They even had television now, courtesy of the company.[66]

The last point was true. The U.S. Air Force had opened a 200-watt television station, AJL-TV, "The Eye of the Desert," in the summer of 1955. The signal could be received in most of the ARAMCO communities and in Bahrain.

It was strictly American fare—*Industry on Parade, Strike It Rich, Ding Dong School, Gary Moore, I Love Lucy, Roy Rogers,* and *Ed Sullivan* on Fridays—but officials were vigilant and made sure to edit all depictions of "alcoholic beverages, symbols of the Christian religion, and references to or symbols of Israel or the Jews."[67] ARAMCO followed the Air Force station into the broadcasting business a year later, aimed primarily at its Saudi workforce. In the aftermath of the strike, government officials pushed the company to provide more entertainment for the Saudis. This was a dream come true, a counterweight to Egyptian propaganda, and the firm began to install televisions in recreation rooms and other spots were workers gathered.

The Public Relations Department surveyed Saudi viewing habits by distributing free sets to Arab homes and asking viewers to rank the shows on AJL-TV they liked best. "Cowboy and adventure films proved to be the favorites." The same kinds of the films were also the most popular among Palestinians and others who attended movies in the intermediate camps. ARAMCO conducted follow-up interviews to make sure that Arabs "did not identify themselves with the Indians, a possibility that had caused momentary anxiety."[68]

OUR MAN IN RIYADH

King Saud Ibn Abd al-Aziz gets a bum rap in every book on the kingdom until now. In Riyadh today you rarely find his picture on display alongside kings Abdallah, Fahd, Khaled, Faisal, and Abd al-Aziz. Even the "Kings of Saudi Arabia" link on the Saudi Arabian Ministry of Culture and Information Website, www.saudinf.com, gives him two short lines (he created the Council of Ministers and started King Saud University) and leaves the space for his picture blank. Yet back in the 1950s the Eisenhower administration had embraced him as a great reformer and leader of the wider Arab and Muslim world. This book is no brief for rehabilitating his reputation, obviously, but it does make clear that historians need to think a little harder than they have to date. For one, he is often portrayed now as creating the conditions for serious dissent and perhaps chaos in the late 1950s and again in the early 1960s, leaving Faisal to clean up the mess each time. Yet, given that it is the militancy of the ARAMCO workers that is the main proof of the "internal" trouble Faisal inherited and dispensed with, the argument ought to be reconsidered. That moment of dissent was already over. Faisal usually gets credited, too, with restoring order to the kingdom's finances. But it was King Saud who invited in the IMF mission and it was the mission that pulled the kingdom back from the brink, a fact that no one until now has thought important enough to mention. It was the Anglo-French invasion of Egypt, which cost the Saudis a few million dollars in lost revenue, that goes far to explain the sudden rise in the kingdom's debts beyond the level that Ibn Saud and sons assumed as a matter of course.

On July 26, 1956, the fourth anniversary of the Egyptian revolution, Gamal Abdel Nasser announced that he had nationalized the assets and was seizing control of the Suez Canal Company. The revenues would be used to finance the building of a new massive hydroelectric project, the Aswan Dam, which would provide power for new Egyptian heavy industries. The Eisenhower administration had promised aid for the dam and then called off the deal in July after hearing of a new Egyptian-Soviet arms deal. Nasser fought back by taking over the canal.

It was definitely not a good time to be a Middle Eastern oil-producing state, especially one that had just signed a joint defense agreement with Egypt. Saudi Arabia's income dropped sharply, and it was being called upon to pay its share of the new Arab defense burden. During the Suez War itself, which began on October 29 with Israel's advance across the Sinai Peninsula, it would have been impossible to do anything but support the Egyptians. Their cities were being bombed. Egypt ordered what remained of the Egyptian air force's Soviet-supplied IL-28 bombers and MIG-15 fighters to fly to Saudi Arabia to preserve them from destruction. Muhammad Alireza, who was in Egypt at the time of the attack, wired the king for permission to enlist in the Egyptian army. The Saudis main contribution, however, came after the fighting was over, in the form of crude oil for the Egyptian refineries at Suez (the Sinai oil fields had been destroyed). ARAMCO's parents reluctantly began shipping oil to Nasser, paid for by the Saudis out of their future (diminishing) royalties, in return for Egyptian pounds.

Still, it was evident in the hastily called summits and deliberations of the Arab League in November, let alone in heated discussions among the princes in Riyadh, Jidda, and elsewhere, that the Saudis were worried about the course of events. In return for all their cash, which is what they brought to the alliance with the Egyptians and Syrians, the Soviets had expanded their influence in the region, Nasser was more popular than ever, the Israelis were more threatening, and at least one monarchy, Jordan, was in danger of collapse. There were signs that the king and his advisors were rethinking the strategy of accommodation with Nasser and looking for ways to chart a course closer to the free world's shore and away from the treacherous currents of neutralism.

Father of the Arabs

Just days after the unveiling of the Eisenhower Doctrine, plans began for dual kick-off state visits to Washington, the first by King Saud, the second by the Iraqi regent Abdul Illah. Both royals would meet with Eisenhower and Dulles and, it was also hoped, with one another. For decades the house of Saud had viewed the Hashemites in Iraq and in Jordan as its main enemy. The Republicans were clearly aiming high, although the plan ran into some early turbulence in the form of a call by the Saudi ambassador to the State Department on January 9. The news, immediately telegrammed to the U.S. embassy in Jidda, was that King Saud "would cancel his visit to U.S. if he [is] not repeat not met at airport by the president." Saudi protocol demanded nothing less, he insisted, although Eisenhower had never done such a thing. In the United States it was American protocol that mattered. The president might send Vice President Nixon to National Airport or Union Station to meet visiting heads of states, but Ike himself never left the White House, that is, until January 30, 1957, noon, in the cold, when he greeted King Saud Ibn Abd al-Aziz, who had been flown from New York on the president's plane, the *Columbine*.[69]

A king whom the CIA described a year earlier as a disruptive force in the West's diplomacy was now getting a makeover. We thought he was a corrupt, dull-witted, inept absolutist who ignored the needs of his people. No, the *New York Times* said, Saud was struggling "to make the transition from a society something like that of Britain in the days of good King Arthur to the twentieth century." The king "appears to have found himself. He has made moderation the watchword of his rule at home and in his foreign relations."[70]

The *Times*'s main Middle East correspondent, Dana Adams Schmidt, filed a front-page story the day the king reached Washington that reported the experts' assessments. The king's visit was "a Mideast key," the talks a possible "turning point in the political life of the Middle East" with a king who was "the spiritual leader of the Arab world." No, the king's counselors said, not just "spiritual leader of the Arabs," he was "coming to Washington as *father of the Arabs*."

The American agenda was obvious. The United States was hoping for a critical realignment, where King Saud would conclude that "he had more in common with King Faisal of Iraq, who is also rich in oil, than with President Nasser, who represents a revolutionary, anti-feudal movement in the Arab world that might one day bring down the Hashemites and the Sauds as well." But, Schmidt said, "what King Saud wants, and what the United States can do for him, is not so clear."[71] He needn't have been so coy. The king wanted money, desperately. He wanted more weapons. He also clearly wanted the Americans to remain on call at Dhahran, although the agreement on the base had expired back in June. He might need them because the maneuvering room for a king trying to turn his foreign policy around was snug and he was "a big man," "blind in one eye and sees poorly with the other. That is why he usually wears dark glasses."[72] Nor did statecraft appear to be his game—witness the chips wasted on dragging the leader of the free world to the airport to meet him.

Eisenhower dictated a memo to cover the gist of the private two-hour conversation that he had with Saud and an interpreter. The king had said he "wanted to talk mainly about very secret, confidential things some of them really personal." The problem was Eisenhower "failed to get any clear understanding of which one he considered absolutely secret . . . and which ones were of lesser sensitivity." Saud ran down the short list at long length: arms ("the British are nibbling at a number of my borders") and aid ("He said he was working hard building new schools, hospitals, roads, and communications. . . . All the money I have received has been wisely spent and for these good purposes."). Eisenhower responded mainly with homilies ("He seemed to be responsive to the idea when I told him that small village and household industries were far more important to a country with very low living standards than were heavy goods industries."). The king even conceded "that Israel, as a nation, is now an historical fact and must be accepted as such."

The secret topic was a proposal, as Ike remembered it, "to invite Nasser and the *King of Syria* to visit me." Saud had consistently warned that the United States was exaggerating the extent of Soviet influence in these two places. Imagine if he knew that Ike wasn't quite sure who ruled Damascus. Earlier, the king had told Dulles that "most of the things that concerned the U.S. were in fact manifestations of Arab nationalism," not communism, to which Dulles had responded with a recommendation that Saud meet his brother, Allen. Saud concluded the private meeting by warning that time was running out. He alone had been able to prevent a conflict from breaking out in the Middle East, but war now loomed.[73]

The American and Saudi sides formed working groups and moved swiftly toward an accord while King Saud, wives, and retainers encamped in White Sulfur Springs, West Virginia, for a week. A Saudi demand for $50 million a year to rent the base was quickly dropped. The Saudis didn't get too far either with the plea for lots of new development assistance. The United States agreed to sell the Saudis more arms and pay about $45 million for training the army and air force, and in return the king agree to extend the Dhahran Air Base agreement through 1961. Add about $5 million to build a new air terminal and improve the port at Dammam, and the Saudis were able to claim a neat $50 million. They haggled over the precise mix of guns, tanks, and planes, on new-versus-reconditioned equipment, and so on, but the basic deal was agreed to months before the Saudis and Americans formally signed the renewal agreement in April.[74]

In Washington in early February 1957 the future of the U.S.-Saudi special relationship looked bright, the president was hailing Saud, the king was praising the Eisenhower Doctrine ("it was a good one"), and all that was left to do was for the two leaders and their aides to meet one last time to prepare a joint communiqué.[75] The irony in the "Eisenhower-Saud Text" is in the repeated stress on the right of the peoples of the Middle East to maintain "their full independence." The third of six points on which the "two heads of state reached full agreement" says: "Any aggression against the political independence or territorial integrity of these nations and the intervention from any source in the affairs of the states of the area would be considered endangering peace and stability. Such actions should be opposed in accordance with the purposes and principles of the United Nations."[76]

Why ironic? Because the easiest way to understand the course of events over the next few months—the eye-dazzlingly fast rise and fall of the king as an alternative to Nasser, the steep climb and then tumble of the Eisenhower Doctrine's share price, and the robust Saud's dramatic collapse—is in terms of three failed covert interventions in the domestic affairs of other states. The first one was by Egypt against the monarchies in Jordan and Saudi Arabia in April. The second was by the Americans against the left-tilting government of Syria in August. The third was by the Saudis, who tried to clean up America's mess in March 1958.

The Man Who Would Be Hegemon

To appreciate the drama, first consider the regional alignment of states as Saud and Abdul Illah arrived in Washington. Iraq was said to be the West's only true ally. Nasser, increasingly portrayed in the *New York Times* as a sub-imperial power, backed by, if not doing, Moscow's bidding, headed a dominant pan-Arab coalition of Egypt, Syria, and Saudi Arabia. The kingdom was said to bankroll the Nasserists. Jordan, ruled by a twenty-one-year-old, untested King Hussein, had been won over as well to the Arab nationalist cause, having thrown out the British military in return for large subventions by the Saudis.

On April 18 or 19, according to the U.S. embassy, "the Saudi authorities uncovered an arms cache" hidden in the king's palace in Riyadh. A Palestinian confessed to have secreted them under instructions by the Egyptian military attaché, Lt. Col Khashaba. The Saudis made an immediate request for U.S. help in reducing its dependence on Egyptian military personnel and identified twenty-five key positions to be filled as soon as possible.[77] Days later, on April 26, a high-level Egyptian delegation led by Nasser's political advisor, Ali Sabri, and including the then head of Egypt's Islamic Congress, Anwar al-Sadat, flew to Riyadh to try to prevent the formation of a full-fledged anti-Nasser bloc. King Hussein of Jordan had just imposed martial law in Jordan, declaring, as the *New York Times* put it, "a fight to the finish against a conspiracy to overthrow him . . . getting its support from Egypt." On May 4, the *Times* ran another front-page story by its Beirut correspondent, Sam Pope Brewer, "Plot to Kill Saud Laid to Egyptians." The Egyptians alleged that the cells were really in place for use against future British targets in Bahrain, but months later the king's aide said the plot against Saud was real, in response to the king's dangerous turn toward America.[78]

What is undeniable is that by June 1957, the Eisenhower Doctrine's six-month anniversary, King Saud was being hailed by some for burying "the long-standing feud between the Sauds and Hashemites" and checking the evil Nasser's "drive for an Anti-Western bloc." Nasser's failed operations had restored the natural order of feudalists versus anti-feudalists and boosted King Saud's standing. "The key figure in the present dramatic upheaval of alignments in the Arab world is King Saud of Saudi Arabia who within the last two months probably has done more than any single figure to halt the march of President Gamal Abdel Nasser of Egypt toward leadership of an anti-Western nationalist alliance in the Middle East."[79]

The one potential complication, which would go unreported in the U.S. press precisely as long as Saud remained loyal to the Washington consensus, was the kingdom's growing financial problems. Foreign banks in Jidda were banding together to resist pressure from SAMA to loan merchants any more dollars against the rapidly falling riyal. Newspaper editorials criticized the ministries for waste and inefficiency, the high cost of living, the riyal's decline, the flight of capital (first and foremost by the princes), and the absence of virtually any productive investment inside the kingdom—Coke and Pepsi Cola did not satisfy this particular thirst.[80]

Under pressure, the king assembled a conference of the country's leading merchants and government officials where he announced the creation of yet another new royal commission under Prince Musaid Ibn Abd al-Rahman charged with reviving the country's economy. The first step was to approve a new import-licensing regime. Members split and action was stalled on a new foreign exchange control law ("that would put the open market money changers out of business"), as well as a plan by SAMA to create a new currency to replace the riyal.[81] Businessmen and bankers hated the import controls, but the latter admitted that SAMA had begun to make inroads on the amount the government owed its debtors. In August, the kingdom concluded negotiations to join the World Bank and IMF. Officials from the latter would arrive in the fall to advise on the implementation of an austerity plan.[82]

The search for funds to make up for the post-Suez decline in oil revenues drove at least in part Abdallah Tariki's new, multifront campaign against ARAMCO. Recall the pressure for retroactive royalty payments on the basis of the profits claimed by the parents but not ARAMCO for selling at the TAPLINE Sidon terminal. The transfer-pricing ploy in turn fed Tariki's argument that the kingdom would realize enormous windfalls if and when ARAMCO undertook its own downstream operations. Finally, Tariki was a month or so away from announcing the first deal that promised Saudi Arabia more than 50 percent of the split, with the Japanese National Oil Company in the Neutral Zone between Saudi Arabia and Kuwait.[83]

In a bumper-crop season for irony, it was Eisenhower himself who would take home the blue ribbon. On August 9, 1957, he submitted a report to Congress on the success of his doctrine, a "milestone in our foreign policy as it relates to the Middle East," in contributing to peace and stability in the "tense region." The successes would "continue to grow as long as the purposes and principles it sets forth are maintained." Three days later, Syrian forces surrounded the U.S. embassy and expelled three undercover CIA agents for trying to overthrow the government.[84]

The kingdom's acting foreign minister, Syrian-born Yusuf Yassin, was suddenly protesting a report on Voice of America that aid to Jordan was on its way from the U.S. base in Dhahran. The Saudis were happy about the aid, he said, but use of the base for this purpose violated the terms of agreement. More announcements followed. Contrary to reports, Dhahran was not a U.S. base at all, the Saudi Arabian government did not accept the Eisenhower Doctrine (*al-Bilad al-Saudiyah*, October 6, 1957), and when the BBC reported that Saudi Arabia had received U.S aid, the government issued an official denial. "The Saudi Arabian Government has not taken financial aid from any country."[85] All references to Eisenhower Doctrine–related funds were to be dropped, and the Saudis also protested when the new chief of the U.S. Training Mission was referred to too publicly by his second title, "Commander, Second Air Division, Europe, with Headquarters in Dhahran."[86] Not that it mattered. Nasser's confidant Muhammad Heikal mocked the king on the pages of *al-Ahram* as nothing but a stooge of the United States.

The retreat from the initial embrace of the Eisenhower Doctrine looked to be turning into a rout. The Saudis hired away an Arab League official, the ex-head of Syria's UN delegation and the future president of the Palestine Liberation Organization, Ahmad Shukairy, to represent the kingdom in the General Assembly, and in his debut as a Saudi in October 1957 he delivered a blistering attack on U.S. Middle East policy. King Saud tried to blunt its impact by sending a "warm message" to Eisenhower. Palace watchers were working overtime: what was happening to Saudi foreign policy? Why did the foreign minister, Crown Prince Faisal, continue to delay his return from the United States after an operation in July?[87]

The White House agreed to a meeting between Eisenhower and Faisal for September 23, the day the Arkansas National Guard was withdrawn from Central High School in Little Rock and a mob tried to storm the nine black students inside. Eisenhower would send federal troops the next day. It is doubtful that Faisal had much to say about desegregation. He was "expected to follow the usual Arab line" on Syria and Israel but the transcript of the entire conversation remains classified. At the press conference afterward, Faisal insisted that Syria did not threaten any of its Arab neighbors, because, if it did, then according to the Eisenhower Doctrine the United States might be obliged to take action. Some said that Eisenhower and Dulles had toned down this kind of talk following a personal appeal by Saud, but the State Department was denying it. Its memo to Eisenhower on the eve of the meeting nonetheless suggested a role for Saudi Arabia in the months ahead. "We hope that the Saudi Arabian government may be able to use its influence to reverse the trend in Syria towards satellization of that country."[88] King Saud tried.

Faisal remained abroad until February 1958, either because he preferred to see his brother continue his stumbling course as America's secret weapon in the war with Nasser while at home rents were shrinking and the riyal continued its fall, or else the king would not agree to terms for his return. By coincidence, around this time ARAMCO tried to increase the pressure on Saud's government. The king's ministers had successfully pressed ARAMCO in November to guarantee another $16 million in new loans with New York banks in November on top of the $25 borrowed in September. This arrangement was an updated version of the one made famous by Ibn Saud. Two IMF experts arrived in November. As might be expected, they wanted to abolish the import and currency controls, a step that would require ARAMCO's cooperation in the form of a large loan to stabilize the riyal. The U.S. embassy reported that they asked for some $90–$100 million. The company had countered with several proposals to produce the necessary funds without resort to raiding ARAMCO's treasury. For example, Tariki might finally agree to ARAMCO's parents' proposal to start new oil trading companies to replace the offtakers system. It was also obvious that resources were being wasted, and the safe bet was that a new $100-million loan would be too. The firm politely turned them down. The work of the IMF limped along. When the first new budget in three years was published in January, the U.S. embassy reported the unprecedented event of a reduction in allocations to the royal family, which was supported by the king but opposed by key princes.[89]

News on the foreign policy front was even more grim. In February 1958 Egypt and Syria merged their two states into one, the United Arab Republic (UAR). Syria's ruler, Shukri Quwatly, handed power to Nasser. Yemen pressed to join. Jordan and Iraq announced creation of a rival Arab Federation. King Saud was stunned, pushed back to square one in his quest to stop the spread of Nasserism. The king's brother, Prince Faisal, was there in Cairo during the announcement of the Egyptian-Syrian union, praising it, denouncing the Eisenhower Doctrine, and attacking the "Zionist propagandists and supporters of imperialism" who claimed Saudi Arabia opposed these momentous developments. He flew home from Cairo on February 3 after an unprecedented eight months away. At the same time the U.S. embassy reassured Washington that the king would not publicly attack the UAR but that he was secretly opposed.[90]

On March 5, the head of Syrian military intelligence accused King Saud of plotting to kill Nasser and destroy the UAR. He was working on behalf of the United States and had paid $5 million in bribes, using his brother-in-law as the go-between. The United States denied it. The Saudis tried to block the news from reaching the kingdom, but Cairo's Voice of the Arabs broadcast the details and attacked the king, relentlessly, over the next weeks. It is hard to establish the facts in the case, particularly when Egypt's one-time journalist-turned-sycophant, Heikal, would published the alleged details side-by-side in *al-Ahram* with "proof" that the king let the Americans stockpile atomic bombs at the Dhahran Air Base, a lie that was repeated for years.[91]

Sorting out truth from fiction didn't get easier over the next few weeks. Kim Philby, the Soviet double agent working undercover for British intelligence as the *Economist* correspondent in Beirut, told the U.S. embassy that a royal commission had been formed to investigate the charge against Saud. After meeting, "three members . . . resigned because they reportedly found evidence that the plot existed," although the king himself allegedly had no knowledge of the plan. Philby's source was his father, residing once more in Riyadh and a fixture at the palace.[92] Not that the truth mattered to events. Supporters of Faisal rallied to return him to power. There were rumors of a military conspiracy. The CIA's Bill Eddy, a longtime admirer of Faisal, believed it, reporting that the king was "terror stricken," but the U.S. ambassador and his own sources in ARAMCO and the air base didn't, and told Washington so.

Some called the king's decree on March 23 turning over responsibility for domestic and foreign affairs to his brother a palace coup. Others saw it as a shrewd move on the king's part, an answer to the Egyptian charge that the kingdom was coming apart and that Faisal was a Nasserist fellow traveler. U.S. ambassador Heath was one of them. As the palace coup against the king went forward, Heath organized a back channel for Saud to send and receive "highly classified information from the U.S. government for His Majesty only."[93]

7 EL JEFE ROJO

*It may be taken as gospel that the strongest antagonism to
government ownership and management will be found among
those who would profit most from their elimination.*
Wallace Stegner

*Hardy said he could not but foresee a stormy time ahead for
ARAMCO no matter what policies the Company adopted.*
State Department memo

KEEPING DISCOVERY SECRET

In July 1955, two weeks after *Island of Allah*'s very short run in New York's
Forty-ninth Street Trans-Lux Theater, the film's producer, Ray Graham, telephoned the writer Wallace Stegner about his latest bright idea. ARAMCO
wanted a book to go along with the movie and proposed that Stegner write
"the story of American oil pioneers in Arabia." He would be paid a minimum
of $6,500 and expenses or what today is around $42,000 for a mere thirteen
weeks of work. Stegner's only worry was that the company would interfere. "I
have no liking either for muckraking or for whitewashing jobs; I should very
much like to make this book straightforward and honest history. . . . To do
that properly I shall need my elbows free."[1]

The two weeks he spent in Dhahran and the surrounding region in December 1955, by coincidence the year of most intense worker struggles in the
camps, weakened Stegner's resolve that he could deliver the book the company wanted in time, and he expressed his doubts in a letter written while he
was still in American Camp.

> It is very clear that there is a great story here, one important not only to the
> company and to Saudi Arabia but to the whole world, and in greatly diverse

ways. It can be told as a simple success story, as the story of a new kind of frontier, as a new alignment of political and economic forces, as a meeting and fusion of cultures, and someday someone will do the whole thing and make a superb book of it. But it is also clear that a good deal of the story can hardly be told now, except for the company files.[2]

What Stegner was beginning to discover in Dhahran was the contradiction at the heart of the firm's self image. If some inside ARAMCO wanted the history told, others feared what it might reveal or what use might be made of such a book by the company's growing ranks of critics.

Stegner wrote *Discovery!* at an astounding pace. On March 3, 1956 he reported

> operations have ground to a limping halt and the offensive has been temporarily called off. Yesterday I sent off to Thompson two copies of a 357-page triple-spaced manuscript. I am positive the smell of it will bring mail inspectors around to 505 but at least it is temporarily out of my hands. . . . And if I ever write 80,000 words in a month again, for any price, take me to Bethesda and have my head examined.[3]

Stegner heard nothing from the firm for the remainder of the year. The vetting of the manuscript by the company's Arabists and public relations people had to compete with other, more pressing concerns facing those on the ground, beginning with a new round of strikes by Saudi workers in June 1956.

News of the death of two of his informants led Stegner to contact ARAMCO once more in January 1957. It was a propitious time. King Saud and the Eisenhower Doctrine were big news, and by April the United States had renewed the Dhahran airfield agreement, and Saud's stock was still rising. ARAMCO finally agreed to move forward on a revised draft. That is when Stegner learned of the fate of the text. Tommy Thompson in the public relations department, reluctantly sent him the company's internal report.[4] Stegner fired off a response to the company's demands for changes.

> As you know, this kind of book may be either of two things: It may be frankly a "company history," written by Company employees according to Company specifications and published with the Company's backing or at the Company's expense. This makes it, essentially, a public relations job. Or it may be a book written by an outside observer, with more or less cooperation from the Company and with greater or lesser access to its records, but representing

his interpretation of people and events and published under his name and at his responsibility. Done on this basis, its aim is the truth of history insofar as its author can attain it, and not the immediate and uncritical promotion of Company purposes and prestige.[5]

The problem appeared to be Stegner's attempt at verisimilitude or "speaking frankly on Arabs and Arab-Company relations." He argued that the "conflicts of culture and personality" were a defining feature of ARAMCO's frontier era. "In these matters it is very hard to know where discretion shades off into timidity on the one hand and into rashness on the other."[6] He had sought to capture the rough-hewn nature of most of the wildcatters, their perceptions of the people they found themselves thrust up against, and the slow emergence of modes of cooperation and accord. As he wrote in one contested passage,

> It is hard for a man who has spent ten years or so of his life drilling holes in a desert, living in a bunkhouse or a prefabricated cottage, and learning to get along with strange and foreign people wearing robes on their backs and rags on their heads to realize the ways in which his daily actions contribute to the mighty pressures that raise and lower nations and empires and great segments of the world's people.

Dhahran objected to his recounting of an incident in the 1930s when a Saudi worker was beaten by one of the Texas roughnecks. Tom Barger wanted the term "native workers" excised ("natives is a bad word respectable only in California and New England. Elsewhere it means 'downtrodden colonials.'"). His own references in his letters to what at the time had been called "coolie camps" were to be changed to "Saudi camps."[7]

Every chapter of the draft seemed to trip an alarm, beginning with Stegner's rendition of the negotiations for the original oil concession. He portrayed it as a "high stakes poker game" that ARAMCO's Arabian Affairs Division (AAD) said made it look as if the oil companies were trying to reap maximum gains, and he was told to change it. But the poker game analogy was the one used originally by the company's own negotiators. "I find it hard to believe that harm can come from showing the negotiation as a game in which clever antagonists each set out to win from the other the best deal possible. To make altruists and disinterested gentlemen of them would be to falsify the process and the personalities both."[8] ARAMCO also told Stegner to remove references from the letters that Hamilton, its chief negotiator, had written home. One described the rival companies' colluding to control production, another to ARAMCO (Casoc then) always operating "in accordance with orders from San Francisco," as the archives showed, and that it had pursued and obtained "a monopoly of virtually all the potential oil lands in the Arabian Peninsula and the coastal regions south of it."[9] These are the precise charges that Abdallah Tariki would make two years later from the floor of the first Arab Petroleum Conference in Cairo in 1959.

References to the southern frontiers of the kingdom or to oil deposits there worried the firm immensely due to an ongoing boundary crisis with Oman and Oman's British allies. ARAMCO had transported Saudi troops to occupy the contested Buraimi oasis, the Arabian Affairs Division under Mulligan provided the research, maps, and other data to support the Saudi claims, and it published the Saudi government's massive memorial for an international arbitration tribunal convened in Geneva in 1955 in order to adjudicate the claims over Buraimi. British intelligence privately reported that ARAMCO had been working against her majesty's government, and other sources insist that ARAMCO conspired with the CIA. In the British Foreign Office, ARAMCO was being branded "the greatest obstacle to Anglo-American harmony in the Middle East."[10]

Little wonder, therefore, that ARAMCO blue-lined virtually every reference to the firm's interest in Saudi Arabian boundary issues and to anti-British actions of the American oilmen, one of the realities of the war years. Stegner clearly was dumbfounded.

> Dhahran has struck out several casual mentions of the Company's inevitable involvement in Saudi Arab boundary problems, but it has left untouched whole sections of the text dealing with the experiences of Tom Barger and Dick Hattrup in the surveying of these disputed boundaries. Is the one permissible if the other isn't? And in fact, aren't both so intimate a part of ARAMCO's frontier experience in Arabia that they can hardly be censored out?[11]

"Finally," Stegner wrote, there was the question of "antagonism against Israel and the Saudi prohibition against Jews." He was right of course that the issue was one of great interest to many in America who would read the book. The firm was then fighting a lawsuit in the New York courts charging it with violations of the law for discriminating against Jewish Americans. ARAMCO's Government Relations Organization (GRO) called for dropping all mention of the company's few original Jewish employees or even of a romance that once blossomed in Beirut in the 1930s between an ARAMCO roughneck and a young woman from Haifa.

Stegner's long response ended by outlining the choices before the parties: end the project because it was too risky, publish it without his name attached, give it to the writers in ARAMCO to rework in a manner acceptable to Dhahran, or let him finish what he had started. "But I would not want my name on it unless I could feel that what I sign is really what I can believe in."[12] His agent successfully concluded a deal, he thought, in May 1957.[13] He promised the redraft in five months, by September, and again turned it in on time while pointedly ignoring most of the requested changes. His contact in ARAMCO's New York office, Gordon Hamilton, wrote in October to apologize for the "mysterious silence," explaining that that there were delays in copying and distributing the revised draft for review, but took pains to reassure him. "From just barely cracking the cover, I would say that this looks like a real 'Stegner job.' Any other words would seem superfluous at this moment."[14] A ten-year gap then appears in the records.

Heaping myths one on top of the other in the 1960s and 1970s, the relations men in Dhahran would explain that ARAMCO pulled the plug on the book ostensibly because Stegner romanticized the Americans far too much and wrote dismissively about the Saudis, as if it were his views that were the problem and not those of the men and women who had built American Camp. Yet, Stegner objected most strongly to ARAMCO's deletion of virtually any passage in the book that appeared to lend weight to the charges of its numerous critics at home and abroad—matters that of course had nothing to do with so-called cultural insensitivity.

Discovery! had become a hostage to the perilous politics of the late 1950s and the regional upheavals that seemed poised to topple the Saudi monarchy itself. Soon after Faisal's palace coup, a second, much bloodier military-led "revolution" modeled after Nasser's overthrew the Iraqi monarchy in July 1958. The United States landed troops in Lebanon in August in response, fearing the collapse of the governments there and in Jordan. The space for criticizing the American firm grew larger, the arguments more threatening, as radical commoners and "liberal princes" began to talk of constitutional monarchy, Saudization of ARAMCO, and the creation of an organization of oil-producing countries to balance the power of the companies. Relations with the United States and with ARAMCO would remain fraught through 1962–63 until the radicals finally were eclipsed, Tariki fled the country, and Faisal, who became king in 1964, began to rebuild the relationship.[15]

The "N" Word

ARAMCO responded to the regional upheavals of the late 1950s and the tremors in Dhahran by circling the wagons. Two kingdoms had collapsed, and a third, led by the impossibly young, sports-car-loving King Hussein of Jordan, survived for the moment with the help, covert and otherwise, of the United States. It was hard not to be influenced by, let alone convincingly refute, the argument heard everywhere from the Casino el Nil in Cairo to the Cosmo Club in Washington that the days of the Al Saud were numbered. Saudis were saying the same thing. Americans heard it constantly in Jidda, among Hijazis, or when they called on Tariki or invited other Young Turks for drinks or dinner.

It would appear, too, that ARAMCO was less able or willing than either Faisal or the late Eisenhower administration, after its Middle East policy crash-landed, to imagine it possible to make peace with Nasser and Nasserism.[16] Nasser, famously, made it easier for Ike in 1959 when he began tossing Egyptian communists into prison and torturing them. Faisal must have felt vindicated that the Americans finally seemed to be following his advice as he tried, it turns out one last time, to repair relations with his "brother, Gamal." Eisenhower once called Faisal "stupid" for holding this positive view of Nasser. The irony is to be savored because a year or so later, Faisal wanted the old order back, arguing futilely with the incoming Democratic administration that Nasser was a communist and that the New Frontiersmen were guilty of appeasement. No, Kennedy's aides would respond, Nasser was really a nationalist. But by then, Nasser had started bombing Saudi Arabia and the royal family feared that the United States might have been plotting its overthrow.

Nationalization of foreign property—the issue that had always concerned ARAMCO's owners most—was by decade's end claimed by all the various rival currents, movements, and personalities in the region, from the Muslim Brothers to the Baath, as a nation's right and a necessity. Imagine the reaction, therefore, when Abdallah Tariki's legal advisor, Frank Hendryx, delivered a paper at the first Arab Petroleum Conference in Cairo in April 1959 that argued the sovereign right of oil-producing states to override a concession's terms where a nation's welfare is at stake. ARAMCO all but declared war on Tariki.[17]

Readers who want to believe the myth that Americans backed a good prince and farsighted reformer, Faisal, against the bad king and sybarite, Saud, may experience some difficulty in making sense of events because Faisal was seen by many in the U.S. administration and in ARAMCO as a less moderate and reliable ally than Saud and the true Nasserite in the bunch. Opinions were split when it came to assessing Faisal's performance as prime minister in the year or so after his brothers brought him back from the United States in March 1958 to govern. He received high marks for staying the course that the IMF advisors had charted, in particular, in stabilizing the price of the riyal and paying back some of the government's massive loans. Austerity regimes though have particular consequences and in Saudi Arabia discontent was growing among merchants as much as among workers, once employment slowed down and spending on public and private works declined. Faisal turned on the press, which had begun to attack the government's failure to pursue meaningful development, and crushed all talk of things like constitutions and elections for Saudi Arabia.

Keep in mind too that Tariki had always been viewed as a protégé of Faisal and that it had long since become routine among Saudi watchers to blame the crown prince for preferring to turn the screws on ARAMCO, which was easy albeit unfair, than to build efficient administrative institutions and to engineer a "take off" of the national economy.

ARAMCO's Late Turn to Uplift

ARAMCO publications portray company policies in the 1950s and 1960s as the voluntary, steady, continuous refinement of the putative partnership-in-progress principle worked out decades earlier with the king, so that one sees in the Jebel School the seeds of the later blooming, advanced training program for Saudis (that, nonetheless, the firm did not want to pay for), or thinks that the home ownership program (that, nonetheless, most Saudi workers were not buying) was the improvement that local managers had always planned to make on the hovels (that the company called the best housing in the Middle East) they built for the natives. The reality is that in 1958 company documents show Barger and others acknowledging that many of ARAMCO's vaunted second-round reforms, which the firm had unveiled in response to the strike wave of the early 1950s, had failed. The Exxon people were again comparing ARAMCO unfavorably with their own venture in Venezuela, Creole Petroleum. Tariki was doing the same. They posed sharp questions in particular about the value of George Rentz's Arabian Affairs Division.

Once again, ARAMCO would accept some change that the Saudis had been demanding for years, always unreasonably, the Americans added, when first forced to confront an issue. So, in 1958, years after Tariki first claimed it as the government's right, he and the king's aging advisor, Yusuf Wahba, were brought onto the company's board. The firm and its allies in the U.S. embassy would then toast this latest proof of ARAMCO's farsighted commitment to the Saudi people, what Barger called "corporate responsibility," even while dismissing Tariki's newest demands as beyond extreme.

A reader might therefore legitimately question my characterization of failure above and argue instead that ARAMCO had evolved a fairly successful strategy, which consisted roughly in discounting the future and "getting what one could get away with" in the short term. As we will see, this was in fact one characterization that ARAMCO's new policy planning group in Dhahran came up with in guessing what were the owning companies' intentions with respect to their Saudi concession.

There is no better place to begin the account of the firm's stumbling in the face of Arab nationalism than with the old Arab Research and Translations Department, renamed the Arabian Affairs Division. The hubris of the postwar Orientalists who marketed themselves—everywhere, not just in Dhahran—as agents possessing expertise that would secure the "full truth" and "intimate knowledge" for firms and states left them vulnerable whenever their principals found themselves surprised by strikes, coups, nationalization campaigns, revolutions, and the like. Hard-drinking George Rentz and his comrades took it on the chin numerous times, from the New York owners and even from the State Department, for example, when the latter accused ARAMCO Arabists of ignoring the growth of communism in the Saudi oil fields.

At the time of the 1953 strike, the firm took the unprecedented step of moving the research department out of American Camp, where, presumably, certain limits to intimate knowledge of the local communities had inhered. It was relocated to Dammam, where the amir of al-Hasa, Saud Ibn Jiluwi, was moving his operations and the would-be state in Riyadh was placing its own experts. ARAMCO convinced a crony of Jiluwi named Abdallah Darwish, a merchant who after being run out of Qatar set up shop in the Eastern Province, to build a modern, air-conditioned office building in town. ARAMCO promised a long-term lease for two of the three floors. Darwish, who was a partner with Muhammad Bin Laden and various princes in the new Riyadh Bank, agreed to the deal. His American backers didn't like him much. "He was gross, walleyed, stentorian, and heavily scented. His peccadilloes (by the standards of the area) were young boys and barely nubile girls."[18]

Bill Mulligan, who became manager of the division under Rentz, once described the Darwish Building as a "beachhead" at a time when ARAMCO "was questioning the colonial aspects of an exclusive, walled compound style of living" and was seeking means "to get closer to the Arab public." Perhaps, but the Darwish Building was sited next door to the local police station, plans to move ARAMCO Public Relations operations went by the wayside, and the "beachhead" was in fact the farthest point reached by the Americans before they retreated, not counting the apartment that one of the CIA agents in ARAMCO's research unit, Homer Mueller, rented for a while with Rentz. Mueller was in fact the first ARAMCO American to actually live outside camp in over a decade, since an early employee, William Lutz, a Muslim convert, was fired and then expelled from the kingdom for moving into Saudi Camp in 1944. Mueller would move on to Riyadh as company representative there. Mueller's boss, Rentz, meanwhile, made scant use of the apartment, but his little extravagance was a bargain in comparison to the Darwish Building itself.

Rentz moved his library and librarian there along with radio receivers, tape recorders, and other gear for the translation unit. He had a fully functional laboratory built for Federico Vidal, the anthropologist, and Bob Matthews, the naturalist, he had hired. The secretaries worked among the "skulls in bookcases [and] frozen lizards in refrigerators," while the various Bedouin paid informers, or "relators" as they were known inside Camp, would wander among the specimen cases (and secretaries, was the joke) before reaching room 319, equipped especially for these visitors. "Please furnish room 319 of the Darwish Building, Dammam in Arab Majlis style . . . to be encircled with appropriate floor and back cushions and provided with an adequate number of elbow rests . . . eight mattresses . . . eighteen bolsters . . . [and] material (washable) to cover above."[19]

The experiment proved costly. The guinea pigs carried on glumly. The researchers and translators had to endure eight hours or more a day in a drab, under-heated, kerosene-fume-filled space with too little water but lots of western-style toilets. The director, Rentz, showed up infrequently himself, preferring to work on his manuscripts from home and, eventually, from Cairo, six months of the year, where he had relocated his Egyptian wife and children away from the racism that pervaded Senior Staff Camp. The company paid Darwish rent for the full ten years of the lease but pulled its grumbling and demoralized employees out of the building in 1958, which saved some $75,000 a year in operating costs, according to Mulligan. The legend lived on, though, and for some years thereafter the Darwish Building would be identified in the dissident press and propaganda broadcasts as headquarters "of the American intelligence service in Dammam."[20]

Pressed repeatedly by the board or various of the owning companies to justify the running of a research center and its own state department, Barger and the others would exaggerate the stakes while ignoring the contradictions in their various representations of the kingdom. Saudi Arabia, they said, was unlike any other operating environment known to the overseas oil industry. It still had an "embryonic" form of government in 1960, according to one briefing paper, where "the only manuals of government organizations are those which our own people have compiled." Yet, back in 1957, it was the fact that the Saudi government had "grown tremendously in size and complexity" that made a research and analysis capacity more necessary than ever.[21] Three years earlier, Barger had framed its mission in the broadest terms imaginable and, yet, insisted that his superbly trained professionals were up to the task.

> The chief function of this organization is to know and understand the Arab and his country—a land that stretches from the Atlantic Ocean to the Persian Gulf, and from the Bosphorus to the Indian Ocean. To know the Arab one must also know and understand his religion which widens the Research Division's area of interest to include large regions east of the Persian Gulf to include Persia, Afghanistan, Pakistan, and Indonesia. . . . The Kingdom of Saudi Arabia is of special interest. The nature of work involves not only knowing and keeping abreast of the contemporary scene in geography, culture and civil affairs but includes the past as well since the present can hardly be evaluated without a knowledge of the history of the Near East.[22]

What is truly remarkable in light of these regular public assertions of their indispensability to the owners and the infallibility of their expertise is the private admission by the Arabists that after years of seeking to know and understand Saudi Arabia they in fact knew very little "about how the Saudi Arabian government operates, and thus how the Company should conduct its business. . . . Who or what is responsible for domestic affairs is not clear. . . . We do not know what channels to follow to try to get business done."[23] In their defense, it was also true that the New York management often saddled staff with work that took them away from their studies. For instance, Matthews had to work with the crew of the ill-fated film *Island of Allah* for almost a year. In 1955, the vice president Harold Minor ordered the team to produce Arabic translations of two Cold War tracts—James Burnham's *The Coming Defeat of Communism* (1950), and Max Eastman's *Reflections on the Failure of Socialism* (1955)—for Saudi Arabia while disguising ARAMCO's role.[24] Warring on the Saudis' side in Geneva over boundary issues took more time still. The would-be professors that Rentz was prone to hire seemed more at home identifying flora and fauna or translating texts in Islamic thought than in collecting political intelligence. ("Do we need this?" some manager scribbled on the memo proposing a comprehensive history of the Shiite communities of the Gulf.) Then there was the matter of Rentz's own perfectionism. He had produced at best a handful of articles and encyclopedia entries in the course of a career.[25]

The rare, unvarnished criticism may explain the hiring of new contemporary affairs analysts Harry Alter, who spent his career in Dhahran, and, once she escaped Iraq after the revolution, the Harvard Ph.D. candidate Phebe Marr, who fled Saudi Arabia two years later. The duo produced the first biographies of Tariki, Ibn Muammar, and other Saudi officials for the revamped GRO in 1958–59. The more immediate outcome, however, was the convening in early 1958 of a cross-department Special Committee to assess the nature of the firm's understanding of and relationship with the Saudi state, and to devise "more effective ways" of dealing with the new governmental agencies. Its brief also included the problem of ARAMCO's industrial relations policies.[26]

Barger, the head of ARAMCO's Government Relations Organization, convened the first meeting of the committee in his office on January 8, 1958. Tariki and his continuing criticisms drove the agenda. ARAMCO would need to deal with the growth of unemployment in the kingdom ("for which the Company must bear major responsibility") by aiding development of local industry. The firm ought to be examining its relationship to the owner companies, the structure of the current board of directors, and the legitimacy of the Saudis' contention that the company was cheating them out of their fair share of the oil rents. ("Does the government suspect our integrity on the ground, for instance, that if we could offer a 50–50 deal in 1950 why could we not have done so before? . . . Should the Company attempt to determine what a fair return would be and decide not to go beyond that point?") At the same time, and underscoring the proximate source of their current difficulties, Barger and his associates considered what alternative—or, at a minimum, complementary— political course they might pursue. "Has the Company given Tariki too much of a build-up? Do we perhaps consider him more important than he really is? Is there someone else with whom we should try to do business?"[27]

The Pendleton Memorandum

One of the committee members, John Pendleton, circulated an incredible memo to management outlining what were, essentially, the three policy currents that governed decision making by the firm's various factions, the owners, the producing departments and, not least, the relations people. First, there were those committed to maintaining the status quo. "Hold[ing] on to what we have, in the minds of the proponents of this objective, appears to require no special measures, only a continuation of what we are now doing." The second current, which I alluded to earlier, was dedicated to maximum exploitation of the concession at minimum cost to the firm.

> This objective reflects a short-term philosophy. It assumes . . . that the concession cannot run its full course [which indeed proved correct]. Therefore the Company should not spend time and money trying to create good will in Saudi Arabia, or, more specifically, seeking to improve relations with the government. Our effort should be directed toward rapid money making without regard to other considerations.

In classic three-position-paper style, Pendleton outlined the Barger group's (and presumably his), for lack of a better term, liberal or accommodationist position last. They believed it possible "to continue to operate successfully throughout the remaining years of the concession," which ran until 1999. To do so, however, required ARAMCO to make numerous specific concessions of its own to the emerging Saudi nationalist and developmentalist forces. ARAMCO should sponsor local industry. It would have to repair relations with the government. Importantly, not least because it would continue to be successfully resisted throughout the following decade, ARAMCO would have to tear down the fences in American Camp. Pendleton called for "integration of the Arab and foreign living communities." Plans to do so would actually be drawn up and then shelved once the nationalist wave subsided in 1964. The firm would also have to provide serious funds for building hospitals, schools, and roads, he said. Taking up the matter that Tariki had hammered on for over a decade, Pendleton argued for "increased hiring of Saudis for Senior Staff jobs, making greater use of them after they are hired, and treating them like any other Senior Staff employee." Stealing a page from the Rockefeller and Exxon program in wartime Latin America, he argued ARAMCO should practice "widespread and somewhat indiscriminate philanthropy in Saudi Arabia," perhaps through "creation of an ARAMCO foundation." Finally, to "get the people behind us," he said ARAMCO should prepare for the possibility of regime change, "drastic or otherwise."[28]

I want to stop and consider the significance of this document for the analysis I have been pursuing throughout the book. First, and most obvious, the memo contradicts the conventional depiction of ARAMCO as somehow pursuing from the start of its operations or from the war or the early postwar period policies that Pendleton was advocating in 1958 as reforms the firm ought to undertake in the future. In other words, it is consistent with the record I have been piecing together of the 1930s, 1940s, and 1950s of the distance, on the one hand, between the firm's treatment of Saudi labor and other of its standard operating procedures in the Eastern Province producing fields and, on the other hand, the images of the partnership-in-progress that the firm marketed abroad.

In contrast therefore to accounts that see policy as originally or exclusively accommodationist, we have a more accurate and useful depiction by Pendleton of a firm being steered by owners and managers who reflected, embodied, and advanced different and contradictory objectives throughout the course of the concession. Most important of all, the Barger group turns out to have gotten it wrong. Their belief that by moving finally to support uplift of the Saudis they could prevent nationalization of the company ("to continue to operate successfully throughout the remaining years of the concession"), proved a costly illusion. The best they might claim, although there is good reason to doubt its truth as well, is that the late turn to uplift meant that the Saudis were more disposed than they might otherwise be, once Tariki was gotten rid of, to negotiate the takeover of the firm in stages rather than by unilateral dictate.

Still, we ought to give credit where it is due. The Barger group kept the GRO and its research arm, AAD, alive for a dozen more years. The Special Committee lived on as well in the form of a new policy and planning staff, which began to function in November 1958, with the objective of providing management "with recommendations for action" with respect to ARAMCO's relationship with the Saudi state.[29] A number of reforms were in fact enacted around this time, along the lines suggested by the Barger group. The Special Committee successfully pressed for the dismantling of the much-heralded, single-skill training regime, which was judged an expensive and misguided failure. In its place, ARAMCO finally approved training in special facilities with full-time instructors, focused not on individual jobs but on broad knowledge areas and skill sets.[30] Its old and studiously ignored commitment to build a vocational education system for the kingdom, which had followed the labor unrest of 1955–56, was brought back to life long enough for ARAMCO and the government to fight over terms, inconclusively, in 1960, but the funds that were offered would ultimately be used to support the government's new College of Petroleum and Minerals seven years later—after many more rounds of negotiations, of course.[31]

Arguably more important, and certainly more striking given the routine praise of ARAMCO's so-called enlightened approach to labor relations throughout the 1950s, was the dismantling in 1958 as well of the company's retrograde system for disciplining and punishing Saudi workers. "These were the shameful days of references to our workmen as 'rag heads' and 'coolies,'" a 1965 report states. The author called the end of the old bunkhouse system of discipline the most dramatic change in the history of the firm. "We moved from punishment and retaliation of the employee . . . to correction and guidance." ARAMCO, it seems, finally discovered that Saudis "responded to fair and reasonable guidance" and "that many . . . disputes were reasonable misunderstandings and not willful violations. . . . We realized that our employees, despite the surface appearances were not essentially different than any other body of workers in their basic motivations."[32]

By now, any impulse to explain these changes in the labor regime, which had been the focus of Saudi demands for a decade or more, as evidence of the exceptional nature of the firm or of the enlightenment of its managerial vanguard, the Barger wing, ought to have been put to rest. But my own emphasis on the "local" political arena—the complex interplay of regime and opposition in the kingdom and the "outside" pressure in the form of Nasserism and its rivals—that compelled management to rethink its interests is probably inadequate. All oil-producing countries confronted similar challenges, from the import quotas imposed by the U.S. Congress to the price cuts for crude that the international firms announced without warning in 1959. A new, now historic wave of expansion and investment abroad by the so-called independents— firms like Getty, Cities Service, Amerada Hess—was just starting to cross the Atlantic, threatening the fifty–fifty formula and other parts of the deals that had been struck in the 1940s–50s. So, Venezuela forced a shift to a sixty–forty split in 1959 and Abdallah Tariki redoubled his efforts to build a front with the Venezuelans. He sent his protégé, the future oil minister Hisham Nazer, who served as his administrative assistant, to Caracas for a year in 1960. Tariki also brought the Venezuelan oil minister, Juan Pérez Alfonso, to the first Arab Petroleum Conference in Cairo. The norms of a new international oil regime were taking shape extremely rapidly.

The question is the extent to which any particular agent aided or resisted this transformation. With rare exception—Pendleton's hint that the relative share of rents might need adjusting—ARAMCO was a reliable redoubt of the resistance, in my judgment. The technical training regime finally put in place with much fanfare in the 1960s was one that British Petroleum pioneered some two decades earlier in Iran, while those who celebrate the introduction of the home ownership program leave the impression that ARAMCO had brought something new to the world oil frontier when it was an innovation developed and in place earlier elsewhere. No one, of course, notes that this modernization scheme allowed the firm to avoid paying the higher costs that equal treatment of the U.S. and Saudi managerial and skilled cadres would require, that it was designed to keep Dhahran itself white, and that many Saudis refused to buy into the program. Then again, ARAMCO could if it wanted claim with pride to be treating Saudi labor better by far than its new U.S. rival, Aminoil (Getty), which was operating in the Neutral Zone between Saudi Arabia and Kuwait, where it built a set of unbelievably squalid worker camps. These reservations reminded some of the old days in Dhahran.

ON THE TRAIL OF THE RED SHAYKH

Abdallah Tariki was a rising star in the Arab world, the Middle East's preeminent oil expert and advocate of collective action against the western monopolies. He was soon so visible and well recognized in the press that, in ARAMCO, where he was hated, some portrayed him as angling to play the role of Nasser inside the kingdom if the House of Saud was ever to be overthrown. The launching pad for Tariki's meteoric ascent was the first Arab Petroleum Conference, which met in Cairo in April 1959. He was a major force in bringing feuding states together for the long-delayed meeting, in the process eclipsing the original promoter, an exiled Iraqi engineer, Muhammad Salman.[33] ARAMCO's delegates watched warily from the sidelines—the strategy that the TAPLINE executive David Dodge convinced the others to adopt—as one or another of Tariki's local rivals tried to cut him down to size.

Tariki's fiercest opponent was a TAPLINE client, the Lebanese contractor and parliament member Emile Bustani, who muscled his way onto Lebanon's delegation. The economist originally appointed to head it left Cairo in protest. Bustani styled himself a champion of the sanctity of contracts and the efficiency of markets—the latter, of course, argued utterly without irony by someone whose fortune was secured through a privileged position of access to the Lebanese state. As ARAMCO's report details, Bustani was so single-minded in his opposition to Tariki that many others at the meeting assumed he was doing the oil companies' bidding.[34]

ARAMCO's agents tracked Tariki's every coming and going during the conference, including the lecture he gave at Cairo University one night ("the first three or four questions tended to be hostile in tone toward ARAMCO but Tariki gave the company's position fairly and frankly"). Rentz was nothing if not melodramatic. Tariki and a handful of other delegates in an "inner coterie . . . held secret daily meetings at the Maadi Yacht Club . . . in the garden to avoid eavesdropping" where "the real decisions were made on resolutions" and where Tariki endeavored to "bring about a rapprochement between the Arabs and the Venezuelans." As the company's captive scholars struggled to make sense of the forces that would lead to the founding of the Organization of Petroleum Exporting Countries (OPEC) one year later, they countered the idea that the firms had invited collective action by policies such as the recent price cuts with what passed for deep insight into culture. "The Latin and the Arab temperaments appear to have a strong affinity for each other," Rentz suggested.

To ARAMCO's growing dismay, in Cairo and in subsequent congresses Tariki pulled the politics of Arab oil down the road that the companies most dreaded. In 1959 Tariki argued the case for integration and Arabization of the concession companies. ARAMCO then viewed him as more or less fronting for the Venezuelan delegation, headed by Pérez Alfonso, in advocating coordinated production controls, but it also believed that the Venezuelans "were a force for moderation" in warning Arab officials away from too confrontational a stance.

Tariki's position at the conference had been identical to the one he had been taking with ARAMCO in Riyadh, and throughout that year his meetings with the oilmen grew increasingly tense and confrontational. When one or another disbelieving executive tried reasoning that such decisions were costing the kingdom millions in new revenues, Tariki answered that he was "not interested in profits but in principles." The only new ARAMCO proposal he would go along with was an offer to create a nonprofit foundation for development, the idea floated in the Special Committee. In this case, Faisal's council of ministers turned it down, on the advice of its World Bank adviser, for seeking to usurp government authority.[35]

ARAMCO tried without success to convince Faisal that the loyal client was running Saudi oil policy into the ground. Its new chief executive officer and chairman of the board, Norman Hardy, went personally to Taif, where the royalty summered, in September in search of a way around the impasse, but after two audiences with the crown prince he concluded that there was "little hope of a change in Tariki's intractable attitude or policy toward ARAMCO."[36] The reason for pessimism was equally obvious to the U.S. ambassador following his own meeting with Faisal a few weeks later. "Faisal has espoused Tariki's animosity toward ARAMCO, accepted Tariki's allegations against the company, and . . . he sympathizes with Tariki's determination to force ARAMCO to become a Saudi company." Faisal had no clue, was the conclusion, and the ambassador warned

> unless Tariki is checked in his accelerating course, relations between the Saudi Arabian government and ARAMCO will become steadily and probably rapidly worse and there will be adverse effects on relations between our two countries. Tariki's policies and aims are not economic but political. He has told ARAMCO's people and members of this Embassy that he believes in the principle of Arabization of the company and does not care whether there would be loss of profits for the government thereby.[37]

The Filth of Riyadh

In essence, ARAMCO executives (and representatives of the U.S. government) found themselves moving, like many other of Faisal's erstwhile supporters, into the ranks of the opposition to him. The first and most famous defector was the crown prince's friend and longtime ally Muhammad Alireza, the Hijazi businessman who resigned his post as minister of commerce in November 1959. He told the U.S. embassy that he was relieved to be through with the "filth" of Riyadh. In Baghdad the new regime had put a number of ancient régime stalwarts on trial, and, Alireza said "he did not wish to be called upon to answer for the Saud family's sins when the day of reckoning came." Faisal had failed to move decisively to "curb the excesses of irresponsible princes" or do much of anything else to move the country forward.[38]

It was a remarkable turnaround. A year earlier, in November 1958, Alireza, who was described in a key U.S. embassy dispatch as extremely close to Faisal (and no particular friend of the United States), was urging the Americans to take the crown prince's side. By April 1959, Alireza was having second thoughts about his patron in the face of continued economic recession and Faisal's deafness to pleas for political reform. The king, he said, was a "weak reed" who would need the Americans' help to succeed. Although the embassy was noncommittal, by November Alireza had cast his lot with the anti-Faisal coalition, reporting to his contact in U.S. military intelligence plans by the king to unseat Faisal and create a new consultative assembly, with Alireza himself as one of its deputy heads, and Saud's outspoken younger brother, Prince Talal, as the other.[39]

Thus, a second breach opened up in the once-formidable walls of the crown prince's fortress with the defection of Talal and other royal family members who would in the next phase of the struggle come to be referred to as the "Liberal" or "Free Princes." The twenty-eight-year-old Talal was, like Alireza, routinely described as someone who despised the king, and members of the family suspected him of actively plotting Saud's overthrow. Perhaps that suspicion explains Talal's sudden transfer from minister of communications to ambassador in Paris in 1955, when plotting against Saud was on the rise, before "retiring" from political life for the next four years and turning to various philanthropic projects in the kingdom and investment properties in Beirut, Cairo, and elsewhere. But in the spring of 1960 he signed on as minister of finance and vice chair of the new Supreme Planning Board in a government led once more by the king. He brought his brothers Abd al-Muhsin (interior), Badr (communications), and Fawwaz (governor of Riyadh) with him.

The third, arguably most surprising set of defectors included virtually all the idealistic "new" men, "modernizers," and "reformers" that the U.S. embassy had been busy tracking and courting over the past few years from Aziz Dia, who had been imprisoned as a suspected communist but was now being given a new magazine to run, to Abd al-Aziz Ibn Muammar. It was the exact moment when the great friend of shahs and oil firms, Berkeley's George Lenczowski, had met and, as I have noted, called Ibn Muammar a "radical and emotionally unbalanced Arab nationalist." Faisal whose guest he was, was merely "uninformed" but, then, Faisal had laughed at Lenczowski's favorite modernizer, the shah of Iran, who "was autocratic while his own house was 'of the people' and 'really democratic.'"[40]

In August, the U.S. embassy reported Faisal's late attempt to court these same outspoken Nasserists. Only Ibn Muammar, one of the oldest of the Young Turks and back from travels in Austria, had not been invited to consult with the crown prince in Jidda. ARAMCO's Terry Duce would later insist that Ibn Muammar's trip to Austria was strong proof that he was a Soviet agent; none of the ARAMCO or U.S. embassy documents speculating about his travels considered it important that one of his daughters had undergone surgery and long-term postoperative care there.[41] The embassy could report, nonetheless that Faisal was at least now speaking "kindly of him."[42] Faisal would soon do more, surprising the Americans by inviting Ibn Muammar to join the government as deputy minister of finance.[43]

Faisal Stumbles

Faisal though was unable to stop the defections of those around him. The king apparently began to look better to these disparate one-time backers of the prince. Even the U.S. embassy had begun to beat the drums: "The optimism of the public following the assumption of power by Prince Faisal in March of 1958 is being rapidly replaced by impatience as government inactivity delays the promised financial and administrative reforms. The feeling is growing that Prince Faisal remains a bottleneck to progress, unwilling to take a strong stand against princely waste and extravagance."[44] Ambassador Heath himself told Washington that he shared the concern "that with the success of his financial stability program nearly complete, Faisal is not actively considering plans for economic development which could meet the rising criticism of his conduct."[45] King Saud and his closest advisors did what they could to encourage the stampede, from holding press conferences, posturing in the still fashionable style ("Israel cannot survive"), quietly repairing relations with the British government, and embracing more of his countrymen. So, late in November, Riyadh announced that Ibn Muammar himself had accepted a position as counselor to the king.

Although not all these factions acted for the same reasons, the Americans' motivation is obvious in retrospect, even if no one until now has bothered to notice. We have already seen the U.S. ambassador describe Faisal as supporting Tariki's "Arabization" policy, and the only real difference between Faisal and Saud, he said, was the king's more reasonable attitude toward ARAMCO. Heath had in fact gone to Saud and warned him that Tariki's policies were taking the kingdom down the road Iran had gone, and the king in turn promised that Tariki would never be allowed to succeed. ARAMCO turned the spigot back on. A year earlier the king's repeated request for loans off the books, use of ARAMCO planes, and the like all had been turned down. The firm confided to the embassy that these decisions had been made "on political grounds." Well, the grounds had apparently shifted. ARAMCO's president approved at least one loan of $2 million, and agreed to the king's request that the government not be informed of the deal.[46]

What did the firm think it might get? The Americans started to speculate, of all things, that Tariki might lose his job. In addition, ARAMCO's records show the Government Relations people referring to a "golden moment" in the early days of the Saud revolution when "it appeared that the government was more disposed than formerly to work for the solution of certain problems."[47]

The historians who perpetuate the myth of Faisal as reformer are wise to ignore the entire period discussed here, since scant evidence can be found to support the idea. The opposite is true. The one claim about the moment that makes it into the brief is that Faisal brought spending into line following the irresponsible behavior of his brother the king. We saw though that it was Saud who brought in the money doctors and began the process of retrenchment, for which Faisal gets the credit, for the simple reason that he had taken over from the king. I suppose he might have reversed the austerity program, and so deserves some credit. Of course, it is his failure to do so that led to the defection of businessmen like Alireza, and if the criticisms of the merchant class are correct, then the problem is that Faisal preferred to shut off the "development" tap while keeping resources flowing uninterruptedly to the royals.

The last is certainly what the Americans began to report as the attacks on ARAMCO grew irksome and Faisal continued to align Saudi Arabia's policies with the most powerful Arab state, Egypt, the main antagonist of the United States in the region. The same un–client-like behavior drove the United States to step up the criticism of the kingdom in 1954–56 when Saud headed the government, and so we find various accounts of his tolerating corruption, spending way too much on palaces, and not doing enough to spur development. When the king briefly aligned Saudi policy with America's, he gained the seal of approval as a genuine reformer committed to bettering the lot of the Saudi people. So now guess when Faisal earned his enduring reputation in the American newspapers and later in history books as a genuine reformer and all-around enlightened ruler?

The answer isn't when he finally endorsed and assisted the efforts of those in the kingdom who began to imagine drafting a constitution that would define and set limits on the royals' power for the first time and opening up governance of the kingdom to its citizens. In all accounts, Faisal consistently opposed all such steps away from absolutism. Those who heralded him as a reformer pointedly ignored his deep-running reactionary streak, either because it doesn't fit well into the story or because they feared political change in the kingdom as much as he did. You'll find the answer to the riddle in the next chapter. Meanwhile, the story gets even better.

THE REVOLUTION THAT NEVER HAPPENED

On April 30, 1960, the *Economist* of London published a dispatch from its Beirut correspondent, under the headline "A Saudi Revolution." *Economist* writers do not sign their names to articles, but all those filed that April on the surprising turn in the kingdom's political affairs were by Kim Philby. Kim was the son of St. John Philby, King Abd al-Aziz's illustrious western advisor whose fame would be eclipsed by the son's eventual exposure as a Soviet double agent inside Britain's Secret Intelligence Service. In his guise as a journalist, Kim Philby had cultivated a wide array of contacts, including many friends inside ARAMCO, which had paid for the publishing of his father's last books and, with Kim's help, bought the father's papers after his death in 1960.[48] The key source for Kim Philby's "Revolution" story, however, turns out to have been Abd al-Aziz Ibn Muammar—"beyond a shadow of a doubt," according to one of ARAMCO's agents.[49]

Philby traced for the *Economist*'s readers the strange twists in the struggle between the king and crown prince, which had turned Faisal into a "reactionary" and Saud into a "proponent of popular reform." These quotes are a U.S. embassy officer's paraphrase. Philby actually said the king "has chosen to join sides with the forces of revolution . . . making a bid for the kind of popular support which the deposed monarchs of the Middle East never enjoyed," and that "his chief opponent is his brother Faisal." The embassy officer added that Philby "was substantially accurate," although the king and crown prince thought more or less alike.[50] That is, the account could not be reduced to a matter of ideological differences. Nonetheless, across a wide ideological spectrum, from men of the left such as Ibn Muammar and Kim Philby to more conservative figures such as U.S. ambassador Heath, there was agreement that the prime minister had failed to mobilize the country's resources to advance the elusive, increasingly talked about goal of "economic development," and that the king had emerged as its amore or less genuine champion. The other problem, at least for some, is that Faisal refused to accede to any type of political opening, notwithstanding his profession (or perhaps because he believed) that his country was already "really democratic." Here too the king appeared to be taking a decisive turn—toward the abyss was Faisal's fear.

Ambassador Heath explained Faisal's failure using ideas that perhaps were new then, but today are evoked as a kind of mantra inside the embassy in Riyadh. The basic problem was, he said, a "conflict of civilizations. . . . Veils with bathing suits, desert sheikhs with oil engineers." Heath called Saudi Arabia "a society in the 'second category' of [Walt] Rostow's recently publicized theory of the five stages of economic growth," and he warned of "a long wait before the country develops the political, social, and institutional framework normally required for passage to the third or 'take off' stage." [51]

The world's best-known expert on the Saudi economy at that time, Anwar Ali, then the acting governor of the Saudi Arabian Monetary Agency, avoided the worst of the mystifications of the moment. The problem was, more simply, he told the ambassador, the "lack of adequate understanding by any influential group in the country of the necessity and the scope of social, educational and economic development." Human capital was about as scarce as water in the desert kingdom. The country's future hinged on systematic and comprehensive planning, which Faisal never committed to nor understood, said Lenczowski. The American who led the World Bank mission to the kingdom said substantially the same thing, adding, however, that Faisal's penchant for handing over concessions to other royal family members did not constitute much of a plan.[52]

The moment, whether one chooses to view it as a principled political advance of the country's most able citizens or a king's cynical manipulation of his "playboy brothers" Talal and Nawwaf and power-hungry commoners like Tariki, produced some of the now famous landmarks of the modern horizon.[53] There were also important and now forgotten institutional reforms put in place that Saudi rulers in 2005 were dusting off and trying to sell as new. So, for example, the king created an economic development committee within the royal diwan or cabinet, headed by Ibn Muammar, and including another of the young Turks, Faisal Hejalan. The spadework done by Ibn Muammar's team provided the foundation for the remarkable December 1960 cabinet installed by Saud, with Prince Talal as finance minister and development agency head, Tariki running a newly created oil ministry, and a contract signed with the UN for the creation of a new Institute of Public Administration.

If the U.S. embassy's account is correct, then Ibn Muammar's real significance to Saudi institutional history is in the political sphere because he is credited in its discussions—and only in its discussions until now—with designing the country's first system of formal representation of citizens' interests. When an ARAMCO executive visited from New York, he passed on to the embassy Ibn Muammar's claim that "one of the real reasons" for his appointment was to devise "some form of representative government in the Kingdom." Ibn Muammar said in fact that he worked harder on this project than on any other. Tariki confirmed it, and said that Prince Faisal "bitterly opposed the whole idea" on the grounds that "once the movement toward representative government was started it could not be stopped."[54]

It is incontrovertible that talk of institutional innovation was in the air and that outside the kingdom the Arab press could actually report it. In early January the editor of the Saudi daily *al-Bilad* confirmed for a member of the U.S. embassy the rash of rumors about plans for a council of deputies that the royal diwan was producing under Ibn Muammar's direction (he of dubious loyalty, Heath opined).[55] The U.S. ambassador fretted that, if true, a parliament might be a first step toward constitutional monarchy, but that at any rate it would be used to strip Faisal of his power. Partisans of reform inside the kingdom turned to the clandestine circulation of pamphlets addressed to the liberal princes, denouncing reactionaries in both the king's (Jamal Hussaini, Yusuf Yassin) and prince Faisal's (Fahd, who ruled until 2005) camps, and urging the creation of a parliament.[56] In Beirut, *al-Nahar* reported on the royal family's fight over a parliament: would it doom or strengthen the monarchy? In Cairo, the Nasserist daily *al-Jumhuriyya* published an interview with Prince Nawwaf, who said that the kingdom was ready to take a "major step forward toward a democratic form of government." Ibn Muammar told the U.S. attaché that Nawwaf was pressured by Faisal's censors to repudiate the interview but that he had refused.[57]

The Americans viewed these events as a campaign by the liberals—Talal, Alireza, Ibn Muammar, and so on—against Faisal, who was scheduled to leave the country for surgery but after a furious round of meetings and maneuvers decided to remain at home. His allies among the princes, since known as the Sudairi Seven (seven brothers headed by the late ruler Fahd and Prince Sultan, who is in line in 2006 to become king following the present ruler Abdallah), were insisting on it after a long, tense meeting among royal factions.[58] Talal pressed his American contact for help in keeping the king on the side of those "determined to carry on the fight for reform as the only means of saving the future of the family." It was a struggle over principles, he said, "constitutional monarchy versus absolute autocracy," and urged that the U.S. ambassador weigh in on the side of democracy or at least not bless the reactionaries. When the attaché said it was U.S. policy not to interfere in Saudi affairs, Talal asked, what then was Colonel Eddy, the longtime champion of Faisal, doing during his recent visits to both the embassy and to the king? The attaché said he didn't know and asked why it mattered. Eddy was an ARAMCO official, wasn't he? Talal said he "believed that Eddy was now working for Allen Dulles [head of the CIA]," which of course was true.[59]

We don't know what Eddy was doing in Riyadh nor what to conclude about the obligatory profession of neutrality toward the royal factions, but we do know that the Americans sided decisively with the royal family when they learned through the head of the military training mission in Taif that Saudi officers were plotting a coup in the summer of 1960. The conspirators planned to kill Faisal and officials of the Ministry of Defense. American officers said their contact among the plotters, Lieutenant Colonel Hazim Sulayman, a Hijazi with deep hatred for the Al Saud, was among the military's most promising officers. Still, he and perhaps others of his comrades were foolish enough to believe that the Republicans might assist their revolution, a result perhaps of the twisted logic of those in the Arab world who imagined either Nasser or Qassim or both had seized power with the help of the United States, and sought to assure his contact that after the takeover the new republic would want ARAMCO's help rather than Nasser's. It didn't matter. Both U.S. military intelligence in Dhahran through the "intelligence liaison" within ARAMCO, identified by the embassy as the Government Relations employee William Palmer, whom we encountered earlier burning all the company's confidential files during the Mossadegh crisis in Iran, and the State Department in Washington sent word to ARAMCO about the existence of a plot against the government. These agencies in turn used the oil firm to deliver the warning to the king and the crown prince.[60]

In a mid-December assessment of current political and military trends inside the kingdom, the U.S. embassy concluded that dissatisfaction with the regime was real and growing, while the reform movement had lost momentum and the promises of liberalization that the king had made to the reformers had "died with the failure of the king's effort to unseat the crown prince." Faisal now stood in the way. The one arguable advance was that local elections were taking place in Qatif, al-Khobar, Dammam, and Hofuf. These unprecedented municipal elections were popular and would continue in other locales, including Riyadh, during the remainder of Saud's reign, before Faisal ended the long since forgotten experiment—the better for those same princes who worked so hard to squash the movement in 1960–62 to offer up a half-century-old idea as their own new, unprecedented step toward democracy in 2004–5. The embassy, however, discounted their importance. Faisal had successfully weathered the storm, it said.[61]

Faisal's survival would have been looked upon warily by ARAMCO, at best, because Tariki had license to go after the oil firms, which had unilaterally announced a second set of oil price cuts in less than a year. Tariki, with Faisal's backing, refused to recognize the reduction for purposes of calculating royalties, adding yet another issue to the long list of disagreements with the company (ARAMCO's president called it a crisis). The price cuts led Tariki personally to step up his campaign for collective action from Maracaibo to Houston, where oil prorationing had been invented in the 1930s and where organized opposition to the oil majors in the form of the Texas Independent Producers Association was strongest. Venezuela, Iraq, Iran, Saudi Arabia, and Kuwait ultimately agreed to create OPEC in response to the July 1960 price reductions.[62]

ARAMCO feared that more trouble was coming, sharing with the embassy its latest, extremely pessimistic analysis of the deteriorating economy around the oil fields of the Eastern Province. Two years of reduced government spending had taken its toll, and goes far to explain the preoccupation with development (or the lack thereof). ARAMCO was no longer quite so easily able to imagine that its routine business operations amounted to a private Point Four plan, as it did in the 1940s, because the boom had ended, production and ancillary facilities were all in place, and the parents had ordered reductions in staffing and production. Reduced demand and increased competition drove decisions in 1959 to begin cutting personnel. Many newly declared "surplus" foreign employees, including dozens of Americans, were let go. The company reduced the number of Saudis as well, but by attrition rather than by firing, trying to hide the drop in numbers for fear of inviting yet more criticism. ARAMCO then put on the best face possible, celebrating whenever it could its putative objective of increasing the overall percentage of Saudis in its labor force. A second success, the gradual elimination of "troublemakers," was only toasted in private.[63]

The effects of the bust went far beyond ARAMCO's decision in November to close down one of its legendary institutions, its own airline—the DC-6 Flying Camel—which flew twice weekly between Dhahran and New York. In the shanty towns that had grown up around the towns, many of the make-shift dwellings had been abandoned, left to "decay and collapse." Other encampments were practically lifeless or taken over at night by single migrants rather than the families that once had resided there. The embassy cautioned, ARAMCO's fears of the size of the unemployment problem might be overstated, because

> many, perhaps most, of those who have lost their jobs during the past few years have left the local labor market and either gone elsewhere in search of work or returned to the desert or the date grove. . . . Such mobility eases the present situation although those who return to their homes tend to spread a desire for change outside the immediate environs of the oil camps and may also be dissatisfied with life at home after having tasted a different existence.

As Tariki continued to campaign for ARAMCO's transformation, some in the firm began to see Ibn Muammar, hard as it is to believe, as a potential counter. The key was a rift that ARAMCO's analysts said was growing between the two friends over what direction Saudi Arabia ought to go. "Tariki is a pure Nasserist while Mu'ammar is essentially a Saudi nationalist." ARAMCO's ex-Navy analyst Carl Barnet reported witnessing a "bitter argument between them on the issue of indigenous versus Nasser-type nationalism." The embassy, wisely, called the whole line of thought into question, although it continued to denounce Tariki as an "opportunist and headline grabber" and caricature Ibn Muammar as "erratic," "emotional," and well versed in communist thought but ultimately a "free enterpriser."[64] The slurs masquerading as analysis aside, there is no evidence of any kind of rivalry shaping polices in the ensuing months.

The judgments of top ARAMCO officials reveal that their prejudices continued to cloud their reading of events. In December, ARAMCO's top Washington official, Terry Duce, denounced Tariki yet again as carrying on an irrational campaign against ARAMCO guided only by his own political ambitions, aimed at bringing down the monarchy. There was a slim hope: The "fact that Tariki has never been elevated to ministerial rank" suggested that the king and prime minister had recognized "Tariki's political menace to the dynasty." Less than a month later, Tariki was given a spanking new oil ministry to run. The embassy got things wrong, too. As I noted, on December 19, 1960, it reported that Faisal's rule was secure. That same day, King Saud officially rejected the budget submitted by Faisal's government, and two days later, Faisal submitted his resignation to the king.

Ibn Muammar Electrifies Riyadh

King Saud took back control of the government in the waning days of 1960 and assembled a cabinet that remains until today the most uncharacteristic and progressive ever seen in Saudi Arabia. Barger, headed to Riyadh to see the king just days after the new government took over, called it "the biggest political poker game in the history of the kingdom."[65] Saud's profligacy, his lack of foreign education and experience, and his failed attempt to topple Nasser hardly seem to explain why to this day his picture is not even displayed in most formal settings. It is this now officially forgotten government that best accounts for his officially forgotten rule over the kingdom.

The king named his brother, Talal, who had by then become Faisal's bitterest enemy, to head the Ministry of National Economy, the start of a journey that would end in exile in Beirut. As finance minister, Talal also steered the new planning initiative. Tariki, who ARAMCO prayed would be eclipsed, instead gained a cabinet position as the kingdom's first minister of petroleum and minerals. He too would lose his post and have to flee to Beirut. Unprecedented at that time, three other commoners were named to ministerial posts. The king announced the creation of a second, new Ministry of Labor. Most amazing of all, for an all too brief moment, Ibn Muammar had come to be seen, in the words of the U.S. embassy, as the most powerful man in the government and the architect of most of the changes. Exile would not suffice in his case. Faisal instead would send him to prison once more, this time for a dozen years.

The embassy's report of a conversation with an engineering consultant who worked in Riyadh and knew Ibn Muammar for over a decade is typical of that remarkable time. Emile George told a visiting political officer that "he was somewhat skeptical and a little afraid about the future course of the new Saudi government," given the "immense power" now in the hands of Ibn Muammar and his ally Faisal Hejalan. George repeated the standard description of Ibn Muammar as having "rather radical ideas." Now, however, the radical was vetting all government appointments. Palace watchers called him "the main power behind the king," and, for that reason, others in the cabinet were treating him with amazing deference.[66] The embassy added that other sources confirmed the account. One was ARAMCO's top official in Riyadh, Ronald Metz, an old CIA recruit, who provided the most detailed reporting of the momentous change under way. The king was fully aware of Ibn Muammar's sway over the intellectuals being given government positions—these followers practically revered him—but Saud nonetheless accepted Ibn Muammar's choices, in some cases in the face of "vigorous protests" by more conservative counsel.[67]

Metz said that Ibn Muammar's ideas about development and state building resembled those of a Baath socialist, although he had no known relationship to the party. He and Arabia's other "new men," as Kim Philby referred to them, had plunged into the job of "planning"—"a key word for the new group"—meeting constantly, drafting new legislation, reorganizing government procedures, and then continuing to meet at the Yamamah Hotel "evenings and often far into the night after the ministry offices are closed." Metz said Ibn Muammar exercised his control over the acolytes through these regular nightly gatherings. Metz also followed the ARAMCO line about a war between Tariki and Ibn Muammar, which the embassy discounted heavily ("Rumors of a split . . . are unfounded though neither dominates the other"). What the Americans thought was probably true was that Tariki had less influence with Faisal in eclipse and with Ibn Muammar, who was more tolerant than Tariki of those who disagreed with him, commanding the loyalty of most of the new appointees.[68]

Another difference between the two, less flattering to Ibn Muammar, is that Tariki never mistook the king's power play for a conversion experience. Metz described Ibn Muammar as a "man who knows very well that he is in possession of power and influence." He might have added who knew precisely from where that power and influence derived, because as a client of the king he paid more reverent tribute to the monarch that had elevated him from obscurity than his own clients would pay Muammar, who was idolized. Thus when Metz asked Ibn Muammar to explain the turn in the kingdom's affairs, "he recalled how 'his majesty said that now that he had all he could possibly desire in the way of palaces and other luxuries, the only further desire in his life is to accomplish something for the people . . . , the mission of developing the country.'" He may have even believed it, but whether he did or not he made sure to register his anger with the cynical views of western journalists writing about the new order in Saudi Arabia.[69]

There was consensus on one point. The atmosphere in Riyadh was charged; "electric," as Issa Sabbagh, the embassy's public affairs officer, put it. Another embassy official said that the new government "has support of a type never before accorded anybody in Saudi Arabia."[70] So the embassy reported on a meeting with Mustafa Wahba, one of those so-called new men, an economist, son of the late king's adviser, Shaykh Hafiz Wahba, and a protégé of Tariki in the General Petroleum Authority. Wahba joined the new government as deputy minister of finance for economic development under Prince Talal. The embassy officer described him in early February 1961 as "brimming with vitality" and "scarcely controllable enthusiasm." Wahba argued "for the first time in the history of the country, people were taking intelligent interest in politics and paying real attention to the promises being made." At the same time he said the kingdom would need the help of outside experts from "the Ford Foundation and Rockefeller interests" to support the development agenda that Saud's new government had begun. A true believer in Ibn Muammar, such as Abd al-Rahman al-Bahijan, made Wahba appear reserved by comparison. As noted in Chapter 6, Bahijan, from Unayza in Najd, was one of the strike leaders in 1953 whom the company refused to rehire a year later and who was instead employed by Ibn Muammar at the Labor Office in Dammam and given a scholarship to study economics in London. He "spoke of Ibn Muammar in terms of awe and reverence," and laughed at the embassy's defense of ARAMCO's commitment to Saudization.[71]

Tariki was even more reserved but nonetheless made clear what side he stood on. He told the ARAMCO envoys to Riyadh, Barger and Brougham, that he hadn't believed the creation of such a forward-looking government was possible in Saudi Arabia "without revolution" but that now "we plan to make the most of it." He was ready to work with the king but would not tolerate treatment as a mere "cat's paw." Barger emphasized Tariki's repeated use of the pronoun "we" when speaking about the changes under way.[72]

In his first interview with the Saudi press in *al-Bilad* (January 9, 1961) after becoming the new minister of petroleum, Tariki called for changing the concession agreement in the interests of the people. If he also spoke of the government's hope for "a friendly agreement" with ARAMCO and TAPLINE then perhaps he imagined the new regime cowing the firm into submission.[73] He rejected the firm's new proposals and during the second half of 1961 used the press to pressure the company, he said, by educating the population. He charged that ARAMCO had evaded payment of some $180 million in income taxes from Sidon pipeline sales. The finance minister, Prince Talal, joined the campaign as well. ARAMCO protested to the king, ineffectually—Saud claimed ignorance about Tariki's activities—and so the oilmen resigned themselves to wage a countercampaign of sorts, using ARAMCO television in the Eastern Province, circulating films about the firm in schools and sports clubs to almost 16,000 Saudis, hosting newspapermen, businessmen, bureaucrats, and professors in Dhahran, and flying the two most influential editors in the country on a coast-to-coast tour of the United States.[74]

Faisal remained the champion of all those who were shaken by the king's surprise defection to the side of history and who fretted about the consequences of the prince's equally sudden retreat. Ghaleb Rifai, a palace official, leavened his fears with a little humor when he invited a U.S. colleague to the American military training mission's weekly movie following dinner at the palace. "The movie has Jerry Lewis. We surely need to laugh in the midst of all this tension." There were many others, however, who were not laughing, including the Saudi contractor who prophesized that "what lies ahead is conflict and confusion," and the Pakistani ambassador, who feared that leftists were taking over the kingdom and wanted Talal and Alireza to outflank them. Indeed, while the Hijazi merchant celebrated his erstwhile patron's defeat ("Faisal is definitely powerless and cannot regain power or position"), Alireza nonetheless dreaded the turn of events. The communists are as "close to the king as my necktie is to my throat," he said, referring to Ibn Muammar, and he, Alireza, was being "forced to withdraw from the business world of Saudi Arabia, against his family's wishes, to enter politics to protect the country against communism."[75] Ultimately, the American ambassador guessed, the new government would be unable to "withstand a concerted effort by conservative elements to bring about its fall."[76] He guessed right.

House of the Reactionaries

Those then known as the Al al-Fahd princes and referred to now as the leading Sudairi Seven amirs, including Fahd, Sultan, Salman, Turki, and Abd al-Rahman, led the opposition to the revolution. As the embassy reported, the princes acted out of different motives but were converging on a single purpose. Some had a fierce loyalty to Faisal. Others feared Ibn Muammar. Others sought payback for losing their ministerial posts. Still others, maybe most, were angered by what the embassy called the "inequitable distribution by the king of favors and bounties."[77] The first to be sacrificed in the war inside the house of Saud was Abd al-Aziz Ibn Muammar.

Another of the Alireza brothers, Ahmad, broke the news to the American ambassador at dinner on February 18. The night before the king had decided it critical to appoint his erstwhile adviser, Ibn Muammar, ambassador to Switzerland and his sidekick, Faisal Hejalan, envoy to Spain. Alireza could hardly contain himself (the "clan is apparently jubilant over the decision"), boasting that "next on the list of removal is the petroleum minister, Abdallah Tariki."[78] The deputy foreign minister, Umar al-Saqqaf, a Faisal loyalist, proved the shrewder judge. The new men had shown themselves unable to adapt to Saudi realities, wanting "to crowd three years of change and progress into one." Prince Talal was the key and he would not last very much longer, al-Saqqaf predicted, and "this would be the end of the King's attempt to rule."

Over the next few months, Saud and Talal would both grow increasingly desperate, which speaks both to the sagacity of al-Saqqaf and, more important, the singleminded purpose of Faisal and the Al al-Fahd. The princes demanded that Faisal be brought back into the government, while Faisal was demanding Talal's head as the price for his return. Perhaps the strongest piece of evidence that the king was beginning to lose his bearings again and had begun improvising, as he had done to disastrous ends in 1957–58, is in the sudden, very public announcement in March 1961 that his kingdom would not be renewing the Dhahran airfield agreement with the United States, which was not due to expire for more than a year. This bit of gratuitous grandstanding, which the U.S. ambassador called "one of the silliest and most disgusting performances King Saud has ever made," was driven by the fear that newspapers in Beirut and perhaps elsewhere in the region were about to publish a story— the palace told the embassy that the opposition princes were the source—that Faisal had been angling to end the American military presence, which was true, and so Saud had driven Faisal out, which was not. Back in December 1960, Faisal had quietly started contact with the Kennedy administration to end the agreement (but not the training mission) ahead of schedule. Talal had done the same two months later.[79]

At that moment, the completion of the new U.S.-funded international terminal at Dhahran, which was part of the quid pro quo for renewing the agreement back in 1956, was years behind schedule. The irony in this case should make one pause. As we have seen, the Saudis had been protesting the American military presence since at least 1955. Riding on the reputation of the Dhahran project commission, the designer of the terminal, the American modernist Minoru Yamasaki, was just beginning his next project and talked of how his work in Saudi Arabia and in particular the monumentality of the Qabba in Mecca—its vast central space—was an inspiration for his proposed design for the new World Trade Center in Manhattan. Decades later, the return of the American troops to the Eastern Province would lead to the destruction of Yamasaki's best-known building.

In 1961, the Americans cared more about the affront of Saudis acting unilaterally than about the continuation or not of a military presence at the base (which was not a base, all sides continued to insist). The main value of Dhahran to the U.S. military was as a peacetime communications hub, both for the Navy, whose regional command (COMIDEASTFOR) relied on it for all its shore-to-ship communications, and the Air Force, for which it served as the single relay station between Wheelus in Libya and Clark Field in the Philippines. Similarly, it served as a transit hub for goods, a fueling station for transport planes, and a rest stop for plane crews. Alternatives could be arranged. The base had no role to play in the Pentagon's war plans because strategists assumed it would either be targeted for destruction or else overrun by the enemy.

The ambassador, of course, cautioned against the king taking precipitous action. It was not the way friendly nations normally resolved their differences, and, when it was clear that the Saudis weren't going to wait to negotiate a mutually agreeable solution, Washington requested time to draft a joint statement—to no avail. Talal asked the U.S. embassy on March 16 to approve an announcement on Radio Mecca of the king's decision to end the (first) U.S. presence on holy Saudi soil. Washington wouldn't be happy, Heath said. Radio Mecca broadcasted it anyway.

By the spring of 1961 the war between the princely factions was out in the open, covered even in the *New York Times*. In Cairo, Prince Talal ("plump, energetic and very wealthy"), on vacation from promoting Saudi development, had been doing a little freelance foreign policy work. He told the Egyptian press that the Saudi government was terminating the Dhahran agreement (mistakenly described as a "lease") because of American aid to Israel! Saud made a more convoluted version of the argument a few days later—perhaps he thought the Americans would appreciate it more. They didn't. Talal asked for a meeting with the ambassador in his peculiarly royal style: his secretary "telephoned from Riyadh with regard to my request to see the Prince," Heath reported to Washington. "I have made no request. . . . Supposedly this is his way of asking me to come see him." The State Department advised that he be given an extremely cold shoulder.[80]

Few in the Arab world would have been fooled by these desperate appeals to "the street" as the contemporary cliché puts it. By May, a number of new clandestine groups—Sons of the Arabian Peninsula, the National Reform Front, and the National Liberation Front—were waging a campaign by broadside mailed from Beirut, Damascus, and Bahrain attacking the royal family and condemning "both Saud and Faisal as 'pro-western'—a term of opprobrium in the vocabularies of nationalists as well as Communists," the *New York Times* explained. Its old resident Gulf expert, Dana Adams Schmidt, reported that demonstrators in the Eastern Province had massed when the king visited in May.[81] A month later, following Baghdad's original claim that Kuwait was in reality a part of Iraq, Shiites in the Eastern Province were supposedly telegramming for Qassim to rescue them as well and bring them into the Arab motherland.

Shiite discontent if not militancy had indeed grown more visible in the Eastern Province and, in particular, in the Qatif oasis in the past six months. A delegation had gone to Riyadh to protest to the king about discrimination and attacks in the press. Shiites in Hofuf boycotted the local elections there after the amir reserved only two out of eleven seats for them, although, as ARAMCO reported, they represented almost three-fourths of the town's residents. The embassy also reported the creation of a number of new voluntary associations to represent Shiite interests. The point always underscored in the embassy's reporting is that, although there were precious few communists in Saudi Arabia, they were overwhelmingly young Shiite men from al-Hasa, where support for Iraq and its populist ruler, Abd al-Karim Qassim, was also strong.[82]

The king and the crown prince met face to face for the first time in six months on August 8, 1961. Saud professed his desire to heal the split for the good of the country, according to the accounts that were leaking out of Riyadh. Faisal demanded Talal's resignation as the price. Some in Riyadh would go on to argue that Talal went to Beirut that next week and gave a series of new and more inflammatory interviews as a cover, attacking Faisal's record of reform at home as well as his failure to stand up to the Americans. The ARAMCO agent in Riyadh, Ronald Metz, probably got it correct, however, when he said that Talal's dismissal early in September stemmed from the deepening rift with the king, and the "struggle for leadership" was "the real issue." A confrontation between the two after Talal's Beirut venture ended with the king ordering him out. Talal's allies among the princes inside the cabinet tried to protect him but failed, and his brothers Badr and Abd al-Muhsin were dropped from the cabinet as well.[83]

Faisal's brother-in-law and his future intelligence head, Kamal Adham, approved the king's action to the extent that it had served to check the danger of liberalization. Adham said "the country is simply not ready for a constitution." He also told his American contacts that Faisal was still not satisfied. It was "impossible for the king and Faisal to get along together and . . . it would take some national crisis to bring Faisal back into power."[84]

A war with Egypt would do it.

8 AMERICA'S KINGDOM

I gather the Chiefs also wonder whether this flea-bitten part of the world is one where we should get involved. I'm afraid we are involved here long since—even though it may have been a mistake in the first place. But remember oil.

McGeorge Bundy, Memorandum to the Chairman of the Joint Chiefs of Staff

King Faisal's visit will depend—more than usual—as much on the tone you set as on the substance. . . . Your first look may make you feel it will be hard to hit it off with this bearded, robed desert king. But Faisal is a lot more modern than he looks. Under those robes, you will find a sharp mind and deep devotion to educational and social progress. I am sure he will warm to your sincerity and frankness. . . . It is worth the effort. Our largest single overseas private enterprise is the Arabian-American Oil Company's $1.2 billion investment in Saudi Arabia. In addition, all our other Middle East interests—from blocking Communism to preserving Israel—depend heavily on gradual modernization under moderate leaders like Faisal who oppose the revolutionary methods of Nasser and Communism.

Walt W. Rostow, Memorandum to President Johnson

After Allah, we trust the United States.

Faisal Ibn Abd al-Aziz,
quoted in Parker Hart, *Saudi Arabia and the United States*

NASSER'S WAR WITH THE HOUSE OF SAUD

On September 28, 1961, rebel Syrian troops seized Damascus, the capital of what was then known as the "Northern Region" of the United Arab Republic, declaring their objective the "restoration of the rights of the nation." The rebellion quickly spread to garrisons in Aleppo and Latakia. The insurgents captured an advance guard of 120 paratroopers flown from Egypt to crush the coup. In the face of this challenge, the UAR's leader, Gamal Abdel Nasser, called off the invasion. His viceroy in Damascus and close comrade Abd al-Hakim Amr retreated to Cairo.[1] The three-year-old union with Egypt was over. Nasser had been dealt a serious blow.

To recover from the Syrian debacle, Nasser began arresting rich Egyptians although they had nothing to do with the succession, seizing their homes and valuables, and promising the nation show trials for the biggest of the businessmen. Officers in his own army who tried to lead a coup to reverse the turn to socialism were much more quietly dispensed with.[2] Nasser also renewed the campaign against the conservative although hardly less repressive Middle Eastern monarchies that he accused of aiding the Syrian "right wing" and "deviationists." Attacks on Saudi Arabia's king and the America oil company—"reactionary Arab rulers, traitors, conspirators, and the forces of imperialism"—on Cairo's Voice of the Arabs kept ARAMCO's translators and analysts busy and entertained. Saud's harem, the clandestine broadcasts said, included "Jewesses from Israel, Yemen and Europe. . . . Has your false faith crumbled before glittering dollars and your American Lords?" The propaganda leaned heavily on poems and parodies of popular songs of the day.

O Slave of Aramco, stooge of Imperialism,
You built Nasriyah on sweat and moving sand.
O, Slave of Aramco, stooge of Imperialism,
You did this with the sweat of the free in your land.
People shall have a hand in exploiting their land
And the day is at hand when they will have revenge;
Yes, they shall have revenge; yes they shall have revenge.
They live in your prisons, tortured and behind bars,
They know naught of justice, are sold throughout the land,
And live in your prisons, tortured and behind bars.
Now you no longer pray, but kneel to the dollar;
America will not from the free protect you,
Nor will Dhahran Airbase when they rise against you,
When the free do revolt and the oppressed avenge;
O Saud, brace yourself for they shall have revenge,
Yes, the people and the free shall soon have their revenge.[3]

Pressed by officials in the State Department on the perennial question of the Saudi regime's vulnerability, ARAMCO executive Tom Barger said that the propaganda broadcasts were "more violent and obscene than ever before," but that the Saudis would likely withstand the heat. The army was divided and lacked the necessary espirit de corps, he said, although to his credit Barger also admitted that no one had foreseen the Nasser and Qassim coups coming in Egypt and Iraq, respectively. What Barger said worried him more was the rising flood of newly educated Saudis returning to the country. After two decades in the kingdom, ARAMCO's most progressive voice was nonetheless still questioning the wisdom of training an intelligentsia in the way that the kingdom's founder had for roughly the same reason two decades earlier.[4]

The more privileged section of the Saudi intelligentsia nonetheless proved its value to the regime in answering the Egyptian propaganda barrage in the pages of the press and on the air once the Saudis launched their own station, Voice of Islam. In December 1961, the Saudi newspaper *al-Nadwah* ran a series of articles criticizing Nasser's mouthpiece, the *al-Ahram* editor Muhammad Heikal (a "stooge" and a "whore"), recalling Egypt's once-important role in advancing Islam, and denouncing Nasser for debasing the religion. While professing to be a believer, he was accused of allowing "taverns, night clubs with naked-bodies that stir the lowest sexual desires, obscure songs on radio and television explicitly urging people to commit adultery, magazines and newspapers with pictures of semi-nude women, gambling clubs, young girls compelled to learn to dance and attend youth-camps conducted at home and abroad, the latter in the capital of communism, Moscow."[5] A little later the Saudi press would run cartoons showing Nasser and Heikal smoking hashish pipes—the point being that the Egyptians must be on drugs given the lies they printed about the king.[6] Saudi propagandists were obviously capable of giving the Nasserists a run for their money. A writer might criticize the regime for torturing members of the Muslim Brothers or for treating Sudanese in Cairo as second-class citizens, but he also might condemn Egypt's role in the non-aligned movement on the grounds that "Castro is a Jew and that two of Tito's closest advisers are Jews."[7]

The King and Kennedy

The escalating war with the newly radicalized Nasser regime in the winter of 1961–62 and all the new anxious handwringing by those who had bet heavily on the Saudis coincided with the sudden deterioration again in the king's health. At his palace in Riyadh in mid-November 1961, Saud became violently ill, hemorrhaging from the mouth while under examination by ARAMCO doctors for a recent heart attack. He was admitted to the ARAMCO hospital, where he was misdiagnosed, the doctors there arguing that his life was in danger and that he needed an operation on what they said was a bleeding ulcer. ARAMCO's political people argued against doing the surgery in Dhahran, fearing the prospect that the king "might not pull through" and that the firm would be blamed. Tom Barger told the State Department that the firm was against transferring him to the hospital at the American University of Beirut for the same reason.

It was a good thing that they hesitated. The U.S. embassy had arranged for a specialist from London to examine the king, and the new doctor ended all talk of emergency surgery. The State Department also politely deflected ARAMCO's request for U.S. military transport for the king. The firm would have to arrange for a TWA plane to fly from New York to Dhahran to take King Saud on what turned out to be a long winter vacation. Doctors at Boston's Peter Bent Brigham Hospital pronounced the king "in surprisingly good condition" save for his high blood pressure. He had work done on his eyes and a month later had minor abdominal surgery, but he and his entourage also took forty rooms at the Boston Plaza after leaving Brigham. He spent time at a resort in Swampscott near the Kennedy family's famous Cape Cod compound as well, and, later, in Palm Beach, Florida, again, within driving distance to the first family's winter retreat.[8]

It wasn't to be all rest and relaxation obviously. When the American ambassador to the United Nations, Adlai Stevenson, paid a courtesy call on the king he found the entire cohort of exiled progressives in Saud's hospital room, including Faisal Hejelan, the Saudi ambassador to Spain, Abd al-Aziz Ibn Muammar, who had arrived from Switzerland, and Saleh Shafan, the ambassador to West Germany. Tariki was expected in Boston soon as well.[9] One concern of those gathered at the king's side emerges clearly from the archival record. The advisors were anxious to arrange for Saud to meet Kennedy and arrest what many saw, hard as it might be to imagine now, as a dangerous tilt away from Truman's and Eisenhower's commitment to preserve the Saudi dynasty and "our" oil.

The Kennedy administration had taken office resolved to improve relations with Egypt, a pivotal Arab state, following what even key figures in the Eisenhower administration had recognized was a failed policy of containment.[10]

The vaunted initiative amounted to a handful of friendly letters between JFK and Nasser and the advisors' magic remedy for all the arch neutralists including Nasser, Tito, Nehru, and Nkrumah, namely dollars for development. Kennedy's team upped the amount of (tied) food aid going to Egypt, known as PL 480 funds, and sent the Harvard "planner" Edward Mason on a new mission to survey socialist Egypt's development prospects.[11]

The truth though is that the effort to woo Nasser never gained much traction. The Saudis wouldn't have known it, but Kennedy was himself highly skeptical of its prospects. The political backlash from liberals and friends of Israel, on the one hand, and the basic and unchanged prime directive in U.S. Middle East policy to secure Saudi oil, on the other, made realignment difficult for both Nasserists and New Frontiersman. The first, more-or-less token gestures, reluctantly agreed to by the White House, triggered protests by Jewish groups and the Israelis, but also lobbying by the oil companies and their allies against the new-look foreign policy. Nasser made it easy for these otherwise antagonistic forces to come together when he took up the cause of revolutionaries in Yemen in late 1962, across the border from Saudi Arabia. As the war in Yemen unfolded, Kennedy's relations with Egypt would grow poisonous.[12]

Saud though had the burden of smoothing relations with the White House early on where the kingdom's UN representative, Ahmad Shukairy, had gained his masters the dubious distinction of being the state that sent the single most "venomous" response to President Kennedy's recent overture to key Arab heads of state.[13] So the king reached out in his characteristically royal fashion. He or, more accurately, his advisors dissembled. Then, when the White House agreed to host Saud for lunch in Washington late in December, he demurred. It would be better for Kennedy to call on him at the hospital given what was claimed to be the Arab custom of visiting an ailing man.

Unbelievable as it might seem, a second U.S. administration bowed—although it pained Kennedy more than it did Eisenhower apparently. The president agreed to see King Saud for a "fourteen minute courtesy call" in Palm Beach in late January 1962—for a "chat" and some "strong Arabic coffee," the *New York Times* wrote—a gesture that would be met by the king's traveling to Washington for lunch early the following month. America's continued support for the monarchy was not in doubt, although Kennedy griped all the way back to the family compound following the klatch with Saud. "Did the oil companies put you up to this?" he snapped at his chief of protocol and heir to the Duke family tobacco fortune, Angier Biddle Duke.

Why had photographers been tipped off to the visit? "There goes New York State," he said. Duke took a lot of flak, but the president even agreed to host a dinner rather than lunch at the White House given that it would take place during Ramadan.[14]

Kennedy played Eisenhower and went with Secretary of State Rusk and others on February 13, 1962, to Andrews Air Force Base to greet the king, who flew up from Florida on the president's plane, *Air Force One*. The two heads of state met together with their aides later that day. Kennedy asked the king to lift the ban on Jews (namely congresspeople) traveling to the kingdom. The Saudis had gotten slicker by then. "Saudi restrictions are only placed upon Zionists and . . . 'many' non-Zionist American Jews have visited Saudi Arabia." The king in turn pleaded for more U.S. aid and a "mission" dispatched to his country too—the Mason mission was then en route to Egypt. He also pressed JFK to explain why America was rewarding Nasser, whose fealty to Moscow was proved by the recent sequestrations. Saud brought up this point repeatedly in fact, and indulged in some very wishful thinking when pressed on this point. "In response to the President's question as to how long the King thought Nasser would stay in power, the King commented that while only God knew, Nasser's days appeared to be numbered."[15]

The Revolution's Last Gasp

Under the pressure of other senior princes, Saud had appointed Faisal as his deputy and acting prime minister during his absence. Faisal had agreed not to engineer any changes in the government during the king's absence. In return, Saud promised to settle the question of Faisal's role in a new government as soon as he returned. Sure enough, four days after landing in Riyadh in March, the king announced the appointment of the crown prince as foreign minister and deputy prime minister. In effect Faisal was being given control over all but security affairs, which Saud insisted on retaining. The last of Talal's allies, Nawwaf, was dropped from the government. Faisal then turned on the kingdom's conscience, Tariki.

The oil minister had made it easy for him to do so. Back in the summer, before Prince Talal's dismissal, Tariki joined the renegade royal in a campaign against the crown prince. The two accused Faisal of profiting personally in a deal with the Japanese owners of the new joint venture Arabian Oil Company (AOC). Faisal had been forced to dignify the charge with a response when the story appeared in print in a Beirut daily. He denied it, of

course, and challenged his enemies to provide proof for such an outland-
ish claim, promising that the government would act swiftly to remedy any
"harm done."

Remarkably, Tariki soldiered on, gathering enough evidence, apparently,
to persuade the Council of Ministers that's Faisal's condition had been met.
On November 14, 1961, the council voted to cancel an agreement after Tariki
revealed how 2 percent of the profits from AOC's operations were assigned
in perpetuity to Faisal's brother-in-law and the kingdom's future intelligence
chief, Kamal Adham.[16] He then hunkered down in Riyadh—the normally jet-
setting minister seemed to be afraid to leave the country.[17] Faisal waited until
March to take his revenge, when he ended Tariki's remarkable run in Saudi oil
affairs and gave the lawyer Ahmad Zaki Yamani Tariki's position.

For the ARAMCO Americans a new day dawned in Dhahran. Yamani
assured them that Tariki was being thrown off ARAMCO's board of directors, and they watched as the ex-minister wandered the region seeking
work. Kuwaiti allies tried but failed to give him a new oil ministry to run.
The *Economist* guessed that he would join the new provisional government
in Algeria.[18] Others said he was destined to head OPEC, but Tariki blamed
Faisal for sabotaging his chances. In the end he settled in Beirut and started
a consulting business, consoling himself with the idea that it wouldn't be
long before he'd be back in the kingdom. "I asked him how would he envis-
age a change in regime. He said that it would be very simple. A small army
detachment can do the job by killing the king and Faisal. The rest of the royal
family will run for cover like scared rabbits. Then the revolutionaries will call
Nasser for help."[19]

Tariki was wrong about the fate of the Al Saud, of course, and as a result,
he spent almost two decades in exile. It was only after his nemesis, Faisal,
was murdered in 1975 that he and other old enemies found it possible to return home.[20] He was far from alone, however, in guessing that the regime's
days were numbered, and it was impossible to predict from day to day who in
Washington or in Dhahran would wake up that morning in a sweat. It is also
clear that the kingdom's rulers assumed that its enemies in Egypt and Iraq
were conspiring with various Saudis at home and abroad.[21] Those who discounted the idea that the royal family's days were numbered were banking on
the fact that there just wasn't much of an opposition. Riyadh was also tightening its grip again.[22] Faisal appointed new editors of his choice to some of
the country's best-known newspapers and periodicals—*Akhbar al-Dhahran*,

al-Bilad, Ukaz, al-Khalij al-Arabi, and *al-Yamamah*—the better to ratchet up the anti-Nasser campaign and clamp down on his critics.[23]

Faisal turned on his brother, the liberal Prince Talal, next for the crime of cabling his congratulations to Nasser on the tenth anniversary of the Free Officer's coup in July 1962. The problem is that Egypt's propaganda ministry rushed to the airwaves with the news that a Saudi royal supported the revolution. Faisal treated it as an act of treason, revoking Talal's passport while he was traveling in Geneva and, putting aside for the moment his objections to Nasser's criminal sequestration program, seizing his brother's land and palaces for good measure. The government decreed that no airplane would be permitted to land inside the kingdom if the now real renegade were on board.

Talal took to the offensive. He flew first to Beirut, where he called a press conference. The campaign against him followed from Faisal's fear of the constitutional reforms that he had been advocating until he had "resigned" from office. He said that he would be traveling on to Cairo and Alexandria, that he was launching a campaign for democracy and social justice in Saudi Arabia, and that "he would return [home] one day to lead the struggle."[24] In Cairo he officially joined the forces of the would-be Saudi revolution. Nasir al-Said, a leader of the strikers back in the 1950s, was already there and so was another progressive ex-minister, Hassan Nassif. Talal's brothers Fawwaz and Badr and his cousin Saud Ibn Fahd, the other so-called Free Princes, joined him, surrendering their passports to the Saudi embassy in solidarity with Talal.[25]

Sides were hardening in what Talal described as the struggle for reform of the absolute monarchy, what Egypt's doctrinaires called a war against reactionaries and feudalists, and what the Saudis' allies in Washington and other species of anti-Nasserists called Nasser's desperate bid for "regional dominance." So if June 1962 represented, in Warren Bass's words, the "apogee" of the Kennedy administration's Nasser initiative, the *New York Times* was weighing in on the other side. While its Cairo correspondent reported on Nasser's increasing isolation and Egypt's lost luster, the paper's longtime specialist on the Gulf was reporting on the "progressive changes" amounting to an "economic and social revolution" inside the kingdom, led by the dynamic combination of King Saud and Crown Prince Faisal who were now working together for the welfare of their people, investing in the development of the kingdom, and even preparing "for eventual alternatives to oil as a source of income"—this last a total fantasy.

Still, take note of the inclusion of Saud because less than a year later the same story would begin to be told with the king omitted or with him treated as the main obstacle to the reforms Faisal was advocating. Dana Adams Schmidt writing in the *Times* even gave some credit to men like Talal, "the ideal of what a Prince should be," and Tariki, "the ideal of what a commoner can become." Although "these men and others like them are on the sidelines . . . it appears that the forces at work must inevitably bring them again into the mainstream of Saudi public life"—in reality, Faisal's worst nightmare.[26] These men, too, would never be mentioned again.

The championing of Faisal required that a number of issues go unmentioned, from the reasons for Tariki's sidelining to the royals' preference for monopolizing power, the ban on unions, Faisal's tightened control over what was already one of the least free presses in the world, and so on. The fact is, U.S. ambassador Hart and others were reporting privately, the threats from Nasser's campaign and later from Talal's defection shook the regime but, paradoxically, also drove Saud and Faisal to cooperate. ARAMCO's Washington representative and vice president of Government Relations, Garry Owen, said the trouble was that Faisal was hardly the progressive ("a pity that he was so negative") that was depicted in the *New York Times* or in other pro-regime propaganda.[27] Just as the Saudi press and radio continued its "total silence" over Talal's flight and the regime's reprisals against him, the *Times* never saw fit to print an account of the rise in repression.[28]

The government raided public cafés and arrested customers in the oil province for tuning into Voice of the Arabs' *Enemies of God* program, which regularly attacked the king and Faisal. By the summer, it was also harassing and arresting teenagers in large numbers in the Eastern Province as well for what was called "political activity" and what the U.S. consulate called, equally vaguely, "Nasserism." Young Saudis would "simply disappear" and then parents would receive permission to visit their sons in custody. As a U.S. official wrote at the time, the new prison that had just opened near Dammam was an "ominous reminder to everyone" of the government's "real means of control."[29] In September, after an Egyptian-backed coup gained control of the capital of nearby Yemen, the ban on Voice of the Arabs was extended to listening in public to Radio Sana, Radio Cairo, and the BBC.[30] Among its objectives, the Saudi government sought to limit the spread of the news of the creation of the Arab Liberation Front in Cairo, the umbrella under which the exiled Saudi communists, nationalists, and Free Princes operated for a while.[31]

If ARAMCO's executives considered Faisal to be a reactionary who had a long way to go before meeting the minimum demands of the liberal opposition, they weren't too troubled by this failing. The opposition was not much of a threat.[32] On balance, most officials thought the prospects for the regime were pretty good. After all, Faisal had cashiered the one man who had done the most to make their lives miserable for over a decade. The Americans also appreciated the tricky situation Faisal found himself in as Nasser turned up the heat against a one-time admirer and ally of a sort. ARAMCO's Washington office began lobbying for a meeting between the crown prince and Kennedy. It lent some additional resources of its own to Faisal's efforts at weakening Nasser. Earlier in the summer the company secretly arranged a $14 million loan to the Syrian government, at Faisal's request, although the company's head, Tom Barger, told his embassy contacts that there was no "solid evidence of Syria's urgent need" of the funds. But ARAMCO did it to build goodwill with Faisal and improve the prospects for negotiations.[33]

Perhaps it was a coincidence, but around the same time the Saudis themselves sent word through Faisal's brother-in-law and agent, Kamal Adham, that the crown prince wanted to meet President Kennedy. Faisal would be traveling to the United States for the new UN General Assembly session, but Adham said

> his real purpose in going . . . would be to see the President and Secretary of Sate and let them see that all Saudis are not like the king. Faisal would like to gain sympathetic understanding in the U.S. for the aims of the new World Islamic Society formed under [the] auspices of the King during the last pilgrimage season. The Saudis feel that Islam forms a significant bulwark against communism.[34]

The *New York Times* had already carried an account back in June of the Saudis' bid "to organize the forces of traditionalist Islam for political purposes." In September 1962, readers would also have found a series of stories for the first but not the last time about Yemen, where a military group had overthrown the imamate. Nasser's Egypt was the first country to recognize the new republican government and to promise aid for the revolutionaries. The crown prince of Saudi Arabia had flown to New York where he called for rolling back the revolution. Faisal met with the U.S. secretary of state on September 25, to be followed by a meeting with the president in early October. A proxy war between Egypt and Saudi Arabia was under way.

THE YEMEN DEBACLE

Yemeni officers, led by a Nasserist, Colonel Abdallah al-Sallal, had mounted a coup in Sana on the night of September 25, 1962, against Muhammad al-Badr, the imam, and his absolutist state. The coup leaders' goal was to end the dynasty's reign and to remake Yemen into a republic. The revolutionaries claimed to have killed the imam, but in reality he had escaped into the sheltering mountains of the north where he began to organize tribesmen for a war to regain the capital. Egypt promised to stay out of the conflict and warned others to do likewise. Nonetheless, within days one of Nasser's generals flew to Yemen to assess the situation and, on his recommendation, a battalion of special forces arrived on October 5 to support al-Sallal. They would be followed shortly by jet fighters and, later, more troops. In Sana, which they controlled directly, the Egyptians quickly set up a police state.[35] The Saudis began almost immediately airlifting weapons and gold in support of the royalists, together with the man they hoped would lead the counterrevolution, the imam's brother, Prince Sayf al-Islam Hassan.

The first days of the Saudi campaign against Nasser and for counterrevolution were a disaster. On October 3, a Saudi Air Force Fairchild C-123 cargo plane took off from Jidda with a crew of three. It was loaded with rifles and bound for Nejran, on the southwest Saudi-Yemeni border. The plane instead turned and flew across the Red Sea for Aswan. The Saudi crew announced that they were joining the Egyptian Air Force. The U.S. mission had trained both pilots, a fact that the embassy thought it wise to bury. The plane's cargo was offloaded and displayed for the press. The crates all had the famous clasped-hands of the U.S. aid administration pasted on them.[36]

Riyadh took emergency measures. Some half-dozen tanks were parked around the king's palace, their guns turned out. Troops blockaded the road leading from Dhahran to the capital. The American commander of the U.S. training mission was ordered by the Saudi commanding general to ground all jet trainers and to cease working with Saudi personnel. These precautions notwithstanding, a second air force crew got away in a two-seat trainer the next day, October 4, landing in Wadi Halfa, at the Egypt-Sudan border. They requested political asylum. Two more planes and their crews crossed over to the dark side on October 8. By then, the king, who at first refused to believe the news, rendered all the heavy weapons surrounding the palace inoperable.[37]

The U.S. embassy report said it was "a blow to the morale" of the vaunted

Royal Guard, among whom only the most trusted of the king's personal sentries carried loaded weapons.[38] In the same way, when the Saudis broadcast news of the defections, they lied and said the men had been bribed, rather than admit that, like many others in and outside the cabinet, the men had opposed the palace's decision to back the royalists.[39] After grounding the rest of the air force, the Saudis used their American TWA pilots to ferry rifles, bazookas, mortars, and land mines.

The American embassy passed on reports from a key contact inside the military that by early October Saudi forces were launching limited attacks across the border. There were two battalions of infantry and two brigades of the National Guard positioned there. The army was opposed to fighting against Egypt in Yemen, and a main factor determining what dissidents would do was the widespread fear that the United States would intervene to protect the king. A good guess, given that just a few weeks later the U.S. Joint Chiefs of Staff specified that one of the main American objectives in Saudi Arabia ought to be to "identify and neutralize subversive elements seeking the downfall of the Government."[40]

In the wake of those first and what turned out to be the only defections by the Saudi pilots, Talal voted, rashly, for what he said was the coming Saudi revolution—hoping it was going to be a bloodless one. In a series of interviews and press conferences in Cairo he predicted the imminent collapse of the monarchy and the birth of a democratic republic. Yemen showed the way, and toward that end he was founding a united front of Saudi dissidents under the name of the Arab Liberation Front. The brothers renounced their titles and took their place among the ranks of the Saudi people.[41]

Talal wanted to go to the States, where his rival, the crown prince, had just met with the president, in order to rally support for his cause, but an embassy officer in Cairo let the ex-royal know that Washington didn't want him. He in turn accused Washington of doing Faisal's bidding in trying to "muzzle him."[42] Egypt turned out to be not much more accommodating, unfortunately. Talal quickly grew wary of Cairo's control over what he could say, let alone over the relish with which his putative comrades in Yemen called for the Saudi subjects to rise up and kill the royals. He chafed even more at the support Nasser's government provided the labor leader Nasir al-Said, a rival for the title of exiled dissident number one and, for at least some in Egypt, a more authentic voice of the Saudi people (who, nevertheless, "welcomed the support of all liberated elements, no matter how recent their awakening," al-Said said about Talal). While al-Said and other exiles banked on Cairo, Talal moved more or less permanently to Beirut, where the newspapers at least would publish what he wanted.[43]

The Free Princes Retreat

By December, as al-Said relocated to Yemen to support his newly proclaimed and quite fanciful Saudi government in exile, which he called the Republic of the Arabian Peninsula, Talal was reaching out to his brother Faisal. The Americans were at least one channel he sought to use. The family council of senior princes had insisted on Faisal reassuming the powers of prime minister and steering the kingdom through the conflict with Egypt. King Saud's decade-long run was essentially over. Talal admitted as much to the Americans in Beirut, in the course of some remarkably fast rewriting of the record.

> Talal and his Arab Liberation Front had not opposed [Faisal] personally but was opposed to reaction, which Faisal represented to date. If Faisal had in fact changed, and was ready to undertake truly progressive reforms, Talal wished him every success. He was prepared to support him. . . . Talal did not wish to see revolution in Saudi Arabia and believed progress via Faisal was much preferred to anarchy and the terrific fragmentation that might result through the efforts of unspecified subversive elements. Therefore it was in the interests of both Saudis and Americans that the latter force Faisal to act quickly. Half measures would not do.[44]

Ever the host, Talal sent his guest off with a copy of his banned book, *Risala ila al-Muwatin* (Letter to a Citizen), and the charter of the Liberation Front. But "truly progressive reform" is a comfortably loose-fitting idea, and Talal's position in support of Faisal was in essence now the same as that of ARAMCO and the Kennedy administration. The crown prince dutifully declared himself a reformer, which gained him critical American support, as we will see. Talal and his brothers petitioned for the right of return and, that summer, rejoined the ranks of princes.[45] One rebellion at least was over.

Nasir al-Said continued to agitate, without success obviously, for revolution inside the kingdom. In December, the U.S. ambassador pressured Egypt's president to close down his operation. "Nasser Said was a known extremist on whom we have full dossier revealing unsavory past history of extremist connections and activity. . . . I urged that Nasser Said be taken out of circulation immediately." He wasn't "taken out"—at least not by Gamal Abdel Nasser, who demurred. Dissidents, however, insist that al-Said was eventually kidnapped and killed at the Saudi government's behest, much later, in 1979.

The Myth of Faisal's Reforms

Nasser's true motives in fighting a long, costly, and ruinous war in Yemen are a matter of conjecture: recover his prestige, test his army's ability, weaken the British position in nearby Aden, foment a revolution in Saudi Arabia, or short-circuit American hegemony in the Gulf. What is indisputable is that over 50,000 troops were tied down in the Yemen highlands over many years, even as he engaged in the brinkmanship with Israel in 1967 that led to the destruction of the Egyptian Air Force and much else. His army used chemical weapons against the Saudi- and British-backed forces in Yemen. Thousands died on both sides. Although Nasser hadn't intended it, as a result of the Egyptian adventure, Faisal secured control of the Saudi state, and that state crushed the remaining dissidents.

The coup and American intelligence's consensus that Nasser was involved in it focused the attention of the White House and State Department. Faisal reached New York just days before Sallal seized power. He met U.S. officials several times, beginning with the secretary of state, Dean Rusk, on September 27, when the prince learned that the president would host him for lunch a few days later. Kennedy's advisor, Robert Komer, outlined what was at stake in the meeting and why Faisal wanted some time alone.

> We've had numerous reports that Saud is rapidly failing. Faisal, next in line, is probably here to find out how much he and his country can rely on U.S. support. You can talk frankly to him, as he is far more intelligent than Saud. . . . Unfortunately, the Yemen revolt has brought to a boil all Saudi fears of Nasserism (the House of Saud well knows it might be next). Faisal wants U.S. backing for the UK-Saudi counter-effort in Yemen. It will be hard to satisfy him on this score.[46]

Some have since surmised that Kennedy and his team forced Faisal to undertake the reforms that are now periodically hauled out as proof of his (second) enlightened turn in office, or that the Americans brokered a deal with Faisal to support him against the king. There has been scant evidence in support of this view until now, and it still doesn't amount to much. Certainly, the handful of declassified documents generated from the many meetings between Faisal and his aides with Kennedy and other top U.S. officials offers no proof of the proposition. In particular, the bland, brief memorandum of conversation of President Kennedy's private meeting with Prince Faisal, with only the ex-ARAMCO-employee-turned-State-Department-officer Issa Sabbagh serving as translator, reveals nothing. Still, there was plenty of opportunity for colluding. Faisal and his top adviser, Rashad Pharaon, a Syrian who had parlayed his position as Ibn Saud's private physician into a lucrative sinecure, remained in the States for close to a month even while the king and his counselors all but began a war with Yemen.

The idea of a "deal" with Faisal rests on the fact that some weeks after finally returning to Riyadh and taking over as prime minister, Faisal announced a vaunted set of reforms referred to as his Ten-Point Program. He did so while negotiating with the Americans for a show of force after Egyptian planes bombed Saudi territory—albeit near the Yemen border where the arms were being transshipped to the royalists. Calls for reform inside Saudi Arabia were all but ubiquitous in the days and weeks following the fall of the imamate. So, the American embassy could provide a representative account of the general response of the "conservative elements" in Saudi society who were arguing for "some measures that will placate people," for instance, movies, some kind of representative assembly, and cutting back on "princely extravagances."[47] For Saudi liberals, the way forward required even bolder, crash programs of military modernization and social and economic development. The early military defections, Talal's rush to Cairo's side, and so on, coming on the heels of what the Saudis viewed as Egypt's stunning success in bringing down the imamate in Yemen, was more than enough to focus the mind of Faisal on the options before him.

In all the archival records I have seen about what Saudi liberals or State Department officers or ARAMCO brass thought Faisal and the Al Saud ought to do to strengthen their position, none focused on slavery in Saudi Arabia, in contrast to, say, making television and other "innocent means of pleasure" available to subjects. This is true as well, by the way, in the transcripts of President Kennedy's meeting, where it is not slavery but, once again, Saudi Arabia's failure to let Jewish Americans land in Riyadh that is the focus of (fleeting) concern.[48] Beginning in the 1970s and until today, however, Americans and other westerners are apt to elevate emancipation to the top of the list of Faisal's achievements. Some Saudis might too, although by and large they are more likely to lead with his putative educational reforms.[49] Manumission was simply not a burning issue in Riyadh nor one around which a domestic opposition was likely to mobilize. Rather, the slavery issue was, as we have seen, much more salient to critics outside the country, from Cairo to New York, who might use it to demonstrate just how reactionary, anachronistic, and noxious was the rule of the Al Saud. Every year at the UN the future PLO leader Ahmad Shukairy would claim that slaveholding in the kingdom was a myth invented by its enemies. Nonetheless, when Talal and the Free Princes reached Cairo they announced the freeing of their slaves. The Yemenis had just done so as well. So Faisal announced that he (and everyone else in the kingdom) would be releasing theirs soon too.

Decades later, Parker Hart published a memoir of his time in Saudi Arabia.[50] Hart had three tours of duty in the kingdom. He served as the first consul to Dhahran in 1944–46, returned as consul general for two years in 1949, moved on to other posts in Cairo, Washington, and Damascus in the 1950s, and was named ambassador to Saudi Arabia in 1961, where he served for four years. Hart says that during the president's meeting with Faisal, Kennedy promised support to the kingdom against Egypt and other threats inside and outside the country in exchange for the prince's taking specific

> early steps to abolish slavery and to institute fiscal and other reforms in the Saudi judiciary and its practices so as to bring about strong public support and greater participation of educated Saudi youth in the administration of the state. The US government would find it hard to justify to its own people a deep commitment to a system of government that was corrupt or bore the stamp of slavery and arbitrary denial of civil rights and personal freedom.[51]

How does he know what transpired in a meeting that he didn't attend and that isn't supported by any documentary source? Unfortunately, he doesn't say. The third person in the room was an old hand in Hart's embassy. "Sabbagh was the best of any bilingual interpreters I knew, a man on friendly personal terms with the prince, and one who would report fully to me on his return." Sabbagh may well have briefed Hart back in 1962 on what transpired upstairs in the White House, and Hart may have recalled that moment accurately, decades later. Then again, the memoir of the eighty-plus-year-old ex-ambassador might have been aided by the impressive pile of books that started to be published about Saudi Arabia following Faisal's ascension to the throne, where the slavery issue serves as the main proof both of Faisal's enlightenment and loyalty to Camelot. Then again, it may be a myth.

Deal or no, reform was, as we have seen, the all-purpose American prescription for the Al Saud any time it faced a challenge in or outside Saudi Arabia's borders from the 1940s to the 1960s. Why not? The notion is infinitely flexible, and, since the royal family survived each crisis in turn where others were convinced it would not, it even appears to have worked. In 1962–63, the Saudis proved wrong a host of prophets who were predicting the imminent end of the monarchy, from a then young Shimon Peres ("the overthrow of . . . Saud-Faisal was unavoidable") to Prince Talal.[52]

The views of ARAMCO's Arabists, which U.S. ambassador Hart sought out at the start of the Yemen crisis, matter here because back then the firm's experts challenged the now-fashionable idea about the time that the Saudis' days were numbered. "Prospects are much better than Talal's one year estimate," the head of Government Relations, Harry McDonald, told him, because, despite the recent defections, the house of Saud "retained strong support" among Bedouin and other conservative elements in society. McDonald could not resist doing a little unregistered lobbying for the monarchy, adding that, nonetheless, the United States could do more to support the king "in his hour of need." To do what? "Implement reform and development," of course.[53]

The subsequent celebrations of Faisal as a successful reformer who headed off a revolution obscure the most important and enduring institutional legacy of his return, namely the takeover of the state by a coalition of princes, the Al al-Fahd. What the British journalist Robert Lacey noted in 1981 about Faisal's return still remains true. "So was formed in 1962 the partnership which continued after Faisal's death [1975] and which still governs Saudi Arabia today."

The fate of Faisal's so-called reforms, announced with much fanfare in November 1962 and embraced as real by his many ardent admirers, thus gives us an early glimpse of the Al al-Fahd's approach to rule. Faisal announced the creation of a constitution, which he called "a basic law" in order to avoid a tangle with those who insisted that the kingdom's true constitution was the Quran, but it was the last that was ever heard about it. Faisal also promised to reform local government. Most commentators, charitably, note that this was yet another of the vague and unfulfilled promises of a, nonetheless, modernizing reformer. The truth is that those of the ten points (drafted by Yamani, according to the U.S. embassy's sources) that looked more like concrete proposals rather than dreams, say, to "raise the nation's social level" or "promote a permanent economic upsurge," all went unfulfilled. Arguably, the one exception was point five, wherein Faisal reaffirmed the state's commitment to the propagation of Islam. The Americans guessed that Faisal meant the new World Islamic Society, but since it did not fit the preferred account of what Faisal was up to, it was dismissed as just the "standard position of the Saudi Arabian government." Nothing new.[54] The reality is in fact quite a bit worse— if one is a democrat, that is. Reform also meant the end of the experiment in local council elections begun by Saud's progressive allies. Citizens would not have the opportunity to vote again for close to fifty years.

Among the journalists of the time, as distinct from Faisal's hagiographers, David Holden did most to set the Faisal-as-modernizer train in motion. Saudi Arabia "was not much changed, administratively or politically from the way he had found it during his first curtailed early office." The only real accomplishment had been, allegedly, to keep to the austerity budget that people like to believe was the crown prince's doing in 1958 (it wasn't). It all began to change in 1962, as Faisal brought Saudi Arabia "backwards into the future," spending on new schools, roads, plans, and ministries.[55] Yet when Saud's government championed these precise kinds of investments, he was condemned for wasting scarce resources. By the 1980s the train seemed to be careening out of analytical control. Thus Gary Samore and his Harvard dissertation supervisor (and the CIA's contract researcher on Saudi Arabia) Nadav Safran credit Faisal's 1962 reforms for publication of a comprehensive five-year plan in 1970 by Prince Talal's old Supreme Planning Board.[56] Faisal obstructed a process that was already under way. Significantly Hisham Nazer headed the organization that wrote the plan, and he was Tariki's old protégé at the Petroleum Ministry. Let's just say that none of the scholars here in the United States worked overtime on making transparent the logic of causality in, let alone the validity of, their arguments.[57]

We can actually account for more of the facts and resolve many of the ambiguities of the moment if we recall that Faisal had been a most serious obstacle to the modernization projects of Tariki and other progressives, and that in the months after taking over as prime minister he set about placing his clients at the head of various institutions. As a result, some projects would finally be permitted to go forward, while others were derailed. For example, the new General Petroleum and Mineral Organization (Petromin) unveiled in November 1962, was the project that Tariki had been about to launch when he fell from power. Faisal had nothing to do with inspiring it save to stay out of the way. Consider in the same light one of Faisal's first real as opposed to imagined acts of planning. He fired the American advisor to the Saudi government, Harold Folk, the World Bank economist. Folk had labored hard for two years in the face of resistance from conservatives. ARAMCO's development people called Faisal the chief obstacle to the planners and the king the better bet for seeing the work of the planners move forward.[58] Folk and his Saudi counterparts had drawn up blueprints for roads, water, power, and light manufacturing.

As the Tel Aviv analyst Sarah Yizraeli points out, "Faisal's government did not carry out any of these plans, modest though they were, even though the cabinet recognized them as necessary."[59] Imagine the feat of logic-twisting necessary to omit this detail of Faisal's efforts. Smarter and more careful than most in her reading of the archival record, Yizraeli concludes that Faisal was a main obstacle to the planners throughout this period. We have seen many examples of his handiwork. The best we can say is that Folk, Tariki, Talal, and others laid the groundwork for future efforts, late in the 1960s, which are remembered now as King Faisal's visionary efforts at transformation.

I'm Your Father Now

Two years of diplomacy failed to end the Egyptian-Saudi proxy war. Early efforts to involve the UN in disengagement talks failed until Kennedy's personal emissary, Ellsworth Bunker, began shuttling between Riyadh and Cairo in the spring of 1963. As a result, UN observers eventually took up positions at the Saudi-Yemen border, where they remained for just over a year. Faisal agreed to stop arms shipments to the royalists in return for Nasser's pledge to begin troop withdrawals. An original token withdrawal of Egyptians was followed by a kind of three-card monte game of rotating troops to disguise new deployments (on the order of 1,000 out, 4,000 in). The bottom line is that 30,000 or more troops remained bogged down until after Egypt's defeat by Israel in June 1967.

Faisal made more of an effort than Nasser to comply with the terms for two reasons. First, the guerillas did not depend solely on the Saudi and Jordanian arms pipeline from the north, as long as they had cash with which they could pay smugglers and as long at the British government continued to supply weapons to the rebels from its colony in Aden and the surrounding protectorates.[60] Second, President Kennedy made closing down the arms pipeline a condition for sending Faisal a squadron of U.S. Air Force F-100 fighters. Air support was the third component of America's promise to preserve the monarchy. The United States had already turned a few training jets over to the Saudi Air Force, stepped up the number of visits by American warships, and sent special forces (Green Berets) to the Dhahran base for schooling Saudis in counterinsurgency techniques.

The Americans ratcheted up their involvement as Nasser's forces began bombing Saudi towns and royalist staging posts along the Saudi-Yemen border. Consolidating the victory of the Arab people against feudalism was not going to be as easy as Nasser thought. Egyptian troops and those Yemenis who could be dragged into battle against the tribes would get chewed up consistently through the next few years. Nasser imitated the Americans in Vietnam by using napalm, and did them one better by gassing the guerillas along with those living in the vicinity of the staging posts. The Egyptians also sent an occasional jet to rattle windows in Jidda. Attacks on the Saudi side of the border declined, however, once the Americans deployed their fighters for a short while in July 1963.

Disappointed that the Egyptians did not just roll over and play dead upon command after a little morsel of AID funds had been tossed their way, the Kennedy administration ultimately gave up on Nasser. Liberals had hammered the White House and State Department for, as they imagined it, ignoring the threat that Egypt, a state now guilty of using weapons of mass destruction, posed to Israel. The oil firms came at Washington from the other direction, for all but inviting Nasser or his followers to overthrow the Saudis. Kennedy decided to bring the F-100 mission home by January 1964, although it would be left to his successor, Lyndon Johnson, to close the chapter on the New Frontier's short-lived Egyptian campaign. The UN would pull out of Yemen too by August 1964.

Republicans and royalists in Yemen stumbled toward a power-sharing accord, driven in part by disenchantment with the new Egyptian overlords and in part by stalemate on the battlefield. Nasser would ultimately give up the long and costly intervention in Yemen, but only after the crushing defeat by Israel in the June 1967 Six Day War. The Americans' main Middle Eastern client, Faisal, was likewise forced to acknowledge that those he financed were unable to reverse the republican turn in Yemen, even after the Egyptians withdrew. The two sides had fought to a draw.[61]

Under pressure by Faisal, ARAMCO, the parent companies, and other oil firms, various congressmen, Joseph Alsop, the most widely syndicated newspaper columnist of the time, and so forth, the New Frontiersmen of the Kennedy administration finally abandoned the short-lived romance with the Nasser regime.[62] The downward slide—or, if you favored the "evolutionists," as Lyndon Johnson's national security advisor Walt Rostow described the U.S. clients in Jordan and Saudi Arabia, then the upward climb—continued after LBJ took office. The Texan agreed to sell the Saudis close to a half billion dollars worth of weapons and services, which included installing the kingdom's first missile defense system and building a string of new bases. The U.S. Army Corp of Engineers, which had developed the Dhahran terminal in the late 1950s, returned to the kingdom to set up a television broadcasting system in Riyadh and, in a massive expansion of its role in the kingdom, became "national manager of Saudi development contracts" in the military and civilian sectors, similar to the role played by Bechtel in the 1940s.[63] Faisal would make a return visit to Washington in 1966. LBJ's secretary of state, Dean Rusk, ordered all "exhortations for reform, which had by then become almost rote and counterproductive, to be discontinued. The Saudi leadership, Rusk believed, was best qualified to judge its own best interests."[64]

Despite America's two-decade-old commitment to preserve the Saudi monarchy and the Johnson administration's renewed, relatively more open, heartier, Texas-style embrace of the Al Saud, Faisal remained distrustful of the United States, uncomprehending of its failure to force Nasser out of Yemen, and afraid of the power America wielded in the Gulf and beyond. In other words, he had come face to face with the kingdom's acute structural dependency on a not-so-merciful and not-so-beneficent protector. His father had made peace with his place in the international order or hierarchy in the 1940s. Faisal did too, in the 1960s, having abandoned his efforts to alter the terms of the bargain, and turning, like his father, to blaming his troubles with America on the Jews.

As far as ARAMCO was concerned, Faisal had satisfied the company that the bad old days were over. His new oil minister, Zaki Yamani, addressed the fourth annual Arab Petroleum Conference in Beirut in 1963 on the very day Faisal had unveiled his reform agenda and told the delegates that the kingdom resolutely opposed the idea of nationalizing the oil industry, a step that would be completely "contrary to the government's policy of maintaining a free economy."[65] In return, ARAMCO was expected to support Faisal by advancing funds to kick-start some more road building and the like.[66] He also pulled the plug on Ahmad Shukairy, the Saudi UN delegate whom the Kennedy administration detested. Shukairy would return to New York, this time sponsored by Egypt as a representative for Palestine before Nasser chose him to head the new Palestine Liberation Organization.

Faisal did some fence-mending in Jidda as well. Through his brother-in-law and intelligence head Kamal Adham he reached out to the long-estranged ally, Muhammad Alireza. The U.S. ambassador described it as an attempt to build political support outside the royal family. Recall that Alireza had quit Faisal's Council of Ministers in 1958 and, wisely, lived abroad until Saud took over. Privately, the ambassador reported, Alireza and his family members continued to criticize Faisal and doubted his ability to guarantee security and stability. Nonetheless, in due course, the rehabilitated Alireza would become Faisal's emissary to Nasser.

Faisal did some more general reaching out as well. In early January 1963, he took to the airwaves to address Saudi citizens directly, the first time any leader had, the U.S. ambassador thought. The prince rehearsed the history of relations with Egypt, lamented Nasser's decision to threaten the kingdom, and called on Saudis to rally in support. It is unclear how effective the speech was. Apart from some royals who vowed to take the war home to Cairo and the like, the people listened, presumably, and many were bussed into the city for war rallies, but since they were never required to do anything but remain loyal it is hard to judge their commitment. The Americans, on the other hand, clearly ate it up. Ambassador Hart wired Washington, modestly, that he and Issa Sabbagh were behind Faisal's "fireside chat" because, they said, they had prodded him, using the example of FDR during World War II. Faisal deserved extra credit for taking the embassy's challenge, since, according to Sabbagh, his command of classical Arabic was "inordinately faulty."[67]

Faisal was, in a word, golden. He had backed off the issues that for many years unnerved ARAMCO, and, after the initial defections of the pilots and the Free Princes, there were few signs of unrest that might unnerve an ambassador's staff. Faisal had even agreed to what amounted to a don't ask, don't tell policy on the issue of Jewish Americans included among the crews and support staff of the U.S. Air Force squadron, which Egypt's propaganda arm seized upon, but so what? For years, stories in the American press about Faisal would lead with an account of how his visionary reforms had averted certain catastrophe. *New York Times*'s editors, among others, lost little sleep trying to solve the mystery of how a regime allegedly on the brink clawed its way back long before any of the magical reforms had been concretely specified, let alone implemented, save if one believes that it was the freeing of the slaves that had convinced all the would-be revolutionaries, Nasserists, communists, and the like to lay down their arms.

The account of how Faisal's modernization efforts preserved the kingdom would be repeated each time the press reported on Saud's sad end and eventual exile, on the checkered course of Faisal's negotiations with Nasser, his crowning in 1964 and follow-up trip to Washington, the sale of missiles to America's hawk-nosed ally, and so on. Perhaps it was better not to confuse readers, because the modernizer had also been busy, as I noted, ending the first experiment with local elections. In July 1964 his agents arrested scores of Saudis apparently a bit too taken with reform.[68] The kingdom sheltered some of Egypt's most violent Islamists, as we would now call them. One of Faisal's first acts after being named king was to abolish the position of prime minister. He dismantled the American-designed Planning Board and turned planning over to his familiar Rashad Pharaon, Ibn Saud's old doctor-turned-profiteer.[69] As I have said, most of the promised reforms were never mentioned again, for good reason. According to ARAMCO's analysts, Faisal was in reality "running the government by himself with no regard for established lines of authority and the theoretical structure of the government."[70]

There were also scores to settle. Sometime in the spring of 1963 Faisal recalled the Saudi ambassador to Switzerland, Abd al-Aziz Ibn Muammar. Friends probably warned him not to go but Ibn Muammar, his wife, and his six children returned to Jidda. The American ambassador remembers Ibn Muammar telling Saud and Faisal to face facts and recognize the Yemen Arab Republic. The journalist Robert Lacey is more vivid. "Ibn Muammar remained a malcontent. He came back from Switzerland to Riyadh, were he criticized Faisal openly and continually, until Faisal lost patience. 'Either you stop stirring up trouble, or you go to prison,' the king eventually cried in exasperation."[71]

Ibn Muammar's youngest daughter, Haifaa, watched when on the night of August 13, 1963, Saudi officials dragged him off to jail. "I remember my mother's screaming protests, begging not to take him, trying to block the arresting soldiers. Father did not resist. Alas, he was taken and not to be seen or heard of for the next ten or eleven years."[72] He was never charged with a crime and never went before a court.

Ibn Muammar's grandfather had hosted Faisal's father back when Ibn Saud was still exiled in Kuwait. Ibn Muammar's father was a loyal counselor to the first king. Ibn Muammar's wife, Umm Muhammad, married him with the king's blessing, and Muhammad, their first-born son, had been nursed with one of Ibn Saud's own sons. So it was possible for Umm Muhammad to seek an audience with King Faisal. The family tells the following story about the meeting. Umm Muhammad brought her eldest daughter Hissah with her to the palace while she pleaded on the children's behalf to free their father. Faisal turned and addressed Hissah directly. "Forget your father. I am your father now."

THE ARAMCO AMERICANS UNDER A MICROSCOPE

Once the White House ordered the U.S. F-100s to the kingdom in Operation Hard Surface, ARAMCO had to scramble, knowing that the spotlight would be shone once more on the American colony in that "flea-bitten part of the world," as Kennedy's advisor put it. A letter written in July 1963 by Harry Mc-Donald, one of the sharpest of ARAMCO's Government Relations people, who learned his craft as part of Eisenhower's intelligence staff in World War II, captures well the mix of superiority and unease that prevailed among the insiders.

> The *Time* magazine Bureau Chief for the Middle East was down from Beirut to do an article on ARAMCO for the business section. We all regarded it as the worst news of the season, and wished they would leave well enough alone. Everyone "wrote" what we knew would be written in advance and even the advance questions which were sent down were a tip off. . . . What is done for the wives to compensate for living in the sweltering desert? What answer can you give that doesn't sound like propaganda if you deny the original point of view. . . . He obviously was working down a list of subjects and the men answered. Every time a woman made some innocuous remark like "dust storms aren't too bad" everyone tensed in case it would come out wrong somehow. . . .
>
> Anyway it would have been marvelous to ask some questions which have been on my mind for years. . . . ARAMCO policy is determined by what is right and good to do in this country. Does *Time*, in its directives, have any element of working for the public good or having a responsibility to the US or anything else, etc.[73]

McDonald meant the letter to be funny, but he was writing privately to another Aramcon and so, presumably, felt no need to dissemble on the company's behalf. The letter is therefore also a testimony to the strength of his conversion to the ARAMCO gospel. Institutions like *Time* were unlikely to get the ARAMCO story right. He doubted that its reporters and editors followed the same higher calling that had moved four of the largest oil firms in the world to exploit (a word that the firm's public relations people never could get their arms around) Saudi Arabian oil. With this late restatement of the idea of ARAMCO as a kind of international development agency or humanitarian mission on behalf of the Saudi people, then the irony in McDonald's other idea is precious, that some fantastic belief on *Time* magazine's part made honest dialogue with the Aramcons impossible. Finally, there is the anxiety about what *Time* might write about the oilmen's wives.

Lifting the Veil

There were 1,250 American men working in ARAMCO in 1966, which, if anything, was a slightly lower figure than in 1963 (Americans were being slowly and steadily retired), and 1,000 married American women living inside the Senior Staff camps in Abqaiq, Dhahran, and Ras Tanura. Another 200 single American women lived inside the camps. Company rules banned married women from working for ARAMCO, and any woman employee who married after moving there was obliged to give up her job. Any women with children had to give them up too, eventually. ARAMCO's schools stopped at the ninth grade. There were "no old people, no young adults, and except in holiday time, no teenagers above age of fourteen."[74]

We have two kinds of accounts of the effects of ARAMCO's paternalism on the women who lived in the camps. There are a few first-person narratives. Nora Johnson, author of the 1960s novel the *World of Henry Orient*, which became a hit film for Peter Sellers, wrote one. She moved to Dhahran at twenty-two, after graduating from Smith and marrying Leonard Siwek, an analyst with Arabian Affairs who had joined ARAMCO out of the Johns Hopkins School of Advanced International Studies (SAIS) in 1949. Siwek gave up his career, though, when his frightened, angry, and lonely wife and mother of a new baby girl forced them to flee Dhahran at the end of his two-year contract.

> The intellectuals of Dhahran put up with the town and cherish the Muslim culture and the barren challenge of the land; the less inspired ones cherish, with equal fervor, the startling efficiency with which the oil company has imported all the comforts and mediocrity of home—the fact that Dhahran is more State-like than the States—and they consider the Arabs repulsive.[75]

Joy Wilson lasted longer in Dhahran than Nora Johnson, a total of eleven years. She was a Radcliffe graduate, class of 1957. Her husband, Robert, a Harvard-trained economist, accepted a position with ARAMCO and they moved to Dhahran with their new son in 1965. They had another son and a daughter soon after arriving in the kingdom. Wilson told the story of these years in Radcliffe's alumni magazine after returning to the United States and settling in New Canaan, Connecticut. What she writes about, and what must have been true about other wives and mothers, is how much she hurt. She wrote honestly, too, about her family wanting to stay and her guilt in uprooting them.[76]

Peter Speers called her a witch who arrived in Dhahran "by first-class broom." He included a copy of the article along with his commentary in a letter from Texas, where he had just retired. Back in 1950 Bill Eddy had called Speers "the most brilliant Arabic language student" in the program at SAIS when Eddy was in charge of recruiting the first group of Arabian affairs analysts and channeling a select few into the CIA. Ellen Speers, his wife, gave up her own career as a Near Eastern affairs analyst in the State Department when he joined the firm and moved to Dhahran. The years may have dulled Speers a bit because his gloss on Wilson missed the basic point. "She nagged Bob Wilson for years and made his life miserable with her demands that they leave Arabia, and here she is carrying on in print about how grand it was living in Dhahran." Perhaps her criticism, however carefully couched, was brushed aside reflexively as all the Policy and Planning staff learned to do over the years.[77]

Two ethnographies of the company towns, one from 1955 and the other from 1963, backed up these first-person narratives. No wonder ARAMCO prepared detailed rebuttals of both. The studies are in fact the only works ever to deal critically with the hierarchy of the camps and the Americans' views of Saudis until now. One was occasionally cited in memoirs from the time. The other has lain unread for decades. The first study, by a leading American anthropologist, Solon T. Kimball, was done on the sly. Kimball, a professor of anthropology and education at Columbia's Teachers' College, pioneered the field of community studies. He worked primarily with the Navaho people and advised the Bureau of Indian Affairs on development issues and on Navaho-government relations. ARAMCO paid him to participate in its 1955 summer institute for teachers at the ARAMCO schools. He spent a week each observing and lecturing at the three oil towns, and later wrote up his notes on "American Culture in Saudi Arabia."[78]

It was a short paper. He didn't waste words but came to the point at once about the contradiction he found in life inside the camps. "I was impressed as all must be who see what has been accomplished, with the magnificence of American technical know-how, somehow coupled with the inability to comprehend or, if understood, to deal adequately with problems of human relationships and emotions." Many of the men and women he met appeared troubled. "Although . . . the phrase 'we've never had it so good' is used often, nevertheless there are deep currents of disquiet and frustration, and perhaps even more serious personal consequences." Kimball's ethnography exposed the hollowness at the core of many of the company's claims about the lack of hierarchy, the equality across races, and the rest. The reality was that physical barriers in each of the towns, "corresponding closely with the structure of bureaucratic hierarchy and ethnic division among employees," made the myriad internal social divisions immediately clear to all. "Even the lowliest" American worker had his or her sense of superiority reinforced through the spatial and racial organization of the camps. "The American position of preeminence is reflected in symbol and fact. It is also a position that imposes a high degree of cultural isolation . . . and contributes to an omnipresent sense of precariousness." When he probed, even those who emphasized ARAMCO's contribution to Saudi development and the Americans' "sense of mission were prone to lapses of pessimism." Many if not most would say, "I'm here for the same reason as everyone else—the money."

Toward the end of the paper, Kimball turned to the way in which hierarchy was gendered, the separate and hardly equal male and female worlds. He said in this respect life in the American colony most resembled the frontier. "The oil industry is man's world and the production of oil under frontier conditions is not favorable to the establishment of stable family and community relations." Rather, he argued, "family and community . . . are, in reality, complicating and disturbing elements" to the main objective of producing oil. From the original decision to provide family housing at the end of World War II, gender was a kind of technical factor of production for the firm's principals to use to arrest the massive and costly turnover in personnel. Many other examples followed, from the rules about when men could bring their families to the field to the decision to ban teenagers from camp after they turned fourteen. He thus underscored at the same time where the frontier model broke down. "The paternalistic control of ARAMCO reaches into every nook and cranny of one's life and gives a security that actually contributes to a feeling of impermanence and precariousness. What is given may be withheld or taken away."

As I said, Kimball's study lends weight to the first-person accounts of Nora Johnson of a world where the men

had an enormous amount of fun, in which they tried to include their wives, and we all sat around and laughed politely and tried to imagine what it had been like out on the pipeline that night in June, or that crazy lunch in Rome the year before or the party in one of the barastis where so-and-so got totally loaded and did this terrific imitation of Banana Nose (King Saud) while wearing only a sheet.[79]

Like Wilson, Kimball showed how the pleasures of simplicity and privilege coexisted with "discontent, anxiety, frustrations," where the institutions of community "are only partially realized and . . . the deficiencies militate against the [married] woman." But Kimball also provides the only account we have of the position of unmarried women in the hierarchy, of how "the social adjustment and, in particular, the sexual behavior of the single female employee is the subject of a great deal of interest, talk, and speculation." ARAMCO records, now at Georgetown University, hint at a problem where company men ruled. Researchers will find William Mulligan's so-called nonsense files, where he kept jokes, parodies of company public relations material, a thick file of women's underwear ads, Joy Wilson's Radcliffe article, the records of "bachelorettes" who alleged they had been raped in camp, and his reminiscence of a female translator whom one of his friends bragged about sleeping with. He must have been lying, but if not, then he "may on occasion be revolted in his drunken hallucinations by the memory of her hideous body."[80]

The firm's reaction to a rare moment when the anthropological gaze its specialists deployed on the world outside the fence was turned back on the senior staff—imagine the al-Murrah or other tribe writing up a point-by-point response to one of Vidal's or Rentz's or Matthews's, et al. reports—is instructive. More to the point, it confirms Kimball's observations about the company's paternalism, the ability to reach "into every nook and cranny of one's life," and simultaneously the sense of precariousness that pervaded life there. So Government Relations prepared a detailed rebuttal to a paper presented at a professional society that no one in Saudi Arabia was ever likely to read. The same contradiction was manifested in Dhahran's obsession at roughly the same point with burying the book that the New York office had hired Wallace Stegner to write.

The Struts and Bolts of Hierarchy

Eight years later, in the fall of 1963, the sociologist Thomas F. O'Dea turned in the final report of the study ARAMCO had commissioned him to produce on "Social Change in Saudi Arabia: Problems and Prospects." O'Dea, then at the University of Utah, was best known for his studies of the Mormons.[81] In Saudi Arabia he led a team that included ARAMCO's own senior anthropologist, Federico Vidal, in an eight-month-long project. The report weighed in at just under 200 pages. The findings, in line with Kimball's with regard to the hierarchy inside the oil province, also went much further. ARAMCO's Arabists prepared an official response, fearing that O'Dea would try to publish the report. Five years later, in response to a request for release of the report, Mulligan argued against it on the same grounds he argued against publishing Stegner's book, which had just fallen into the hands of *ARAMCO World*'s new editor in Beirut. O'Dea was wrong on matters of fact, despite his having worked with a member of Mulligan's organization, apparently, but, then Mulligan never said what any of these errors were. Readers without the requisite "specialized knowledge" would be unable to "make the necessary qualifications and distinctions" needed in the case of those of O'Dea's valid conclusions. And it would be offensive to most Saudis, he insisted. O'Dea portrayed ARAMCO as the solution to all of the kingdom's woes and "continually underestimated the resilience of Saudi society and its capacity for self-generated change." O'Dea in fact argued the opposite.[82] What, needless to say, scared ARAMCO was the report's point-by-point rehearsal of the story we have told here at some length.

O'Dea's central claim, framed in the terms made famous by the development economist and Kennedy administration official Walt Rostow a few years earlier, was that the transitional Saudi society was poised to begin a "take off" marked by "modest development in agriculture and light processing industries." ARAMCO was only one engine that had set the train of modernization in motion over the preceding decade. Most important, the "epoch of the tremendous impact of ARAMCO's innovation *has closed*." ARAMCO's position would increasingly be subordinated to the state or as O'Dea preferred, the government. "The initiative for change . . . has shifted from the oil industry to the government." The third key factor, Arab nationalism, was responsible for the "atmosphere of urgency" surrounding development efforts and government actions.[83] O'Dea would return to this key point in discussing the implications for ARAMCO's policy planning. "In every way except as a source of capital, in the form of oil revenue, ARAMCO's participation in the future will be a subordinate one." The real problem he said was the extent to which a government headed by Faisal was up to the challenge.[84]

O'Dea argued as Sarah Yizraeli would decades later in the 1990s that in the Saud years government agencies had been built that were managing to a greater or lesser degree various aspects of development. The Council of Ministers had started to get a handle on overall coordination of policy. If there were agencies that hardly functioned, notably, the Ministry of Health and the Ministry of Agriculture, there were others that could be relied on, beginning with the Saudi Arabian Monetary Agency, but also the Finance Ministry and Labor Office, however much local firms griped about the latter. One ministry stood out from all the others. "Our researchers suggest that the best functioning ministry is the Ministry of Petroleum and Mineral Affairs. While having its share of common problems, it would seem to be nearest in approach in the entire government structure to a genuine civil service."[85] This was the organization built by Abdallah Tariki.

O'Dea's study reported accurately the sense among many Saudis and outside observers in the aftermath of the Yemen revolution that the kingdom's prospects had been set back rather than advanced as Faisal consolidated his grip on power. To his credit, O'Dea ignored all the noise about reform and the like. The government ruled "not through any widespread support" but instead by keeping the military divided and unarmed, buying the loyalty of key groups, thus dispensing with the fantasy that paying off clients was something Saud had invented, "and suppression of political freedom, now said to involve an expansion of secret police activities." Then there were the Americans. "One man who we interviewed summarized his reaction to the recent arrival of American troops in Saudi Arabia in this complex situation by saying, 'You (meaning the Americans) are backing the wrong side (meaning the royal family), but where is the right side?'"[86]

ARAMCO's records back up O'Dea on this point. One of the firm's sources in Riyadh, a Palestinian named Abd al-Hamid al-Dirhali, who had transferred to the Ministry of Petroleum from the Council of Ministers, complained of the growing problems inside the oil ministry. Al-Dirhali said that Tariki's protégé, Hisham Nazer, struggled valiantly to keep up standards. Tariki had staffed the organization with a cadre of professionals, many of them Palestinians who were given Saudi nationality, designed to parallel ARAMCO's organization. "His reason, according to al-Dirhali, was to be able easily and smoothly to accomplish a transition from ARAMCO management" to government control.

Dirhali reported a serious decline in morale after Tariki's flight, particularly among the Palestinian cohort. He placed the blame on the new minister, Yamani, who he said had reversed Tariki's commitment to hiring and promoting men on merit in favor of the "more traditional approach." Al-Dirhali went further. The one-time protégé and member of Faisal's staff denounced the prince as a dictator, "and the only dictator who would be acceptable is Gamal Abd el-Nasser." He blamed Faisal for blocking constitutional reform, "in which . . . the rights of people would be clearly defined." It was Faisal's "greed for power" that set back the changes that had been under way, he said.[87]

O'Dea argued that ARAMCO faced a dilemma. The firm had to cooperate with Faisal's government but in a way that did not alienate what he called "advanced opinion," because—and here O'Dea echoed Shimon Peres and many others—"sooner or later, gradually or abruptly, advanced opinion will possess the future. Such enlightened opinion, which ARAMCO's presence did so much to bring into being, often sees ARAMCO's relations policies as anachronistic in the present circumstances, and suspects them of compromising with the forces retarding progress." The main proof O'Dea offered for the proposition was the weakness of the government that Faisal had empowered against the one that ARAMCO had struggled with for so long. "There are many who feel that the earlier cabinet with Jamjoun, Nasif, Tariki, Dabbagh and others showed much more promise, and that Faisal fears to appoint anyone other than yes men. . . . The government doesn't seem to be responding to challenge with anything remotely resembling adequacy."[88] While this reads now like a breath of fresh air after all the overheated, stultifying testimonials to Faisal's great leap forward, we can begin to see the real problem if "advanced opinion" got hold of the O'Dea report. It gets worse.

O'Dea wrote a long section on the experience of ARAMCO "on the ground"—this is my term, not his—both in terms of its undeniable catalyzing effects in the Eastern Province but also in terms of "all the usual problems that accompany culture contact between two different kinds of people." This latter topic was the area of O'Dea's expertise after all. Admirably frank, succinct, and, importantly, right, he noted that although "Americans were foreigners and the Saudis were citizens of a politically independent country, the Americans were in an effective sense those who occupied the positions of power." Saudis resented it. For a long time,

> the relationships between Americans and Saudis were openly expressive of this distinction. The separate residential areas of the three main employee

groups . . . rendered visible an operational distinction with far reaching social consequences. Today, because of some important improvements, especially in their personal relationships between Saudis and Americans, but also the residence of a few Saudis in Senior Staff camps and the presence of some 100 Saudis in Senior Staff positions, the situation has improved. Yet there is reason to believe that attitudes formed in the earlier period, both among Americans and Saudis, have not been dissipated. Moreover, there is a general resentment of Americans all having Senior Staff status, and of their residence in a fenced-in area, items which probably have more symbolic than real significance. It appears to symbolize to Saudis what are still felt to be inequitable aspects of Saudi-American relationships.[89]

O'Dea reported the results of a survey of Saudi employees and others in the Eastern Province, which illustrated the ambivalence of those working for the company. Some workers, for instance, singled out the company's schools for praise, but O'Dea argued that the Americans generally had trouble in relations with "blue collar Saudis," while even "white collar Saudis sense that Americans feel superior to them." One respondent reported "at present there is a lot of discussion among Saudis of the race problems in the U.S. Every Saudi has a radio now. Some Saudis feel that Americans say they are democratic, but they discriminate in their own country." Another said, "Americans do not understand Saudis. They should be more integrated socially. Americans don't go outside their own homes and they do not go to the homes of Saudis. I feel that many Americans resent Saudis living in Senior Staff camp. Why do they have to feel superior? Why do they want to discriminate?" When pressed for details, the Saudi employee pointed to the use of "mister" for Americans but first names for Saudis, or, worse, the use of the word "boy."[90]

When employees, business people, and others were pressed on what kinds of changes they thought were necessary in Saudi Arabia, they noted several things that Prince Faisal ought to do: remove his family from government, reduce the princes' allowances, create an elected parliament, give people their freedom, develop the non-oil sectors of the economy, build roads (Faisal listened to this one), and commit resources to neglected regions of the country. O'Dea mined one sentiment that holds true until today. "The dislike of the present increase of American military forces in the Kingdom was also expressed. The United States Air Force was brought here to protect the royal family, not to protect 'us.' Most Saudis think this is a bad thing. There have been pamphlets distributed about this."[91]

O'Dea ended his report with a series of recommendations for the near term. He pressed the company to ratchet down its "lofty" claims about a Saudi being eligible to "rise to any position in the company . . . providing only that he demonstrates the necessary capacity." Saudis knew this to be patently untrue. ARAMCO's managers all made "mental reservations in cases of jobs considered 'sensitive' for one reason or another," and thus off limits to Saudis. The real problem was the "considerable lack of clarity about what such jobs really are and why a particular job may be considered sensitive." It was, in short, a device that worked to lock non-Americans out of far too many positions and to keep decision making in the Americans' hands. Saudis resented the intact institutions of the hierarchy inside the camps, including Americans' higher status, the large salary differentials, and the century-old device of separate payrolls, dollars (or gold) for some, riyals (pesos or silver, etc.) for others. The firm needed to address at least some of those practices of hierarchy that remained after having ended the worst of the abuses "of the older imperialist kind."[92]

O'Dea made exactly the same observation as Kimball about the Americans' superiority leading to their isolation, not only from Saudis but from the non-Saudi Arabs, Pakistanis, and European residents in nearby towns, all of whom resented the Aramcons (the westerners, not least, because they had to live less luxuriously along side Arabs). Like Kimball, he said the isolation explained the deep ambivalence the Americans displayed toward their situation and the precariousness they often felt about life inside the company town. He even used the same example, namely the easy profession of faith in ARAMCO's mission—as we saw in the letter McDonald wrote to Mulligan—and the just as easy rejection of it.

> These Americans often express pride in the benefits which ARAMCO's presence and acts conferred upon the region. . . . But as soon as someone raises any point of criticism, an extreme defensiveness is displayed. This is accompanied by a denial of any American or company responsibility for the development of the region. One is told, "ARAMCO is here to make money." Or, "Let's face it, we're not do-gooders, we're here to get oil out of the ground."[93]

It would seem that ARAMCO's managers or some of them were at least as smart as the professors and had recognized that the symbols of American privilege fueled the resentment and backlash. Why else, as O'Dea's report revealed, would ARAMCO's planning team have launched an effort at the end of 1961 to prepare for pulling down the fences and moving Americans into the "natural Saudi communities" such as Dammam, that, the curious nomenclature notwithstanding, the firm had developed in the 1950s? The only other explanation is that it anticipated being forced to by Tariki or other agents of the more powerful and assertive state-in-formation.

ARAMCO's planners came up with some different strategies, and O'Dea and the special study group were to assess the feasibility of going ahead with one or another strategy. O'Dea came back with a qualified but nonetheless negative recommendation. He said the key question was would integration "improve relations between Saudis and Americans by removing suspicion and resentment?" He argued that there were already Europeans living in close quarters with Saudis, and the "mere physical presence did not necessarily make for a high degree of interaction." To dismantle the various American camps would be expensive, since he presumed the housing stock and infrastructure would not only have to be expanded but also upgraded. It would still mean, at least for a few years, a less luxurious style of life than those in senior staff camps currently enjoyed. Social science also showed that proximity often led to the "development of antagonism" rather than "friendship." Large numbers of Americans "would make those aspects of American life which go against the grain of Saudi culture more visible and obvious." They would probably not learn "to speak Arabic with any degree of competence." Why expect them to be any less "insensitive to local feelings and prejudices" or less "psychologically isolated" than they already were inside the camps? Arguably, integration "would increase rather than minimize present problems." Certainly it would not end them. Finally, O'Dea said, it was likely that if the company forced the issue, it would lead many to resign. "In the long run, mass resignations could have a beneficial effect upon the company's work, and if handled wisely could improve functioning with younger blood and even reduce costs," but the initial effect would be to cause considerable dislocation. ARAMCO might encourage some symbolic "small-scale integration" and it might also, more easily, simply move more Saudis into the American quarters, but integration was unlikely to solve the firm's problem.[94]

Bill Mulligan basically misrepresented the O'Dea report, egregiously, it turns out. Were one to want to give him the benefit of the doubt there is the unfortunate fact that he used the exact same strategy at the exact same time to try to keep Stegner's 1956 book from being published. There he had failed, fortunately. O'Dea (or Stegner) would offend Saudis, he insisted, through his exaggerated portraits of Americans as the kingdom's salvation. O'Dea wrote the opposite, as we have seen, and there was only one Saudi the company worried about, and that was Faisal, who became king in 1964. The O'Dea study reported accurately the retreat from the progressives' state-building program, the fears of many that Faisal would respond to Nasser's challenge (and they were correct) not by liberalization but by repression.

Most obvious, the report was plainspoken about the racial hierarchy that operated inside the camps more or less intact a decade after the workers' uprisings. As we have seen, the strikes had been the lever that nationalists (no, power-hungry fanatics, ARAMCO said, driven by irrational fears and personal slights) such as Tariki had used to move the company slowly along history's path. But, as we have also seen, unfortunately, those who had tried before, there and elsewhere, found the resistance to be tenacious.

Mulligan's brief got one point at least half right. It was undeniably true, with the vantage of five years' hindsight, that O'Dea's November 1963 report underestimated Faisal's ability to weather the storm by marrying the old technologies of clientelism with the newer ones of repression, the latter supplied by America and paid for by rising oil revenues. The firm's enemies had been defeated. Men like Tariki had been forced to flee the country. The man who would later write a novel about the rise and fall of the workers' movement in ARAMCO's camps, Abdelrahman Munif, was stripped of his citizenship. Meanwhile, Ibn Muammar, who ARAMCO said made Tariki sound reasonable, would rot in one of the miserable prisons that have served the house of Saud so well for so long.

Ayyam Jamila (Beautiful Days)

A few weeks before O'Dea sent in his report, the U.S. ambassador wrote to Mary Eddy. Her husband Bill had died in Beirut late in 1962 and was buried near his birthplace in Sidon. He had suffered a stroke in 1959 and made his last trip to Dhahran for the company in 1960. He was writing his memoirs when he died, but they were never published. His patron at ARAMCO, the Texaco official who had gone from running the camps in Colombia to become ARAMCO's Washington operative, had paid tribute to his longtime friend and tutor: "I have a feeling of personal loss—he has so long been a pillar in the councils of ARAMCO, and of State and CIA. We shall miss him greatly."[95] Now the embassy in Jidda was paying him tribute as well. "The Embassy's 33-foot cabin cruiser has just undergone a complete overhaul and refitting and is now in the water. It was unanimously agreed that the name for the boat should be the 'Colonel William A. Eddy' as a memorial to the first U.S. Minister Plenipotentiary to Saudi Arabia."[96]

In June 1964, Mary made her own final trip to Dhahran, as a guest of the Bargers. Kathleen and Tom were traveling in the States while Mary stayed with the three young Barger girls. She wrote, as always, about the changes taking place in camp.

> You may have heard how now dogs are allowed in Dhahran and Theresa [Barger] is hoping her mother is going to bring her one. Poor Kath! Bill Owen arrived with a German Shepherd puppy and I wonder how it will stand the heat.
>
> It seems sad to see the Davies house and the Jeyes house empty. Two new tennis courts have been added to those near the school. There are many more Arabs in Senior camp but they seem to integrate with no trouble.[97]

It was the summer that Malcolm X made his famous pilgrimage to Mecca. When he reached New York he promised that he would be bringing the case of African Americans before the United Nations General Assembly and "would compel the United States Government to face the same charges as South Africa and Rhodesia."[98]

That same summer, a group of concerned parents in the Dhahran camp warred with others over a new proposed code of behavior for the seventh, eighth, and ninth grades drawn up by the junior high school and the local Parent Teacher Association. Parents complained about the rise of thievery, vandalism, and petting, but also of disrespect for Saudis, from those in the nearby towns to workers servicing the camps. "Our Saudi bus drivers who handle the school traffic must be direct descendents of Job for they take considerable abuse. They are screamed at, called names, among the lesser epithets is 'Dirty Arab' and on occasion physically abused. The school knows this situation, the community knows, but it still continues."[99]

There were a few others signs of trouble that summer. Saudi workers were beginning another boycott of company facilities. A Government Relations official, Malcolm Quint, reported on what was driving the unrest.

> Running through my discussion with these employees was the constant theme that the company was not concerned with the welfare of its employees. One complaint repeated most often concerned wages. All indicated that they felt wages had not kept pace with expenses and the benefits they received—which

the company publicizes to the entire world—do not compensate. . . . Mr. Barger is a Jew, in the sense that he is tight-fisted and bakhil (a miser).

The men protested that they were increasingly given "coolie work," and contrasted their miserable conditions with those of the Americans and their many privileges. The average Saudi worker had no future with the firm. "The company is not interested in them and their lot."[100]

The ARAMCO Americans could hardly control the future or even the story that Saudis would begin to tell about them. In al-Rass, a town in al-Qasim, the province north of Riyadh, a poet spoke of the hatred for the men "with hats who smoke cigarettes that blacken their teeth." Another, who once worked for ARAMCO in Dhahran and Ras al-Mishab, tells how "he did not like to live in the time of which even the wolf expressed his hatred and in which crooks and women are considered superior, while valiant men have lost respect."[101]

Then there is the tale still being repeated today, in the United States, of Saudi Arabia, struggling to make the transition to the twentieth (now twenty-first) century. The one I have told here is of an oil company, ARAMCO, struggling to do the same.

AFTERWORD

"So tell me what you're doing here?" Renee asked again.

*"I'm researching a book," I said. "I'm an historian, Renee. I
specialize in relatively esoteric subjects. Modes of Diplomatic Deceit,
Methodologies for a Late Millennium Corporate Self-Question, stuff
like that."*

*As I'd hoped, her eyes had grown glassy with the effort of feigning
interest. . . .*

"What I'm researching is Kurash in 1958."

"Whatever for?"

*"To understand what it was like to be a part of the American
overseas subculture during the cold war."*

"You expect people to buy this book?"

Henry Bromell, *Little America*

When the assistant secretary of state, Dean Acheson, briefed the new presi-
dent, Harry Truman, for the first time in May 1945 on the importance of
Saudi Arabia to America's national security, he came armed with a visual aid.
Acheson spread out a map on Truman's desk. He had used the same one to
fill in key senators less than two weeks earlier. The map showed the United
States with the oil-producing countries of the Middle East superimposed over
it. Acheson's pitch was that Saudi Arabia was not as distant or different from
America as imagined. The United States had a vital interest in Saudi Arabia's
vast reserves and hence in Ibn Saud, who had racked up debts and needed to
be bailed out. Truman bought the argument and agreed to go ahead with the
Dhahran Air Base.[1] A few years later, ARAMCO began to use a variant of its
own in a new publication, *Middle East Oil Developments*. A map of the physical features of the Middle East, stretching from Libya across to Iran, had a

transparent sheet attached with an outline of the United States that the reader could lay on top of the Middle East. San Francisco fell between Bengazi and Tobruk. Denver lined up with the Saudi-Jordanian border. Chicago was close to Bushire. Mecca was somewhere in Texas.

This book does something like those maps did. It brings the history of the two places together, telling them both as part of a single, bigger story, truer in the end than the ones that are typically told about either place on its own, the ones that treat such places and moments as special, autonomous, or as we started out saying, exceptional. The ones invented at the start of the Cold War. The story of the ARAMCO concession runs about fifty years, from the time Karl Twitchell opened negotiations with Ibn Saud and his finance minister to the phased nationalization of the owning firms' assets beginning in the early 1970s and ending in the late 1980s. I covered the first thirty years here. It was only about fifty years earlier that large mining companies first became powerful agents of incorporation into the world economy. That's the story told in Part 1, the Nearest Faraway Place.

The setting of the typical extractive industry differs from that of the factory. Copper, like oil, was found in remote, isolated places. Firms both had to import labor and build, alone or in cooperation with others, the basic infrastructure that we associate with modern municipalities—housing, streets, power, water, security, and so on—in order to bring the ore to market. A firm that survived the market booms and busts—a handful among the hundreds that spent their capital and disappeared—typically will commemorate this pioneer developer role when it commissions histories and publishes anniversary issues of its in-house magazine. In the setting sun of memory it might even appear as if the fact of a once-thriving camp or town or way of life made the firm special, different from other firms.[2] Don't let the heft of a coffee table book fool you.

Manufacturing rather than mining firms built some of the best-known examples of the turn-of-the-century "new" company town ventures. Pullman, Illinois, was founded by the railroad car manufacturer, George Pullman, and Indian Hill in Worcester, Massachusetts, by the Norton Company, then the country's largest producer of industrial abrasives. In the South, mill owners updated the classic New England village model of the previous century, setting their factories close to fast-running streams and supplying low-rent housing for an agricultural labor force that needed to be turned into workers.[3] The paternalism that characterized early New England milling villages was echoed as well in the emerging welfare work movement—the effort by large, capital-intensive firms to defeat the labor movement in pre–New Deal America. Cleveland Dodge and his Arizona *compañero* Walter Douglas among many other businessmen believed, or wanted workers to believe, that all of them were "members of a family," as another titan, Walter Teagle, often said about his own tribe, Standard Oil of New Jersey, now known as Exxon Mobil.[4] What ultimately was a choice or strategy for manufacturers was more nearly a necessity for mineral producers.

The argument does not clear much ground for a defense of the idea that there exists a unique American tradition of welfare-work mining enterprise, with firms like Phelps Dodge, Doheny's Pan American Petroleum, or ARAMCO occupying a position somewhere between the Board of Foreign Missions and Point Four, the forerunner of the Agency for International Development, on a timeline of international humanitarian assistance programs. The two pieces of the production process that some historians seem genuinely mystified by—a company's need to lay power lines and build roads, along with the so-called benevolent stewardship of the camp's denizens—were found across the frontiers of the world mineral market. Ideas traversed the Atlantic alongside capital. Engineering and other professionals were vectors. To underscore what should be most obvious, extractive industries confronted and so are identified by a common set of structural features. Firm size likely explains more of the variance in worldwide workplace conditions than does the patrimony of the absentee New York, Boston, Paris, Brussels, Edinburgh, etc., owners. And what, in the end, might distinguish American from other mining enclaves is that here "we" came to call them camps.

Then there is the paradox. Building clubhouses and golf courses, running power plants, organizing parades, and staffing hospitals all purportedly show that a company like Phelps Dodge stood apart from most other global mining enterprises, and so these elements of the social landscape of "old" Arizona and New Mexico are memorialized. The general office of the Copper Queen mine is now the Bisbee Mining and Historical Museum. Needless to say, the dual racial wage structure and other stark features of the Jim Crow system in place in the camps are not commemorated there, and the company-bought professor is always careful to ignore these parts of the story. But the alternative strategy, the recourse to "custom" to explain (away) the firm's heavy investment in racism, would seem to work against the idea of Phelps Dodge as special or unique. Indeed, it makes little sense.

What was true about the world of work in the copper mines of late nineteenth-century Arizona was true about the oil fields of Dhahran in the mid-twentieth century. The best history of mining and manufacturing "model" towns and camps, *Building the Workingman's Paradise*, argues that the Jim Crow settlements together with their welfare work norms started to disappear in the 1920s. The argument though fails to consider the famous expansion of U.S. enterprise abroad in the same decade and thus the spread of these "Little Americas" across the world mineral frontiers. The war to end the racial wage system would be "reenacted" in dozens of places, including in Saudi Arabia.

The work it took and the costs men and women paid in the course of challenging injustice and inequality in the mines and camps from Sonora to Dhahran is the even bigger, unresolvable paradox in the tale the company historians tell. The benevolence of the corporate Great White Fathers went only so far, apparently, and each strike exposed anew the border beyond which benevolence and oppression, paternalism and authoritarianism, are no longer distinguishable. An analytical boundary parallels this experiential one. Ultimately, one stands on the side of producing a causally adequate account of the historical process or one crosses the line to a place where artful rehearsal of the beliefs, fears, anxieties, hopes, justifications, and rationales of the company and its townspeople is the time-honored custom.

The dilemma for those who do cross over becomes clear with just a little reflection. No company official *ever* admitted to workers acting justifiably in their confrontations with the institutions of hierarchy. Defenders of hierarchy never in fact recognize themselves as such. Men and women on the other side of the barriers it seems are *always* selfish, wrong, misguided, the ones really grasping power and exploiting privilege. The list of outside, controlling interests is long. Wobblies; German or Bolshevik projects; "foreign" ideas; politicians; Mexican or Venezuelan, Chilean, Persian, and Arab nationalists are the explanation for a company's troubles.

Strange, no, that these bundles of beliefs in particular are routinely reached for in any post–World War II "balanced" history of the great corporate power? The very same agents also usually held quite striking beliefs about the superiority of their own white race, about mixing, miscegenation, and the threat of Anglo-Saxon decline, and about the naturalness of privilege. The rich vein of beliefs about the inferiority of the men and women whose lives they controlled ran even deeper. It is, however, the rare company historian who brings this archive to the surface or who is willing to tell us much about the technologies of hierarchy on which the mining firms depended.

During the long decade of the 1950s, the Saudi ruling family, members of the political class, and, above all, the workers in different and at times competing ways pushed at the margins against the hierarchies of the camps, the world oil cartel, and American hegemony. That's the most basic story told in Part 2, Desire's Empty Quarter.

Faisal Ibn Abd al-Aziz had conducted foreign affairs in a government that tried to test the power of the international petroleum cartel and to capture the rents from transporting oil, and then it hadn't been the Jews that worked to thwart him. His brother, King Saud, expelled the American Point Four mission and, later, the U.S. Air Force, moves that Faisal had long been agitating for. More generally, the family started to challenge American preferences for the region's security regime and certainly its place in the international petroleum order. Faisal joined the campaign against the Baghdad Pact and built Saudi Arabia's alliance with Egypt precisely when Egypt was challenging its own position in the American-dominated order. For the first but by no means the last time the policies pursued by the house of Saud led to the branding of the government by the CIA as a disruptive force in world affairs and threat to vital U.S. interests.

Then, as now, Americans swung back and forth between two parables about the Saudis. Recall how in 1957 Saud had pledged allegiance to the Eisenhower Doctrine, which gained him his *Time* magazine cover, a trip by Ike and Dulles to National Airport to greet the bold, brave modernizer whose wise policies promised to take Saudi Arabia out of the . . . well, you know the rest now by heart. ARAMCO's vice presidents swore that it was all true. The *New York Times* reported it. In fact, it was the same *Times* reporter, Dana Adams Schmidt, who might as well have pulled the old story out, changed some names, and refiled it when Faisal arrived in Washington to meet John Kennedy in 1962.

The Saud-as-reformer talk was just as remarkably and quickly forgotten. He had, foolishly, taken to heart the Republicans' talk about promoting him as the Arab world's natural rival to Nasser, a role that, not surprisingly, some in the Democrats' camp imagined that Faisal was poised to play. The real problem was that Saud tried to act like a little regional superpower and undermine the Syrian-Egyptian union. The American-backed coup in Damascus failed, and it almost led to war. Saud's follow-up plot did not fare much better. Yet historians hardly remember the former debacle, not least because no U.S. newspaper at the time would give the time of day to the Syrians' revelations. Saud's plot, by contrast, would become a main proof that he-who-in-1957-was-making-Saudi-Arabia-modern was one year later a misguided, dull-witted drunk. The racism that licensed this type of simple-minded contrast might be a thing of the past, but the fact's half-life is long, and books are being printed now that repeat it like a mantra.

We might consider too the reinvention of Faisal as the true voice of reform in the kingdom. First, of course, the reforms that he was originally credited with spearheading were his putative efforts to rein in spending, force his family slightly off the dole, and so on. Saud's celebrated modernizing reforms in 1957, ministry building, road and school construction, the kingdom's first university (a total waste), industrialization plans, and so on, were a year later America's main proof that the king's wastefulness was threatening the solvency of the state! It was Faisal's opposition to development spending that led to the break with many erstwhile close clients, beginning with Alireza, and that drove the developmentalists to support Saud. Needless to say, it was Saud's short-lived 1960 government, even now the most progressive ever seen in Saudi Arabia, that put forward the plans and projects that Faisal later and cynically permitted the sycophants, including some western journalists, to claim as his own vision for Saudi Arabia's modernization.

Abdallah Tariki, who did more than any other Saudi to develop the state's administrative capacity and, against all odds, bring the American oil company under a modicum of Saudi control, exposed the fault lines in the Americans' wild reinvention of the crown prince as modernizer. ARAMCO's local Cairo representative had invited him to lunch in December 1962 together with George Rentz, the firm's Arabist who had fled Dhahran for Cairo, and a U.S. embassy officer, who sent a report of Tariki's views back to Washington.

> Saud was weak and corrupt but not basically malevolent. Faisal was the great deceiver. He had deceived Nasser into forgiving Saudi Arabia for the assassination attempt 1958. . . . Faisal has now achieved the supreme deception of all. He had convinced President Kennedy that he, Faisal, was genuinely interested in reform and was the best hope for Saudi Arabia. Faisal was as corrupt as Saud, and was interested not in reform but only in persuading the U.S. that reforms were actually underway while he pursued his usual do-nothing policy. The U.S. is in

for a rude awakening. It was time that the U.S. government tried to get in direct contact with "the people" of Saudi Arabia.[5]

I am not as compelled as Tariki was by the idea that Kennedy and his advisors had been taken in. Still, we might follow his lead and reconsider the earnest testimonies to Faisal's austerity, abstemiousness, and overall wise stewardship of the country's wealth, as when he tried to compel Nasser to release the hundred or so properties owned by the royal family that had been taken over by the Egyptian state, including his own Garden City villa, which Nasser gave to the Algerian FLN for its new embassy. It may be less fun but even more useful to consider the one distinction that matters most between the late 1950s and early 1960 governments, on the one hand, and the one that Faisal headed in 1963, on the other, when he made his peace with the American-dominated order.

In 1959, as Tariki, Ibn Muammar, and other would-be developmentalists were growing bolder, the great University of Wisconsin historian William Appleman Williams had just come out with his famous study of early twentieth-century U.S. foreign relations, *The Tragedy of American Diplomacy*. He argued that across the many places that American capital and the American state had begun to penetrate and influence, the boundary between reform and the challenge of privilege and hierarchy was sometimes hard to make out. Each time American officials found themselves at that border, they retreated, compelled apparently to defend the privileges that firms or local clients or both depended on. It was the threat to privilege that mobilized the royal family, ARAMCO, and the U.S. state in the 1950s, and it was Saud's failure to recognize the danger that did him in. Faisal proved himself more adept and, needless to say, much more vigilant.

Defense of privilege is of course never viewed as such, save in retrospect, through dogged detective work in private records and the like. Instead, across the twentieth century, as we have seen, workers might be dismissed as captives of foreign ideologies and agents, or out to aggrandize power for themselves. Their supporters in government were power hungry too. Americans called them dangerous radicals, and not least because racism was so entrenched, they often added ignorant to the list, insisting that someone like Tariki remained unschooled in the technical operations of markets and hierarchies. ARAMCO's officials always knew better what the Saudis wanted or needed. Reporters readily wrote what the firm's agents or State Department desk officers insisted was true. The oil company also liked to argue that the international oil cartel was a myth propagated by its enemies at home and abroad. And, of course, its executives ritualistically asserted to every visiting congressman and in every hearing room in Washington that every Saudi ruler was wisely leading his people out of the past and into the future.

I have tried to tell an alternative history here, one that I uncovered over many years working in the archives of the company, the diaries of its employees, the records of the American agencies with which ARAMCO and its parents were entangled for so long, the letters of the historian the company hired, and documents rescued from Saudi Arabia by the family of a patriot, one of many whose lives Faisal destroyed. In retrospect that story seems so obvious, predictable, banal even. Yet, for many, it was and is unimaginable, to judge by what they began to say and write about the place and time just a few years later and continue to say and write today. What was amnesia for some, artful public relations campaigns for others, and advanced theory for a handful of professors is also a part of the Cold War. Above all, it is a testimony to the power of the norm against noticing.

There is one final thing to say about that time and place when and where one U.S. firm exercised a kind of power over the lives of the people of the kingdom that is hard to imagine now. They are long gone, although some still believe that the oil firms operate in Saudi Arabia, decades after ARAMCO's nationalization, as if it were still 1950, with U.S. companies "dominating" Saudi Arabia. The truth is that production and refining of oil have been in the hands of the Saudi state for almost three decades and closed to foreign investment.

It is not that the entanglement between the two countries ended (or that inequality had been transcended); far from it. Arms sales, bases, military assistance, petrodollar recycling, petrochemical projects, construction boondoggles, joint covert ventures across Africa and Asia, and negotiations over oil prices are a few of the hallmarks of U.S.-Saudi relations from the 1980s until now.[6] The oil industry itself is just another national sinecure where the Saudis are free to rebuild hierarchy anew in Dhahran. One day we might have histories of the Shiites in the Eastern Province and of the migrant laborers from Yemen, Egypt, and elsewhere who worked in the oil and construction industries during the second wave of expansion, after 1973–74.

Outside the Gulf it may be a different story, but we need to know more than we now do. On the newest oil frontiers of global capitalism and inside its oil camps, we are apt to find not only echoes of the ideologies first invented in places like Dhahran (and Madison Avenue), but the din of a building frenzy, with firms pressed between the competing demands of getting production going and supplying a labor force drawn from a half dozen or more countries. We know that mining firms use ethnic cleavages to weaken labor organizing. There will be visible gross disparities in pay, benefits, and amenities among classes of employees that just happen to divide along national and ethnic lines. The innovation that ARAMCO gradually adopted over the 1950s and 1960s, once its nascent labor relations division recognized that the ordering effects of "skills" and "races" were more or less identical, is now the convention. The Americans' quarters will still look strikingly different from those occupied by Azeris in Baku.[7]

The secular shifts in the world economy since decolonization and the first wave of oil nationalizations in the 1970s mean that oil companies have few illusions about the security of these newest concessions and so little incentive to add to their labor costs or to act other than to get what they can get away with. The "maximum exploitation" option that Pendleton identified in his policy memorandum back in 1958 has lost none of its force, even if the history of ARAMCO in Saudi Arabia reminds us that politics can undo, at least in part, the order that ideas about markets, merit, culture, and race represent and sustain. In Dhahran, some of the very same Saudis who, according to the firm's experts, in the 1940s allegedly did not want western creature comforts are not only living in company-built villas but are now in charge of the company.

I spent a day in 1996 in the old American Camp, which is still called "Camp" in fact by the residents who live and work there. I was desperate to try to tie past and present together, like those maps that try to depict the United States and Saudi Arabia as if they were one place. There are, needless to say, few signs indicating that the events that I was then just beginning to piece together from the archives had ever taken place. No plaques mark the spot where the workers began the first demonstrations or where they met King Saud's caravan with the banners hailing him and condemning U.S. imperialism or where they tore down the fence posts in Saudi Camp. The government had Saudi Camp itself torn down in 1979 to allow for the expansion of the University of Petroleum and Minerals at Dhahran. ARAMCO's other names for the Saudi reservation, General Camp and al-Salama, had never caught on.[8] Presumably, Americans ceased arguing that universities for Saudis were a waste of resources.

Along one of American Camp's winding roads, in a little bungalow built in the 1950s and now partially hidden by tall trees and hedges, an American

Heritage Center has been set up, just like in Bisbee, but on a much more modest scale. It is a little sad, in fact. A few showcases display such memorabilia as ARAMCO employee badges, twenty-five-year pins, school yearbooks, old photographs, and the like. One part of the house is outfitted in *majlis* style, with rugs on the floor and cushions running along the walls. I sat there and watched the ARAMCO film, *Island of Allah*, for the first time. The shrine is a volunteer effort if I recall correctly. Let's just say that it was not a high-traffic spot that day. Very little else exists to mark the era and the events that I have written about here. True, women can still drive cars inside Camp. The fence still surrounds it.

The other Malone Fellows and I had dinner at another landmark, Steineke Hall, on our first night in Dhahran. I talked quietly with one of Saudi ARAMCO's top officials, telling him about what I had been learning in the State Department records and the Mulligan papers. He nodded, smiled, and said that when he had first began working as an office boy the toilets and drinking fountains in Camp all had signs designating some American only, others for Arabs.

Harry Alter, the spanking new Government Relations analyst in 1958, was our guide on the trip. It was a piece of luck for me. He was retired and living in Washington, D.C. He stood up in front of the bus as we wound our way from Camp to Dammam. "ARAMCO learned early on," he preached, "about how to operate as a good citizen in the kingdom, building roads, schools, hospitals, housing, training Saudis." I pressed him to be clearer about what he meant by "early on." The firm began exploring for oil in 1933, while the strikes and other challenges of the mid-to-late 1950s that ARAMCO was responding to had taken place a quarter century later. The reforms advocated as a result, as we have seen, and implemented only partially and haltingly over the next years, came at a point when the company had, it turns out, roughly a dozen years to go before the Saudi government began the takeover of its assets.

In the years since, the myths that Alter and his colleagues invented about ARAMCO have continued to be conjured when needed. The destruction of the Twin Towers in September 2001 guaranteed that the true believers would recite them all once more as if they were an incantation for use against Bin Laden and those that friends of Saudi Arabia often believe stand just behind Al Qaeda in the ranks of enemies of America's real national interest, namely the Israelis and their myriad supporters.

I don't know where Nicholas Lemann, the *New Yorker* writer and author of, among other books, the prize-winning *The Promised Land: The Great Black Migration and How It Changed America*, stands on these matters. He is not a captive of the norm against noticing, certainly. Yet, in April 2002 he published a two-page "album" titled "The Way They Were" in the *New Yorker*. It included five photographs by the renowned documentarian Harold Corsini, who had begun photographing Standard Oil operations during World War II. The images, from 1947, were of the swimming pool in American Camp, the Saudi government railroad in Dammam, the office doorway at the experimental farm in al-Kharj that Lemann says "Aramco constructed in the vain hope of introducing agriculture to Saudi Arabia," a market street in al-Khobar, and Saudis walking home from work in the fields. Lemann, who has since been appointed dean of the Journalism School at Columbia University, wrote the captions and a short text.

After I read the article, I sent Lemann a letter, pointing to the errors of fact and more in virtually every caption and every paragraph of the text. We have read them all before. ARAMCO did not introduce agriculture to Saudi Arabia. The firm had been forced to run the model farm for the king. It charged the palace for this service. It was not ARAMCO's rail line. The Saudi government owned it, and had forced the company to fund it out of future royalties, after ARAMCO failed to convince Ibn Saud to forget the idea. The Saudis worked as office boys and the like, but Americans did not trust them as servants. Jim Crow ruled in Dhahran just like in the South from where all those migrants Lemann wrote about had fled. The first "demonstrations" were in 1945 not in 1967. The firm did not introduce modern medicine. Missionaries did. You get the point.

Lemann sent back a nice note one month later.

> Thanks for your letter. I relied on the captions written at the time by Harold Corsini, the photographer, and also on two outside experts, Daniel Yergin, author of *The Prize*, and Teresa Barger, an economist at the World Bank who grew up in Saudi Arabia as daughter of the president of Aramco. But each of these in his or her own way would be inclined to take a rosy view of Aramco. I wish I'd known about you beforehand. But I appreciate your letting me know about this after the fact. Thanks again.

Everyone knows Daniel Yergin. Teresa is the little girl who was hoping for the puppy when Mary Eddy was visiting Dhahran in 1964. For the rest of us, nonexperts, it is still not too late.

ACKNOWLEDGMENTS

The book cost a lot of money to write, so thanks go first to the funders, the American Council of Learned Societies, the American Heritage Center in Laramie, Wyoming, the Alice Higgins Center at Clark University, the Humanities Forum at the University of Pennsylvania, the International Center for Advanced Study at New York University, the National Council on U.S.-Arab Relations, and the Rockefeller Foundation.

My editor, Kate Wahl, and her colleagues Richard Gunde of UCLA and Judith Hibbard, made the book better and made finishing it a pleasure.

I am grateful to Aramcons who agreed to talk with me over the years, including Harry Alter, Rebecca Copeland, David Dodge, Baldo Marinovic, Barbara and David McDonald, Mary Norton, Brock Powers, and Peter and Ellen Speers.

I am grateful even more than I can express here to Maggie Browning for her years of love and mercy.

I have learned so much from so many during the time I have been working on *America's Kingdom*, but the problem with taking a while is that some of you may have forgotten and will be surprised to find your names here. I haven't. Thank you Steve Aaron, Jeremy Adelman, Cary Akins, Lisa Anderson, Abd al-Muhsin Akkas, Shiva Balaghi, Joel Beinin, Cathy Boone, L. Simon Bromley, Carl Brown, Jason Brownlee, Tom Callaghy, Nate Citino, Donald Cole, Simon Davis, Eleanor Doumato, Abdelaziz Ezzelarab, Abd al-Aziz al-Fahad, Alexa Firat, Tom Ferguson, B. J. Fernea, Kevin Gaines, Greg Gause, Irene Gendzier, Kim Gilmore, Michael Gilsenen, Linda Gordon, Ellis

Goldberg, Marie Gottschalk, Janette Greenwood, Bassam Haddad, Abdellah Hammoudi, Debi Harrold, Geof Hartman, Vicky Hattam, Steve Heydemann, Nubar Hovsepian, Anita Issacs, Sana Jaffrey, Toby Jones, Bill Jordan, Paul Kaiser, Ibrahim Karawan, Persis Karim, Resat Kasaba, Alex Kitroeff, Philip Khoury, Yossi Kostiner, Lynne Lees, Michael Libertazzo, Walter Licht, Doug Little, Zachary Lockman, Miriam Lowi, Anna McCarthy, Karen Merrill, Joel Migdal, Tim Mitchell, Fareed Mohamedi, Haifaa Muammar and family, Mary Nolan, Anne Norton, Gwenn Okruhlik, Roger Owen, Marsha Posusney, Eve Trout-Powell and Tim Powell, Dan Raff, Haggai Ram, Regivam Ram, Madawi Al-Rasheed, Adolph Reed, Ashley Salisbury, Mustafa Kamil al-Sayyid, Bill Sewell, Peter Sluglett, Susan Slymovics, Rogers Smith, Joe Stork, Ted Swedenburg, Josh Teitelbaum, Mary Ann Tetrault, Bob Tignor, Peter Trubowitz, Ed Webb, Lisa Wedeen, Mary Wilson, and Marilyn Young.

Although many have read parts or all of it, it is Chloe Silverman's reading the book that matters most. I am grateful for the deal we have, trading her wanderlust for my little bitty fire. "It's a harsh world, sure can be a cold place / If you don't have someone to kiss your face / When it doesn't make sense."

I wrote this book in lots of places over the years, and need to thank Apple Computer for building the G4 Powerbook and the iPod, and for writing iTunes. The playlist has changed some from 1995. Thank to Miles, Ferron, Charlie Haden, P. J. Harvey, Aimee Mann, Laura Nyro, and Lucinda Williams, but above all to Brian Wilson and Neil Young, who are the two inspirations for the book.

New Year's Day, 2006

NOTES

Foreword

1. Hartz, as first quoted and then described by Daniel Rodgers, "Exceptionalism," in Anthony Mohlo and Gordon Wood, eds., *Imagined Histories: American Historians Interpret the Past* (Princeton, NJ: Princeton University Press, 1998), 21–40: 29.

2. See Robert Vitalis, "The Closing of the Arabian Oil Frontier and the Future of Saudi-American Relations," *Middle East Report* 204 (1997): 15–21.

3. Quoted in Sean Wilentz, "America Made Easy," *New Republic* (July 2, 2001): 3.

Chapter 1

1. I pledged fealty that day to "Chatham House rules," meaning statements or views could be reported but identities could not.

2. Two smart social science versions of the conventional story are Nadav Safran, *Saudi Arabia: The Ceaseless Quest for Security* (Cambridge, MA: Harvard University Press, 1985), and Iliya Harik, "The Origins of the Arab State System," in Giacomo Luciani, ed., *The Arab State* (Berkeley: University of California Press, 1990), 1–28. I am not alone in challenging the ideas that a causal logic links the "first," "second" and "third" Saudi states, and that empire did not matter to the outcome of twentieth-century wars of conquest and to the institutions that Ibn Saud inherited. I would cite the Tel Aviv school (my term) as exemplary. That a host of scholars, starting with Chaudhry and ending with Bronson, don't see fit to reference the works is noteworthy but a bit depressing as well. See Joseph Kostiner, *The Making of Saudi Arabia, 1916–1936: From Chieftaincy to Monarchical State* (New York: Oxford University Press, 1993); Joshua Teitelbaum, "The Rise and Fall of the Hashimite Kingdom of the Hijaz, 1916–1925: A Failure of State Formation in the Arabian Peninsula," Ph.D. dissertation, Tel Aviv University, 1996, revised and published as *The Rise and Fall of the Hashimite Kingdom of Arabia* (New York: New

York University Press. 2001), and Sarah Yizraeli, *The Remaking of Saudi Arabia: The Struggle between King Sa'ud and Crown Prince Faysal, 1953–1962* (Tel Aviv: Moshe Dayan Center, 1997). Two other new and valuable, needless to say non–Tel Aviv–based sources are by Madawi al-Rasheed, *History of Saudi Arabia* (Cambridge, UK: Cambridge University Press, 2002), and Alexei Vassiliev, *History of Saudi Arabia* (New York: New York University Press, 2000).

3. Robert H. Jackson, *Quasi-States: Sovereignty, International Relations and the Third World* (Cambridge, UK: Cambridge University Press, 1990).

4. See G. John Ikenberry, *After Victory: Institutions, Strategic Restraint, and the Rebuilding of Order after Major Wars* (Princeton, NJ: Princeton University Press, 2001).

5. Rodgers, "Exceptionalism." The subsequent cites to Rodgers in the text are all from this essay. Google "exceptionalism" and you will find a story about Alexis de Tocqueville first using the term in the early nineteenth century, which is true, but that fact tells us more about the fondest hopes of the believers than about the emergence of this particular paradigm after World War II and its embrace by professional history writing and those social science frameworks that look like history writing. The latter "theoretical schools" have different names in different disciplines and in different sections of the professional associations, including American political development, historical-comparative and historical-structural analysis, dependency theory, historical sociology, and so on. These schools and approaches are the ones that wrestled with the problem of absences, why America had no socialism, why the Third World had no bourgeois revolutions, why Germany was not democratic (then), or why the Middle East is not democratic (now).

6. See Robert Vitalis, "Birth of a Discipline," in David Long and Brian Schmidt, eds., *Imperialism and Internationalism in the Discipline of International Relations* (Albany: State University of New York Press, 2005), 159–82.

7. Rodgers, "Exceptionalism," 22–23.

8. We might add to this list a young historical-oriented political scientist, Ralph Bunche, and a young historical-oriented sociologist, Oliver Cromwell Cox, who wrestled with the dialectic of race and class in ways that prefigured the world systems analysis of the 1970s and 1980s, using terms that we have not yet escaped. We ought to add as well, though, that Bunche, Cox, and, most remarkable of all, Du Bois, would be dropped from the professors' reading lists for at least a generation. Barrington Moore could publish his *Social Origins of Dictatorship and Democracy* in 1966 on America's distinctive road to modernity, rooted in an argument about the Civil War and its consequences, and yet not cite Du Bois's masterful *Black Reconstruction*, written thirty years earlier.

9. Dorothy Ross, "The New and Newer Histories: Social Theory and Historiography in an American Key," in Molho and Wood, eds., *Imagined Histories*, 85–206, but see in particular, 89–90.

10. Rodgers, "Exceptionalism," 29.

11. The oil firms' changed their corporate identities many times during the twentieth century, and I am using the names they are best known by rather than the ones that were in use when the concession with Ibn Saud was first signed. Chevron was then Standard Oil of California (Socal). It sold a share of the concession to the Texas Company (Texaco) after it discovered commercial quantities of oil, because Texaco needed crude and Chevron needed Texaco's outlets around the world. Their operating subsidiary in Saudi Arabia was originally known as the California Arabian Standard Oil Company (Casoc), which was renamed the Arabian American Oil Company (ARAMCO) in 1944. Three years later, Exxon, then known as Standard Oil of New Jersey, and Mobil, then known as Standard Oil Company of New York, provided the capital needed for expansion in return for ownership shares in ARAMCO. ARAMCO is now wholly owned by the Saudi government. Chevron and Texaco merged their operations in 1999 under the name Chevron. The other two partners merged their operations too and now operate under the name Exxon Mobil.

12. Control of oil at a reasonable price is frequently described as a key component of U.S. foreign policy long since the firms themselves were nationalized. See for example F. Gregory Gause, *Oil Monarchies: Domestic and Security Challenges in the Arab Gulf States* (New York: Council on Foreign Relations, 1994).

13. As Patty Limerick, who is probably the best known of the historians in the now old controversy, said, the "show is about as revolutionary as if you had a Southern history exhibit, hung romantic paintings of plantations, and then said slavery was a rough business—not a very wild proposition, and the same kind of proposition this show offers about the West." This account is found in Eric Foner and Jon Wiener, "Fighting for the West," *The Nation* (July 29, 1991): 163–66. Readers are welcome to judge for themselves. See the catalog by William H. Truettner, ed., *The West as America: Reinterpreting Images of the Frontier, 1820–1920* (Washington, D.C.: Smithsonian Institution Press, 1991).

14. See the review essay by Michael Adas, "From Settler Colony to Global Hegemon: Integrating the Exceptionalist Narrative of the American Experience into World History," *American Historical Review* 16, 5 (2001): 1692–1720, which examines the progressive exceptionalism tradition among others.

15. Gregory Nowell, *Mercantile States and the World Oil Cartel, 1900–1939* (Ithaca, NY: Cornell University Press, 1994), 6.

16. Fred Halliday, "A Curious and Close Liaison: Saudi Arabia's Relations with the United States," in Tim Niblock, ed., *State, Society and Economy in Saudi Arabia* (London: Croom Helm, 1992), 125–47; Helen Lackner, *A House Built on Sand: A Political Economy of Saudi Arabia* (London: Ithaca Press, 1978), 131–34.

17. One that I learned in graduate school in the early 1980s and credit to the UCLA sociologist Richard Sklar, the then UCLA political scientist Jeffry Frieden, now at Harvard, and the Harvard economist Raymond Vernon, but others no doubt have canonical citations of their own. See Raymond Vernon, *Storm over the Multinationals: The Real*

Issues (Cambridge, MA: Harvard University Press, 1977); Jeffry A. Frieden, "Oil and the Evolution of U.S. Policy towards the Developing Areas, 1900–1950: An Essay in Interpretation," in R. W. Ferrier and A. Fursenko, eds., *Oil in the World Economy* (London: Routledge, 1989); and Richard Sklar in David Becker et al., *Postimperialism: International Capitalism and Development in the Late Twentieth Century* (Boulder, CO: Lynne Reinner, 1987).

18. David Painter, *Oil and the American Century: The Political Economy of U.S. Foreign Oil Policy, 1941–1954* (Baltimore: Johns Hopkins University Press, 1986), 148.

19. Douglas Little, *American Orientalism: The United States and the Middle East since 1945* (Chapel Hill: University of North Carolina Press, 2002), 67–69.

20. So, Douglas Little has Iranian nationalist Mossadegh agitating for "legislation forcing AIOC [British Petroleum] to split its profits with Iran fifty–fifty, as ARAMCO had recently done across the Persian Gulf in Saudi Arabia," but "the British firm refused to budge." *American Orientalism*, 56. The problem is his chronology. The Iranians pushed the fifty–fifty deal first. British Petroleum refused. ARAMCO officials then moved forward in Saudi Arabia, aided by the promise of a tax break inside the United States and fearing what Mossadegh actually started to agitate for, not profit sharing but nationalization.

21. Irvine Anderson, *Aramco, The United States and Saudi Arabia: A Study of the Dynamics of Foreign Oil Policy 1933–1950* (Princeton, NJ: Princeton University Press, 1981).

22. Petter Nore and Terisa Turner, eds., *Oil and Class Struggle* (London: Zed Press, 1981), 72.

23. Toni Morrison, *Playing in the Dark: Whiteness and the Literary Imagination* (New York: Random House, 1993), 9–10.

24. Here we follow the lead of the historians and sociologists in the 1990s who rediscovered what black radical thinkers in the 1960s were arguing about the relationship of the Cold War to the course of American civil rights policies. C. Eric Lincoln, "The Race Problem and International Relations," and Charles Cheng, "The Cold War: Its Impact on the Black Liberation Struggle within the United States." Lincoln's essay is reprinted in George Shepherd, Jr., ed., *Racial Influences on American Foreign Policy* (New York: Basic Books, 1970), and Cheng's in Ernest Kaiser, ed., *A Freedomways Reader: Afro-America in the Seventies* (New York: International Publishers, 1977). Sadly, they both go uncredited by the new wave, which includes Mary Louise Dudziak, "Desegregation as a Cold War Imperative," *Stanford Law Review* 41 (1988): 61–120; Brenda Plummer, *Rising Wind: Black Americans and U.S. Foreign Affairs, 1935–1960* (Chapel Hill: University of North Carolina Press, 1996); Frank Füredi, *The Silent War: Imperialism and the Changing Perception of Race* (London: Pluto Press, 1998); Azza Salama Layton, *International Politics and Civil Rights Policies in the United States 1941–1960* (Cambridge, UK: Cambridge University Press, 2000); and Mary Louise Dudziak, *Cold War Civil Rights: Race and the Image of American Democracy* (Princeton, NJ: Princeton University Press, 2000).

25. The first article on labor struggles in the kingdom by ARAMCO's Italian workers (but only the Italians!) was published in 1986. See Ian Seccombe, "A Disgrace to American Enterprise: Italian Labor and the Arabian American Oil Company in Saudi Arabia, 1944–56," *Immigrants and Minorities* 5 (1986): 233–57.

26. Toni Morrison, "Unspeakable Things Unspoken: The Afro-American Presences in American Literature," in Henry B. Wonham, ed., *Criticism and the Color Line: Desegregating American Literary Studies* (New Brunswick, NJ: Rutgers University Press, 1996), 24, reprinted from *Michigan Quarterly Review* 28 (Winter 1989).

27. Compare the details provided in Thomas Lippman, *Inside the Mirage: America's Fragile Partnership with Saudi Arabia* (Boulder, CO: Westview, 2004), 88–89, with my original account in "Aramco World." Lippman requested the article and later wrote to thank me for it ("Hello again from Washington. I have finished my book about Americans in Saudi Arabia; it is due to come out around Christmas time. I'm indebted to you for alerting me to the memoir of Nora Johnson, which I would not otherwise have found. I'll see to it that you get an early copy of the book if you'll send me a mailing address.") My study meanwhile is never referenced. The norm against noticing apparently trumps the one about a journalist's responsibility. Robert Vitalis, "Aramco World: Business and Culture on the Arabian Oil Frontier," in Karen Merrill, ed., *The Modern Worlds of Business and Industry: Cultures, Technology, Labor* (Turnhout, Belgium: Brepols, 1999): 3–28, reprinted in Madawi al-Rasheed and Robert Vitalis, *Counternarratives: History, Society and Politics in Saudi Arabia and Yemen* (New York: Palgrave / St. Martins Press, 2003).

28. Robin D. G. Kelly, "'But a Local Phase of a World Problem': Black History's Global Vision, 1883–1950," *Journal of American History* 86 (1999): 1045–77; Howard Winant, *The World Is a Ghetto* (New York: Basic Books, 2002); Thomas Borstelmann, *The Cold War and the Color Line* (Cambridge, MA: Harvard University Press, 2003).

29. Sanford M. Jacoby, *Modern Manors: Welfare Capitalism since the New Deal* (Princeton, NJ: Princeton University Press, 1997), 3–5 and 11–34 for the nature of the emerging system of welfare work capitalism designed to thwart both government intervention and union building.

30. On race development, see Vitalis, "Birth of a Discipline," and Jessica Blatt, "'To Bring Out the Best That Is in Their Blood': Race, Reform, and Civilization in the Journal of Race Development (1910–1919)," *Ethnic and Racial Studies* 27 (September 2004): 691–709.

31. Raymond Leslie Buell, "Panama and the United States," *Foreign Policy Reports* 7 (Jan. 20, 1932): 409, cited in Walter LaFeber, *The Panama Canal: The Crisis in Historical Perspective*, expanded edition (New York: Oxford University Press, 1979), 66. LaFeber argues for seeing the Zone as a colony.

32. Michael L. Conniff, *Black Labor on a White Canal: Panama, 1904–1981* (Pittsburgh: University of Pittsburgh Press, 1985), 6.

33. William R. Scott, *The Americans in Panama* (New York: Statler, 1912), 189. Woodrow Wilson's program to enforce Jim Crow in the District of Columbia was still to come.

34. Gary Gerstle, *American Crucible: Race and Nation in the Twentieth Century* (Princeton, NJ: Princeton University Press, 2001), 271. The other quotes in the preceding paragraphs are general enough not to require specific page references. I have extracted the account from a richer, more complex narrative developed in the first seven chapters.

35. Lippman, *Inside the Mirage*, 84.

36. Robert Norberg, "Saudi Arabs, Americans, and Oil," *Saudi-American Forum*, Essay Series 10 (Mar. 20, 2003), www.saudi-american-forum.org/Newsletters/SAF_Essay_10.htm, accessed on Apr. 16, 2003; Lippman, *Inside the Mirage*, 79.

Part 1

1. Phyllis B. Dodge, *Tales of the Phelps-Dodge Family: A Chronicle of Five Generations* (New York: New York Historical Society, 1987), 142–46, 328–29.

2. Photostat of application found in Box 1, Folder 16, William E. Mulligan Papers, Archives and Special Collections, Georgetown University, Washington, D.C. [hereafter cited as Mulligan Papers with the filing information].

3. William Eddy to James Terry Duce, Apr. 5, 1949, Employment of David Dodge, Box 1, Folder 16, Mulligan Papers; my interview with David Dodge, 1995.

4. See, for instance, Andrew I. Killgore, "David Stuart Dodge," *Washington Report on Middle East Affairs* (September 9, 1985): 10.

5. From the conclusion of Eddy's unfinished memoir, "Adventures in the Arab World," ms., Box 1, Folder 1, William Alfred Eddy Papers, 1859–1978, Seeley Mudd Library, Princeton University, Princeton, New Jersey [hereafter cited as the Eddy Papers with filing information]. See as well, "Point $ Without the Taxpayers," editorial, *Los Angeles Times*, Mar. 20, 1951; and Eddy, "Impact of an American Industry," Aug. 13, 1952, Folder 19, Box 15, Eddy Papers.

6. Wallace Stegner, *Discovery! The Search for Arabian Oil*, as abridged for *ARAMCO World Magazine* (Beirut: Middle East Export Press, 1971), v–vi. Both Eddy and his father were born in Sidon, a part of what was called Syria (now south Lebanon).

7. Richard Slotkin, "Buffalo Bill's 'Wild West,'" in Amy Kaplan and Donald E. Pease, eds., *Cultures of United States Imperialism* (Durham, NC: Duke University Press, 1993), 166.

8. For Dodge's doubts about the CIA's involvement in TAPLINE, see William R. Chandler to David Dodge, July 19, 1991, Box 9, Folder 7, Mulligan Papers. As Chandler reports, "Yes we had contacts with [the] intelligence community, but as for having anyone who was on CIA payroll" it was most unlikely. The correspondence was the result of Dodge sending Chandler Doug Little's "Pipeline Politics: America, TAPLINE, and the Arabs," *Business History Review* 64 (Summer 1990): 255–85. "TAPLINE spent freely dur-

ing the 1950s to win the friendship of pro-Western leaders from Riyadh to Beirut, and rumors persisted down through the 1960s that some of its personnel were actually on the CIA's payroll." Dodge expressed surprise when I asked about Eddy and then told him what I had found in the archives.

9. See the interview with Dodge in Robert Kaplan, *Arabists: The Romance of an American Elite* (New York: Free Press, 1993).

Chapter 2

1. The description "American oil colony" is from the *New York Times*, Jan. 26, 1947.

2. Preparations had begun over a year earlier in San Francisco, where the ARAMCO board authorized $100,000 for the visit, although the local planning committee eventually spent over three times this amount, the equivalent of about $2.4 million today. See "Report on the Visit of King Abdul Aziz ibn Saud to the Arabian American Oil Company, January 1947," Box 6, Folder 6, Mulligan Papers.

3. Duce to Sanger, Nov. 14, 1946, enclosing memorandum, Cypher to MacPherson. Dhahran, Oct. 20, 1946, Gifts for the King and Crown Prince, General Records of the Department of State, Record Group 59, 1945–49, 890F.6363/11–1946, National Archives, Washington, D.C. [hereafter cited as RG 59, with filing information, series years omitted].

4. Notes by Mulligan, Folder 39, Ohliger, Floyd W., Box 1, Mulligan Papers.

5. Round robin letter from Eddy, Washington, Jan. 25, 1947, Folder 2, William A. Eddy, 1947, Box 6, Eddy Papers.

6. Memorandum titled "Crown Prince Saud's Official Visit to America, Notes on the Period Monday, January 13, through Wednesday, January 22, 1947," RG 59, 890F.0011/2–747.

7. Eddy to Merriam, Oct. 24, 1945, "Comment on the Draft of a Proposed Treaty of Friendship, Commerce, and Navigation between the USA and the Kingdom of Saudi Arabia," RG 59, 711.90F2/10–2445. For ARAMCO's role in paying the costs of the Saudi delegation between 1945 to 1947, see ms. marked "confidential," "Arabian American Oil Co., Donations, Contributions, and Assistance to Saudi Arabia 1933–1970," Folder 10, Donations, Contributions and Assistance to Saudi Arabia, Box 5, Mulligan Papers.

8. Jidda to State, Dispatch 186, Mar. 11, 1947, ARAMCO's Relations with Saudi Arabian and United States Governments, RG 59, 711.90F/3–1147. On Norris, whose most famous novel, *The Octopus*, appeared in 1902, see Don Graham, "Frank Norris," in *A Literary History of the American West* (Fort Worth: Texas Christian University Press, 1987), 370–80.

9. Lebkicher to Sanger, Oct. 25, 1949, Folder labeled "Africa," Box 8, William H. Sanger Papers, American Heritage Center, Laramie, Wyoming [hereafter Sanger Papers].

10. Memorandum from Robert Thompson, Chairman, Committee on Unofficial Publications to Awalt, Nov. 17, 1950, "Clearance of manuscript entitled 'Arabian Frontiers, An Introduction to the Arabian Peninsula,'" Folder "Africa," Box 8, Sanger Papers.

The book was published under the revised title *Arabian Peninsula* in 1954 and reprinted in the 1970s.

11. "Crown Prince Saud's Official Visit to America, Notes on the Period Monday, January 13, through Wednesday, January 22, 1947," RG 59, 890F.0011/2–747; "A Partnership in Oil and Progress," *Standard Oil of California Bulletin*, Autumn (1946): 3, 8; "Middle East Junior 'Marshall Plan' Costs Taxpayers Nothing," *Oil and Gas Journal* 47, 18 (Nov. 1948): 54; and editorial, "Point $ without the Taxpayers," *Los Angeles Times*, Mar. 20, 1951.

12. Anderson, *Aramco*, 3–4, emphasis added.

13. For debates since the West-as-colony case was first argued in the 1930s, see William G. Robbins, "The 'Plundered Province' Thesis and the Recent Historiography of the American West," *Pacific Historical Review* 55 (Nov. 1986): 577–97. For the turn to dependency theory, see Richard White, *Roots of Dependency: Subsistence, Environment and Social Change among the Choctaws, Pawnees and Navajos* (Lincoln: University of Nebraska Press, 1983). The most influential of the new western histories of empire is Patricia Limerick, *The Legacy of Conquest: The Unbroken Past of the American West* (New Haven, CT: Yale University Press, 1987).

14. "Mining set the pace and direction of western development." Limerick, *Legacy*, 99, 108, 124; also see Philip J. Mellinger, *Race and Labor in Western Copper* (Tucson: University of Arizona Press, 1995), 1.

15. Oil may have helped free the region from its utter dependence on "foreign" that is, private Eastern, finance by creating the first local zones of capital accumulation in places like Texas. World War II hastened the pace of industrial diversification and, in a sense, "autonomy" vis-à-vis the East Atlantic core. William G. Robbins, *Colony and Empire: The Capitalist Transformation of the American West* (Lawrence: University of Kansas Press, 1994), 14.

16. Ibid., 23.

17. This summary is based on Robert Glass Cleland, *A History of Phelps Dodge* (New York: Alfred Knopf, 1954); and Dodge, *Tales of the Phelps-Dodge Family*.

18. Leah Dilworth, *Imagining Indians in the Southwest: Persistent Visions of a Primitive Past* (Washington, D.C.: Smithsonian Institution Press, 1996) for the location of various southwestern peoples.

19. As Phyllis Dodge says about her ancestors, it "was a way for merchants like Phelps to salve their uneasy consciences without antagonizing their cotton-growing suppliers." Dodge, *Tales of the Phelps-Dodge Family*, 70.

20. Donal F. Lindsey, *Indians at Hampton Institute, 1877–1923* (Urbana: University of Illinois Press, 1995), 11.

21. Dodge, *Tales of the Phelps-Dodge Family*, 197–98.

22. Lindsay, *Indians at Hampton Institute*; David Levering Lewis, *W. E. B. Du Bois: Biography of a Race, 1868–1919* (New York: Henry Holt, 1993). Washington's famous speech in September 1895 at the Cotton States International Exposition in Atlanta is

now referred to as "the Atlanta Compromise." Lewis paraphrases Washington's biographer, Justice Louis Harlan, about Washington's purpose in the Atlanta Compromise. "In exchange for black acceptance of restrictions on the franchise and no further demands for 'social equality,' the South's white rulers were to allow gradual progress in agriculture and business and to rein in the rednecks," 174–75.

23. Adolph Reed, Jr., *W. E. B. Du Bois and American Political Thought: Fabianism and the Color Line* (New York: Oxford University Press, 1996).

24. W. E. B. Du Bois, "Education in Africa," *Crisis* (June 1926), 86–89; John H. Stanfield, *Philanthropy and Jim Crow in American Social Science* (Westport, CT: Greenwood, 1985), 25–42; Kenneth James King, *Pan-Africanism and Education: A Study of Race Philanthropy and Education in the Southern States of America and East Africa* (Oxford: Clarendon Press, 1971); David Levering Lewis, *W. E. B. Du Bois: The Fight for Equality and the American Century, 1919–1963* (New York: Henry Holt, 2000), 70, 190–91; Dodge, *Tales of the Phelps-Dodge Family*, 205, 211.

25. David Roberts, *Once They Moved Like the Wind: Cochise, Geronimo, and the Apache Wars* (New York: Simon and Schuster, 1993), 13. For the locations of the Phelps Dodge properties and the siege by "the most cowardly of the Indian tribes," see Cleland, *History of Phelps Dodge*, 77–80, 82, 84–85. Phelps Dodge employees named an early ore train's engine *Geronimo*.

26. Ramón Ruiz, *The People of Sonora and Yankee Capitalists* (Tucson: University of Arizona Press, 1988), 10–14. By 1916, the railroad company was capitalizing in the new market for ethno-tourism in Indian country. See Carlos A. Schwantes, *Vision & Enterprise: Exploring the History of Phelps Dodge Corporation* (Tucson: University of Arizona Press, 2000), 99, 101; Dilworth, *Imagining Indians in the Southwest*, 16–17, 78–91.

27. Margaret Crawford, *Building the Workingman's Paradise: The Design of American Company Towns* (London: Verso, 1995), 29–30.

28. The only careful work I have found on wage rates and working conditions in the Phelps Dodge mines suggests, in light of the racism, inequality, and labor activism, that such claims may not withstand close scrutiny. See Mellinger, *Race and Labor*, 33–43, 154–73, and below.

29. Mellinger, *Race and Labor*, quotes company executives and engineers in the 1910s who describe the mixing of nationalities as the weapon against unions. "As long as little dissensions and rivalries keep the men busy, they are fairly amenable," 6–8.

30. F. Remington Barr, "Integrated Results of Sixty Years' Operation, Phelps Dodge Company, Morenci Branch," manuscript, Sept. 1940, chapter on the Detroit Copper Mining Co. of Arizona, 1875–1919, 22, 72–74, 79–81, enclosed in Cleveland E. Dodge to Mrs. Frank Ayer, Pelham, NY, Nov. 7, 1940, Box 22, Frank Ayer Papers, American Heritage Center, University of Wyoming, Laramie.

31. Crawford, *Building the Workingman's Paradise*, 136.

32. Chinese, who were considered worse than Mexicans, were entirely barred from

Bisbee. See Carlos Schwantes, ed., *Bisbee: Urban Outpost on the Frontier* (Tucson: University of Arizona Press, 1992), 16, 61.

33. Phelps Dodge eventually tore down the decrepit remnants of the once-remarkable Beaux Arts camp in the 1960s when it developed an open-pit mine in Tyrone. Crawford, *Building the Workingman's Paradise*, 150–51; Schwantes, *Vision and Enterprise*, 119, 122–24.

34. Mellinger, *Race and Labor*, 18–19, 21–22.

35. Ibid., 30.

36. Ruiz, *People of Sonora*, 7–40, 84; Miguel Tinker Salas, *In the Shadow of the Eagle: Sonora and the Transformation of the Border during the Porfiriato* (Berkeley: University of California Press, 1997), 92–98, 194–97. For a contemporary account of the racial hierarchy as viewed from "the American Colony" by a Phelps Dodge engineer at Nacozari, see Ralph Ingersoll, *In and Under Mexico* (New York: Century Co, 1924). "The attitude of the larger companies is necessarily paternal; the people dealt with are children and must be looked after and guarded, to insure any production whatever," 116.

37. James Byrkit, *Forging the Copper Collar: Arizona's Labor-Management War of 1901–1921* (Tucson: University of Arizona Press, 1982), 38–62.

38. Mellinger, *Race and Labor*, 85–86; Byrkit, *Forging the Copper Collar*, 44–55. The *New Republic* (Nov. 6, 1915): 4, mounted a defense in part of Arizona: "We should not overlook the fact that states like Arizona have a very serious problem in the mining or industrial camps financed by absentee capital, manned by alien labor, and governed despotically by a little group of higher employees. In such camps nothing like an American form of local government is possible. The anti-alien law attempted to substitute, for an exploited body of aliens, citizen laborers, by education and temperament fitted to check the arbitrary tendencies of mining camp capitalism." Cited in Byrkit, *Forging the Copper Collar*, 54.

39. "The informed U.S. reading public had only recently read about the Ludlow Massacre in the great Colorado Fuel and Iron Company strike of 1913–14. A detachment of Colorado state militia had wantonly attacked a strikers' tent settlement along the railroad tracks near CFI company property, deliberately setting the tents afire and killing over a dozen women and children. Vengeance was swift, and western Las Animas County became a battlefield. Miners, militia detachments, and company guards fought a small war in southern Colorado, until U.S. Army troops managed to separate them. . . . Retrospective accounts of the Arizona strikes of 1915 credit Governor Hunt with having avoided 'Ludlow Massacres.'" Mellinger, *Race and Labor*, 157.

40. Byrkit, *Forging the Copper Collar*, 58; James R. Kluger, *The Clifton-Morenci Strike: Labor Difficulty in Arizona, 1915–1916* (Tucson: University of Arizona Press, 1970), 36–38.

41. Byrkit provides evidence for Phelps Dodge's control of the *Arizona Gazette*, the *Arizona Daily Star*, the *El Paso Herald*, the *Douglas International*, the *Copper Era*, and the *Bisbee Daily Review*. See *Forging the Copper Collar*, 110. These investments are another subject left untouched by Schwantes, the company's newest historian.

42. Byrkit, *Forging the Copper Collar*, 63–93. The *New Republic* (Jan. 22, 1916) saw

labor as in confrontation with "a process by which a handful of owners in New York, Boston or Edinburgh can impose upon ten or fifteen thousand men and women the choice between surrendering their liberties or starving." The governor referred to Douglas, who became a vice president of Phelps Dodge in 1916, as "the consort of the queen of Arizona copper mines," 64. Hunt lost the election in 1916 by thirty votes (of 56,000 cast). Byrkit describes how the elections were rigged.

43. Byrkit, *Forging the Copper Collar*, 187–235. This book provides the only detailed account of the 1917 deportation.

44. Mellinger, *Race and Labor*, 185–87.

45. Ibid., 190, quoting the *Copper Era and Morenci Leader*, Oct. 19, 1917.

46. The commission "settled the strikes but could not alleviate the basic issues. Short-term gains were made in both organization and wages, then were lost to widespread blacklisting and postwar depression. In Arizona, too, copper unionism was dead until the Wagner Act and the Second World War." George Hildebrand and Garth Mangum, *Capital and Labor in American Copper, 1845–1990* (Cambridge, MA: Harvard University Press, 1992), 141.

47. Doheny was and remains notorious in Mexico from the time of the revolution, where he epitomizes a brand of brazen interventionist politics engaged in by investors during the revolution. He made good copy in the American press in the 1920s and appeared in one guise or another in a host of novels, including Upton Sinclair's *Oil* (1926), B. Traven's *The White Rose* (1929, original German edition), and Carleton Beals's *Black River* (1934), but in the intervening years was all but forgotten. New full-length studies on Doheny have appeared in the past decade, alongside an account of the oil industry and politics in Mexico, in which Doheny is a featured player. See Dan LeBotz, *Edward L. Doheny: Petroleum, Power, and Politics in the United States and Mexico* (New York: Praeger, 1991), Jonathan C. Brown, *Oil and Revolution in Mexico* (Berkeley: University of California Press, 1993), Martin R. Ansell, *Oil Baron of the Southwest: Edward L. Doheny and the Development of the Petroleum Industry in California and Mexico* (Columbus: Ohio State University Press, 1998), and Davis, *Dark Side of Fortune*.

48. Nancy Quam-Wickham, author of one of the very first accounts of oil workers and politics in California in a generation, opens with a pointed reflection on Daniel Yergin's massive history of world oil, *The Prize*. This 800-page, self-styled epic is all but silent on the role of labor both in development of the oil industry and in the grand political conflicts spawned throughout the world by expansion of private oil investment. Leading voices in the new Western history movement have been calling for work on the oil industry for almost two decades. Nancy Lynn Quam-Wickham, "Petroleocrats and Proletarians: Work, Class, and Politics in the California Oil Industry, 1917–1925," Ph.D. dissertation, University of California, Berkeley, 1994, 1–2.

49. Gerald T. White, *Formative Years in the Far West: A History of Standard Oil Company of California and Predecessors through 1919* (New York: Appleton-Century-Crofts, 1962),

chap. 20, quote from 520. Even more expansively, White concludes that Socal led "most oil companies *everywhere*" in wages, hours, and working conditions, 527, emphasis mine.

50. Quam-Wickham, "Petroleocrats," 63–64, 78–79, 81.

51. The most detailed account available is Nigel Anthony Sellars, *Oil, Wheat and Wobblies: The Industrial Workers of the World in Oklahoma, 1905–1930* (Norman: University of Oklahoma Press, 1998).

52. Quam-Wickham, "Petroleocrats," 12–14, 159–60; White, *Formative Years in the Far West*, 526–27.

53. Carl B. King and Howard W. Risher, Jr., *The Negro in the Petroleum Industry*, Report No. 5 in the series the Racial Policies of American Industry, University of Pennsylvania, Wharton School of Finance and Commerce, Industrial Research Unit, 1969; Quam-Wickham, "Petroleocrats," 69.

54. Quam-Wickham, "Petroleocrats," 65–66.

55. Sellars, *Oil, Wheat, and Wobblies*, 63, emphasis mine.

56. Ibid., 108–9, 163–67.

57. Mira Wilkins, *The Emergence of Multinational Enterprise: American Business Abroad from the Colonial Era to 1914* (Cambridge, MA: Harvard University Press, 1970), 62–64, 82–87.

58. See the speculative and unsourced discussion in Gene Z. Hanrahan, *The Bad Yankee/El Peligro Yankee*, vol. 1 (Chapel Hill, NC: Documentary Publications, 1985), 6.

59. Friedrich Katz, *The Secret War in Mexico: Europe, the United States, and the Mexican Revolution* (Chicago: University of Chicago Press, 1981), although the case could be made for going back a decade earlier to Lorenzo Meyer's *México y los Estados Unidos en el conflicto petrolero (1917–1942)* [1972], second ed., translated by Muriel Vasconcellos, *Mexico and the United States in the Oil Controversy, 1917–1942* (Austin: University of Texas Press, 1977).

60. My discussion is based on Brown, *Oil and Revolution*, chapter 5, 307–65. This is the only discussion I have been able to find of conditions inside the Mexican oil camps.

61. Ibid., 315–17.

62. Ansell, *Oil Baron*, 149–50. To be more precise, the company magazine was originally called the *MexPet Record*, first issued in March 1916, around when other oil companies began to publish magazines, and it was replaced by the *Pan American Record* in March 1917.

63. Ansell, *Oil Baron*, 171–73.

64. Some scholars have since taken virtually all the most fantastic charges of foreign machinations against the revolution at face value. Suffice it to say, as a recent, generally skeptical account notes, "all that can be said for sure is that there was enough intrigue among business interests in Mexico to support any number of conspiracy theories." Ansell, *Oil Baron*, 153. The best study of the intrigues of firms and states during the war years in Mexico is Katz's *Secret War*.

65. Dodge, *Tales of the Phelps Dodge Family*, 318–34; Byrkit, *Forging the Copper Collar*, 276–80; LeBotz, *Edward L. Doheny*, 47–48.

66. Anderson, *Aramco*, 22–25, drawing on St. John Philby's *Arabian Oil Ventures* (Washington, D.C.: Middle East Institute, 1964), a book ARAMCO paid for.

67. Stegner, *Discovery!*, v.

68. Ibid., 102.

69. Anthony Cave Brown, *Oil, God, and Gold: The Story of Aramco and the Saudi Kings* (Boston: Houghton Mifflin, 1999), 59.

70. Lloyd Hamilton to Airy Hamilton, Hufuf, Sunday Dec. 30, 1934, Folder 11, Lloyd Hamilton Letters, 1934–35, Box 28, Letters, Journals, Interviews, Wallace Earle Stegner Papers, MS 676, Special Collections, Marriott Library, University of Utah, Salt Lake City [hereafter cited as Stegner Papers with filing information].

71. Stegner, *Discovery!*, 23–63. In writing the history of the pioneer era in the mid-1950s Stegner worked with a company researcher and a three-volume in-house history she had compiled from archival sources that have not been available to scholars.

72. For this discussion of Venezuela, I have relied on Edwin Lieuwen, *Petroleum in Venezuela: A History*, University of California Publications in History, vol. 47 (Berkeley: University of California Press, 1954); Wayne Taylor and John Lindemann, *The Creole Petroleum Corporation in Venezuela*, United States Business Performance Abroad (Washington, D.C.: National Planning Association, 1955); Stephen G. Rabe, *The Road to Opec: United States Relations with Venezuela, 1919–1976* (Austin: University of Texas Press, 1982); Laura Randall, *The Political Economy of Venezuelan Oil* (New York: Praeger, 1987); and, most helpful of all, Miguel Tinker Salas, "Venezuelans, West Indians, and Asians: The Politics of Race in Venezuelan Oil Fields, 1920–1940," in Vincent Peloso, ed., *Work, Protest and Identity in Twentieth Century Latin America* (Wilmington, DE: Scholarly Resources, 2003).

73. John Douglas, Oct. 6, 1924, File of John Douglas's Letters from a Wildcat Well Venezuela, 1924–1925 [accompanying a photo album], Box 1, John G. Douglas Collection, American Heritage Center, Laramie, Wyoming.

74. Darlene Rivas, "Like Boxing with Joe Louis: Nelson Rockefeller in Venezuela, 1945–1948," in Peter Hahn and Mary Ann Heiss, eds., *Empire and Revolution: The United States and the Third World since 1945* (Columbus: Ohio State University Press, 2001), 217–41; Randall, *Political Economy*, 6–17; Lieuwen, *Petroleum in Venezuela*, 103–15.

75. See Larry Barnes, *Looking Back Over My Shoulders* (Peterborough, NH: Private edition, 1979), 54–57. I have scare quotes around the word Persian both because Barnes makes a point of noting that it was an identity "claimed" by all the "nubile young ladies" and because the identification is a flexible one in that time and place. One cannot conclude much from it.

76. See Michael Field, *Merchants: The Big Business Families of Saudi Arabia and the Gulf States* (Woodstock, NY: Overlook Press, 1985), 209, emphasis mine; and manuscript

by Mulligan, untitled, on Bahrain, written Feb. 1985, Folder 2, History Project, Box 8, Mulligan Papers.

77. See Henrietta Larson, Evelyn Knowlton, and Charles Popple, *New Horizons, 1927–1950* (New York: Harper and Row, 1971), 836, questioning (I think) the argument in the case of Standard Oil (NJ) in Colombia: "It would be easy to conclude that the British colonial tradition was at work in Colombia, but the situation apparently was far too complex to allow of any single explanation."

78. See *Organized Labor*, official organ of the State and Local Building Trades Councils of California, San Francisco, Apr. 21, 28, and May 5, 1906, online excerpt at www.sfmuseum.org/1906.2/invasion.html.

79. Barger to Mom and Dad, CampTarfa, Sunday Feb. 26, 1938, Folder 2, Barger Letters, Jan.–Mar. 1938, Box 28, Letters, Journals, Interviews, Stegner Papers. The long excerpt is close to being, but is not precisely, a word-for-word reproduction. In transcribing the letters I sometimes left out words, prepositions and the like.

80. Thomas C. Barger, *Out in the Blue: Letters from Arabia, 1937 to 1940* (Vista, CA: Selwa Press, 2000), 47–49. The copy I quote from is found in Folder 2, Barger Letters, Jan.–Mar. 1938, Box 28, Stegner Papers.

81. See Folder 2, Master Copy II The Explorers, Box 26 Discovery, First Draft, Reader's Copy, Stegner Papers; and compare with *Discovery!*, 74–75. Barger's notes and suggestions are found throughout the various drafts as well as in correspondence.

82. For use of the term *coolie*, see, for example, the letter, Mary to Mother, Feb. 17, 1953, "This is coolie payday Thursday, and I am going to Bahrain Thursday PM and staying over Friday." In Folder 9, Hartzell, Mary Elizabeth–1953, Box 11, Mulligan Papers. Hartzell was the librarian in the Arabian Affairs Division. For Mulligan's description of Barger as later leading the campaign against racism in the camps, see his note on Barger (handwritten yellow sheet), Folder 2, "History Project," Box 8, Mulligan Papers.

83. See Notes of Stegner Interview with Phil McConnell, 1956, Folder 19, Interview, Phil McConnell, 1956, Box 28, Letters, Journals, Interviews, Stegner Papers; and Stegner, *Discovery!*, 84.

84. Letter by Florence Steineke, recipient unclear, Dec. 16, 1937, Folder 16, Max Steineke, Excerpts from Letters, 1937–38, Box 28, Letters, Journals, Interviews, Stegner Papers.

85. See the discussion in Tinker Salas, "Venezuelans, West Indians, and Asians."

86. Barger to his parents, Dec. 15, 1937, Folder 1, Tom Barger Letters 1937, Box 28, Letters, Journals, Interviews, and Folder 4, IV Growing Pains, Box 27, Discovery, Stegner Papers; Barger to Kathleen, Dec. 15, 1937, *Out in the Blue*, 4–7; Stegner, *Discovery!*, 81–82; Nicholas Lemann, "The Way They Were," *New Yorker*, Apr. 15, 2002: 72–73.

87. Mimeograph, Summary of Operations of California Arabian Standard Oil Company for Years 1938 and 1939, Folder 42, ARAMCO Annual Reports, 1938–43, Box 3, Mulligan Papers. The "army of young Arabs" quote is from Barger's letter home, Dec. 15, 1937, Folder 1, Tom Barger Letters 1937, Box 28, Letters, Journals, Interviews, Stegner Papers.

Chapter 3

1. Daniel Yergin, *The Prize: The Epic Quest for Oil, Money, and Power* (New York: Simon and Schuster, 1991), 305–56.

2. Painter, *Oil and the American Century*, 11–12.

3. Figures in Anderson, *Aramco*, 110.

4. U.S. Congress, Senate, Special Committee Investigating the National Defense Program, *Petroleum Arrangements with Saudi Arabia, Part 41, Hearings before a Special Committee Investigating the National Defense Program*, 80th Congress, First Session, 1948, 24707–24710.

5. Aaron David Miller, *Search for Security: Saudi Arabian Oil and American Foreign Policy, 1939–1949* (Chapel Hill: University of North Carolina Press, 1980), 36–45.

6. "Oil and the Near East," *New York Times*, Feb. 25, 1944: 16.

7. Ibid., 38, 44.

8. For evidence of the role played by events in Iraq in shaping ARAMCO's decision making in Dhahran, see James Terry Duce, "The Near East Today," Paper for presentation at a Production Group Session during the 31st Annual Meeting of the American Petroleum Institute, Stevens Hotel, Chicago, Nov. 7, 1951, Box 1, Philip C. McConnell Papers, 1937–63, Hoover Institution Archives, Palo Alto, California [hereafter cited as the McConnell Papers with filing information]; see as well Stegner's notes of interview with McDonnell, Folder 19, Interview, Phil McConnell, 1956, Box 28, Letters, Journals, Interviews, Stegner Papers.

9. Again, see Duce's paper for a discussion of the grave questions posed for the firms by war-driven inflation, cost of living questions, and political turmoil, particularly in the Iraqi market. "The Near East Today," 1951, Box 1, McConnell Papers.

10. John Blair, *The Control of Oil* (New York: Pantheon, 1976), 39. This discussion also relies on Painter, *Oil and the American Century*, 3–9, 35–47, and Yergin, *The Prize*, 396–99.

11. This last quotation is culled from Robert Norberg's talk at Duke University in March 1989, "ARAMCO and the Saudi Arabian Government: The Special Relationship." Folder 7, ARAMCO, 02/11/39–09/23/91, Box 7, Mulligan Papers.

12. See Barger's letters home, Feb. 18, 1940, Folder 6, Barger Letters, 1940, and July 19, 1941, Folder 7, Barger Letters, 1941, Box 28, Letters, Journals, Interviews, Stegner Papers.

13. "Government Relations: How It Is Organized and What It Does" [n.d.] Folder 1, Box 7, Mulligan Papers.

14. Quotation from Mulligan to Greg [Dowling], Feb. 21, 1989, enclosing "Notes on Development of Government Relations," Box 8, Mulligan Papers.

15. Writing about Iran in the 1920s, before ARAMCO was created, British Petroleum's official historian says that the borders dividing municipal and corporate responsibilities at Abadan were nonexistent. J. H. Bamberg, *The History of the British Petroleum*

Company, Volume 2, The Anglo-Iranian Years, 1928–1954 (Cambridge, UK: Cambridge University Press, 1994), 63–76.

16. Stegner, *Discovery!*, 168–71 (Point Four quote); Lorania K. Francis, "Arab Farms Boom under Americans," and Editorial, "Point $ without the Taxpayers," *Los Angeles Times*, Mar. 20, 1951; Lemann, "The Way They Were." Barger himself had disparaged these kinds of gross exaggerations. He told Stegner in 1956, "Al Karj was primarily the creation of that man of great energy and imagination, Abdulla Suleiman—we were simply his tools." Barger's notes to Stegner's draft manuscript, Folder 8, Master Copy, VIII [penciled in, IX], The Time of the Hundred Men, Box 26, Discovery, First Draft, Reader's Copy, Stegner Papers.

17. See Jidda to State, A-487, Dec. 19, 1949, RG 59, 890F.9111 RR/12-1949, and Jidda to State, A-506, Dec. 29, 1949, RG 59, 890F.9111 RR/12-2949.

18. From Stegner's files of excerpts from McConnell's diary of the war years, Folder 13, Phil McConnell Notes, Box 28, Letters, Journals, Interviews, Stegner Papers. "From the very beginning" quote found in ARAMCO's retrospective account of all agricultural assistance to the kingdom in ms. marked confidential, ARAMCO, "Donations, Contributions, and Assistance to Saudi Arabia 1933–1970," Folder 10, Donations, Contributions and Assistance to Saudi Arabia, Box 5, Mulligan Papers. The most referenced accounts on these matters against which I am writing are Stegner, *Discovery!*, 170, and Anderson, *Aramco*, 111–12. See as well George A. Lipsky, *Saudi Arabia: Its People, Its Society, Its Culture* (New Haven, CT: Human Relations Area Files Press, 1959), 214–15; Carleton Coon, "Operation Bultiste: Promoting Industrial Development in Saudi Arabia," in H. Teaf and P. Franck, eds., *Hands Across Frontiers: Case Studies in Technical Cooperation* (Ithaca, NY: Cornell University Press, 1955), 349–50; and Nathan Godfried, *Bridging the Gap between Rich and Poor: American Economic Development Policy toward the Arab East, 1942–1949* (New York: Greenwood, 1987), 160–62.

19. Sanger, *Arabian Peninsula*, 60. An Arizona farmer, David Rogers, and his successor, Kenneth J. Edwards, a Texas county agent who had been recruited by the U.S. Extension Service, were the primary architects. With the dissolution of the Middle East Supply Center in 1945, the project apparently fell for a short while to the Foreign Economic Administration, the government's first postwar development agency, the forerunner of A.I.D.

20. Nils Lind, "A Summary of Views on United States Relations with Saudi Arabia," RG 59, 711.90F/5-147.

21. Godfried, *Bridging the Gap*, 160.

22. See ARAMCO, Dhahran, Aug. 28, 1954, Condensed Executive Audit Report Al Kharj Farms 1954, Folder 10, Al-Kharj, Box 8, Mulligan Papers.

23. Mildred Montgomery Logan, "The Arabs Call Me Madam Sam," *Cattleman* (Jan. 1952): 22.

24. Mildred Montgomery Logan, "I Like Being the Garden of Eden's First Lady," *Cattleman* (Oct. 1957): 30; typed sheet with handwritten title, "Summary on al Kharj

Farms by Sam T. Logan," Feb. 1986, enclosed in Mildred Logan to William Mulligan, Feb. 17, 1986, Folder 10, Al-Kharj, Box 8, Mulligan Papers.

25. The original principals were Hussein Aoueini and Ibrahim Shakir, who were joined by Muhammad Alireza, a rival of theirs and of their patron, the finance minister.

26. Dhahran to State, 50, Aug. 2, 1946, RG 59, 890F.77/8–246; Jidda to State, 38, Aug. 14, 1946, RG 59, 890F.70/8–1446; Outgoing Telegram, State to Jidda, Nov. 29, 1946, RG 59, 890F.77/11-2946; Sanger, *Arabian Peninsula*, 118.

27. Dhahran to State, Dec. 18, 1946, A-73, RG 59, 890F.77/12-1846.

28. John H. Crider, "Saudi Arabia's Oil Looms as Vital," *New York Times*, Oct. 3, 1943: 4.

29. Editorial, "Arabian Oil," *New York Times*, Feb. 23, 1944: 18.

30. "Topics of the Times," *New York Times*, Mar. 3, 1944: 14.

31. Editorials, "Oil and the Near East," and "American Oil Policy," *New York Times*, Feb. 2, 1944: 16, and Mar. 10, 1944: 14.

32. "Moffett Attacks Arabian Oil Plan, Asks Ickes Ouster," *New York Times*, Mar. 2, 1944: 1; Editorial, "The Oil Controversy," *New York Times*, Mar. 4, 1944: 12; Edwin L. James, "Arabian Oil Argument Serves Signal Purpose," *New York Times*, Mar. 19, 1944: E3; and Jane Wyeth Knight, Letter to the *Times*, "Arabian Oil Deal Opposed, Plan to Seek Reserves Abroad Held Move toward Imperialism," sent March 1, *New York Times*, Mar. 10, 1944: 14.

33. See manuscript, "The Saudi Arabian Partnership" and the cover letter by Ohliger, Oct. 7, 1944, Folder 12, The Saudi Arabian Partnership by Phil McConnell, Box 28, Letters, Journals, Interviews, Stegner Papers.

34. Mary Eddy to Family, Oct. 31, 1944, Folder 13, Correspondence, William A. Eddy, 1944, Box 5; and Mary to Dora, Feb. 28, 1945, Folder 4, Marcy Garvin Eddy, 1945, Box 3, Eddy Papers.

35. Mary Eddy, Apr. 13, 1954, Folder 4, Mary Garvin Eddy, 1945, Box 3, Eddy Papers.

36. Eddy to State, 55, Jan. 13, 1945, enclosing document titled "Saudi Arabia," RG 59, 890F.00/1-1345.

37. See Memorandum from Under Secretary of the Navy to Acting Secretary of State, [n.d], with attached Memorandum for the President, June 26, 1945, RG 59, 890F.245/6-2645.

38. Eddy to State, Dispatch 162, Aug. 8, 1945, Signing of Agreement for a United States Military Airbase at Dhahran, RG 59, 890F.248/8–845; Henderson to Acheson, Dec. 29, 1945, Background on Airport at Dhahran, Saudi Arabia, 890F.248/12-2945.

39. See for example, State, Division of Near Eastern Affairs, Oct. 11, 1945, "Interest of Representative Philbin in Construction of Airfield at Dhahran, Saudi Arabia," RG 59, 890F.248/11-1045; letter from A. S. Hediger, San Anselmo, CA, to Secretary of State James Byrnes, Nov. 10, 1945, RG 59, 890F.248/11-2645.

40. Merriam to Lichter, Dec. 28, 1945, RG 59, 890F.248/11-2645; Henderson to

Maslow, [William Maslow, Director, American Jewish Congress], Jan. 8, 1946, RG 59, 890F.4016/1-846; Eddy to State, Airgram a-33, Apr. 19, 1945, RG 59, 890F.4016/4-1945.

41. Jack Winocur, president American Communications Association, CIO, to State, May 8, 1945, RG 59, 890F.4016/5-845; Memorandum, Parker Hart to Near East Department, July 24, 1945, RG 59, 890F.4016/7-2445. Hart reported that Sheets of ARAMCO "did in fact discourage inclusion of personnel of the S.S. *George Bellows* to be presented to Saudi authorities for the purpose of obtaining shore leave. He even ventured the belief that there was a law prohibiting the entry of Jews, and he felt that negroes might be taunted by the Arabs, who had learned the expression 'Sambo.'"

42. See Tuck (Cairo) to State, 854, May 16, 1946, RG 59, 890F.4016/5-1646.

43. See Typed Notes, entry for Nov. 7, 1955, Folder 11, Notes for Book, Box 27, Stegner Papers.

44. The best work to date is Simon Davis, "Keeping the Americans in Line: Britain, The United States, and Saudi Arabia, 1939–1945—Inter-Allied Rivalry in the Middle East Revisited," *Diplomacy and Statecraft* 8, 1 (Mar. 1997): 96–136.

45. Wm. Roger Louis, *The British Empire in the Middle East, 1945–1951: Arab Nationalism, the United States, and Postwar Imperialism* (New York: Oxford University Press, 1984), 191.

Chapter 4

1. ARAMCO, Planning Committee, Minutes of Planning Committee, Meetings of Jan. 4 and 5, 1944, Untitled File, Box 3, McConnell Papers.

2. McConnell, Diary, Entry dated Nov. 29, 1944, Box 3, McConnell Papers. The wording here is a close but not exact transcription.

3. McConnell Diary, Entry dated Dec. 29, 1944, Box 3, McConnell Papers, emphasis mine. Again, wording is close but not exact.

4. The myth began with the very first strike. In the postmortem to Washington, the consulate wrote "Labor organizations and strikes are against the law in Saudi Arabia." See Dhahran to State, 5, Aug. 30, 1945, RG 59, 890F.5045/8-945.

5. Department of State Instruction, No. CA-3384, Dec. 29, 1953, "Comment on the October–Nov 1953 Strike at the Arabian American Oil Company Installations in Saudi Arabia," section titled "The Legal Roots of the Strike Itself," 10, RG 59, 886A.062/12-2953.

6. We have a record of the incident because as a result the government pressed ARAMCO unsuccessfully to reduce the working day for the rest of Ramadan from seven and a half to six hours. See Memorandum, Barger to Davis, Labor Strike at Riyadh, Sept. 26, 1942, Folder 2, Correspondence, Tom Barger, 1957–58, Box 29, Discovery Correspondence, Stegner Papers.

7. See J. B. McComb [Supervisor, Employee Relations Department, ARAMCO] to State, Aug. 10, 1945, RG 59, 890F.504/8-1045. We have the data because of an inquiry to

Washington seeking assurances that the company would not be in violation of government regulations by forcing these employees to pay their own ways home. See as well, Phil McConnell, Journal, entry for Apr. 2, 1945, Box 3, McConnell Papers.

8. Phil McConnell, Journal, entry for June 11, 1945, Box 3, McConnell Papers.

9. Phil McConnell, Journal, entry for July 16, 1945, Box 3, McConnell Papers; Birge [Vice Counsel, Dhahran] to State, Airmail 2, July 28, 1945, Strike of Two Thousand Saudi Arab Employees of the Arabian American Oil company at Dhahran, RG 59, 890F.5045/7-2845.

10. See, Dhahran to State, Incoming Telegram 41, July 21, 1945, RG 59, 890F.5045/7-2145; Birge to State, Airmail 2, July 28, 1945, RG 59, 890F.5045/7-2845; and Dhahran to State, Airmail 5, Aug. 30, 1945, General Strike of All Arabian American Oil Company Saudi Arab Employees, RG 59, 890F.5045/8-945, the source of the extended direct quotation.

11. Dhahran to State, 3, Aug. 3, 1945, RG 59, 890F.5045/8-345; Stefano [Italian embassy, Washington] to Villard [Deputy Director of Near Eastern and African Affairs], July 4, 1947, RG 59, 49/4 890F.504/7-447.

12. Memo by Owen for the attention of Spurlock, Oct. 27, 1945, SZ-251, ARAMCO, San Francisco, "Subject: George Rentz"; and Memo by Osborne [ARAMCO, Cairo], Nov. 27, 1945, to ARAMCO, Bahrain Island, 270, "George Rentz," Folder 52, Rentz, George S.—Correspondence re, 1945–52, Box 1, Mulligan Papers.

13. See the account of American consul Parker Hart when, later, MacPherson too got the boot, the changes once again taking place during a time of increased labor strife. Dhahran to State, 99, July 2, 1949, RG 59, 890F.6363/7-249.

14. See Owen to Duce, Nov. 28, 2945, SZ-366, Identification System, included in RG 59, 890F.6363/12-2945.

15. Details found in Dhahran to State, 46, undated [received Aug. 13], RG 59, 890F.5045/8-1345; Dhahran to State, 49, Aug. 15, RG 59, 890F.5045/8-1545; and Dhahran to State, Airmail 5, Aug. 30, 1945, Subject: General Strike of All Arabian American Oil Company Saudi Arab Employees, 890F.5045/8-945.

16. See Dhahran to State, 54, Aug. 27, 1945, RG 59, 890F.5045/8-2745. In 1942, the government had requested the oil company to reduce the working day during Ramadan from eight to six hours. When the company declined, the government in Riyadh passed a new royal decree. The company then spent the next six months negotiating with the government over the terms of these provisions and gaining exemptions from at least some of them. I have limited records and there are no easily identified secondary sources on the early labor legislation. I also have not read the State Department materials for the war years. For all these reasons, this discussion may well require revision. I have relied on a memorandum from Barger to Davis, Labor Strike at Riyadh, Sept. 26, 1942, Folder 2, Correspondence, Tom Barger, 1957–58, Box 29; as well as a letter from Garry Owen to Najib Salha, Oct. 18, 1942, and a memorandum by Lebkicher, June 26, 1943, on discussions in Riyadh June 14 and 15, 1943, concerning general relations with Saudi Arabian

government labor regulations, Folder 8, Discussion in Riyadh, 1942–1943 Regarding Labor Relations, Box 30, Stegner Papers.

17. Confidential memorandum by Ohliger, 765, Oct. 10, 1943, Folder 23, Correspondence Regarding Disloyal Employees, 1943, Box 29, Stegner Papers.

18. *Standard of California Autumn Bulletin*, 1946: 8.

19. See Dhahran to State, Airmail 15, Sept. 15, 1945, Saudi Government Representatives Suggest Fair Solution to Grievances of Arabian American Oil Company Saudi Laborers, RG 59, 890F.6363/9-1545.

20. Ibid.

21. See memo by G. Owen to W. Spurlock, Dhahran, Oct. 27, 1945, SZ-251, ARAMCO, San Francisco, subject: George Rentz, Folder 52, Rentz, George S.—Correspondence re, 1945–1952; and ms. by Mulligan, Folder 53, Rentz, George S.—Correspondence re, 1963–1987, Box 1 Mulligan Papers.

22. See Mulligan's handwritten notes: "heavy drinking, affection for the girls (pinch/pat) . . . loved . . . drinking-partying." Folder 57 Rentz, George S.—Notes re, Box 1, Mulligan Papers.

23. Mulligan biographical sketch of Rentz, Folder 53, Rentz, George S.—Correspondence re, 1963–1987, Box 1, Mulligan Papers.

24. See Kemper Moore, volume titled "Developments, 1939," pp. 10–15, Folder 13, Research Material, Box 30, Discovery: Personnel, Articles, Research materials, News clippings, Stegner Papers.

25. Ibid.

26. Rentz to Barger, Jan. 12, 1948, "Supervisor's Report on Arabic Research and Translation Section," Folder 13, ART, AAD, Etc., 1947–1948, Box 2 Mulligan Papers. The account of Matthews is found in Mulligan to Donna Drake [ARAMCO Services, Houston], Dec. 26, 1988, Folder 1, Ameen, Michael M., Jr., Box 1, Mulligan Papers.

27. See Jidda to State, 342, Aug. 18, 1947, Treatment Accorded Italian Employees of Arabian American Oil Company at Dhahran and Ras Tanura, Saudi Arabia, RG 59, 890F.6363/8-1847. For conditions of Italians in the Bechtel camp in Jidda, see in addition Dhahran to State, 131, Enclosing Memorandum from Bergus, Jidda, to Childs, Nov. 15, 1948, Italian Workers Employed by ARAMCO, RG 59, 890F.6363/11-848.

28. Jidda to State, 233, Oct. 25, 1948, Some Observations on the Position of the Arabian Am Oil Company with Regard to American Policy in Saudi Arabia, RG 59, 890F.6363/10-2548.

29. "Aramco Insultingly Behaves towards Pakistanis," *New Orient*, Mar. 27, 1949, enclosure in Karachi to State, 99, Mar. 31, 1949, Dispute between Arabian-American Oil Co and Its Pakistan Employees, RG 59, 890F.6363/3-1349.

30. "Where Yankees Misbehave," *Freedom*, Mar. 25, 1949, enclosed in RG 59, 890F.6363/3-1349.

31. Incoming Airgram, Karachi to State, Apr. 14, 1949, RG 59, 890F.6363/4-1449.

32. State Department, Memorandum of Conversation, Treatment of Indian and Pakistani Workers by the Arabian-American Oil Company in Saudi Arabia, n.d. [but probably Apr. 14, 1949], RG 59, 890F.6363/4-1449, emphasis mine.

33. State Department, Memorandum of Conversation, June 7, 1949, ARAMCO and Pakistani and Indian Labor, RG 59, 890F.6363/6-749.

34. See the articles by Patrick Clawson and Joe Stork in Nore and Turner, eds., *Oil and Class Struggle,* 151, 175.

35. Lind, "A Summary of Views on United States Relations with Saudi Arabia."

36. "Improvement of ARAMCO Relations with the Saudi Arabian Government," enclosed in ARAMCO, Relations Department, Monomonock Inn, Mountainhome, PA, Oct. 25, 1948 to Oct. 28, 1948, Folder 25, Relations Department, 1948, Box 5, Mulligan Papers.

37. See Freda Kirchway to Clark Clifford, June 18, 1948, and enclosures, RG 59, 890F.6363/6-2248.

38. William Eddy to Mary Eddy, letters dated July 8 and 15, Folder 3, William A. Eddy, 1948–1949, Box 6, Eddy Papers.

39. MacPherson is portrayed by both Anderson, *Aramco,* and Brown, *Oil, God, and Gold,* as on the losing side of a battle to make ARAMCO a fully integrated oil firm to rival rather than serve the interests of Exxon. He was U.S. consul Parker Hart's primary source of information about the firm, and it is these self-serving conversations with Hart and the latter's reports to Washington on which the historians' portraits are based. None has considered the materials I present here. See Hart to State, 99, July 2, 1949, Outline of Factors Leading to Retirement from Aramco of James MacPherson, Vice President and Resident Administrative Officer, 890F.6363/7-249; and Mulligan to Harry McDonald, 22 Apr. 1991, Folder 7, Correspondence, 1991, Box 12. In the letter, Mulligan revealed that he was a source for Brown's book but had changed his mind about MacPherson along the lines I suggest here.

40. I have used BP for convenience's sake. The firm was then known as the Anglo-Iranian Oil Company, and was owned jointly by investors and the British government. The Iraqi concessions were owned by a consortium, including BP/AIOC, Royal Dutch Shell, Jersey Standard (Exxon), Socony-Vacuum (Mobil), France's Compagnie Française des Pétroles, and an individual, Calouste Gulbenkian. Specialists know the story of these ventures. Others can get up to speed using Yergin's *The Prize.*

41. Harold Hoskins, Mar. 17, 1948, Report to Coordination Committee, Standard Oil Company (NJ), Trip to Middle East—Jan. 10 to Mar. 9, 1948, Folder titled Professional File, Standard Oil Reports and Memoranda 1947–1950, Box 3, Harold B. Hoskins Papers, American Heritage Center, University of Wyoming, Laramie. See as well, Mark Crinson, "Abadan: Planning and Architecture under the Anglo-Iranian Oil Company," *Planning Perspectives* 12 (1997): 341–59.

42. See the account in Harry Roscoe Snyder, "Community College Education for

Saudi Arabia: A Report of a Type A Project," Ph.D. dissertation, Teachers College, Columbia University, 1963. Snyder headed training and directed the institute that ARAMCO and the Saudi government ultimately built some twenty years later.

43. ARAMCO, Relations Department, Monomonock Inn, Mountainhome, PA, Oct. 25 to Oct. 28, 1948 [T. C. Barger copy], Folder 25, Relations Department, Box 5, Mulligan Papers, emphasis mine.

44. Ibid., "Pressure for Improved Housing."

45. See Bamberg, *The History of the British Petroleum Company*, 99, for planning of a garden-city-type housing development for Iranian families along with acceleration of Iranization. The first mass actions by workers dated to the 1920s.

46. For the turn-from-paternalism argument, see Taylor and Lindemann, *The Creole Petroleum Corporation in Venezuela*, 42, 55; and Indonesia Project, Center for International Studies, MIT, *Stanvac in Indonesia*, Sixth Case Study in a National Planning Association Series on United States Business Performance Abroad (Washington, D.C.: NPA, 1957), 54–55, 83–90; but Larson, Knowlton, and Popple, *New Horizons*, also make clear that costs drove the change in Standard Oil's Venezuelan and Colombian camps in the 1940s. For the parallel movement inside the United States, see Roger and Diana David Olien, *Life in the Oil Fields* (Austin: Texas Monthly Press, 1986), 122.

47. See "Agenda for Second Day," discussion of "Community Development," ARAMCO, Relations Department, Monomonock Inn, Mountainhome, PA, Oct. 25 to Oct. 28, 1948 [T. C. Barger copy], Folder 25, Relations Department, Box 5, Mulligan Papers.

48. See "The Growth of Training Policy (1949–1959)," Folder 32, Box 5, Mulligan Papers.

49. See ms. marked "confidential," Arabian American Oil Co., Donations, Contributions, and Assistance to Saudi Arabia 1933–1970, Folder 10, Donations, Contributions and Assistance to Saudi Arabia; document headed "Examples of Representation Indicating Conflict of Company Interest in Our Approach to Saudi Arabian Government," Folder 13, Government Relations Organization Presentations; and document headed "Historical Survey ARAMCO Schools for Children of Its Employees (1952–1967)," Dhahran, 1967, Folder 14, Mulligan Papers.

50. Memo, Sept. 25, McGuire to Ness, General Policy Saudi Arabian Development, RG 59, 890F.51/9-2546.

51. Ibid.

52. Snyder to Sanger, Dec. 27, 1946, enclosing memo on proposed education survey in Saudi Arabia, RG 59, 890F.42/12-274.

53. Robert Norberg, "Saudi Arabs, Americans, and Oil," *Saudi-American Forum*, Essay series 10 (Mar. 20, 2003), www.saudi-american-forum.org/Newsletters/SAF_Essay_10.htm, accessed on Apr. 16, 2003.

54. In California and elsewhere oil was a white man's preserve, and it remained one of the most segregated industries across the United States through the Second World

War. See King and Risher, *The Negro in the Petroleum Industry.*

55. Ohliger to CASOC, June 30, 1941, S-220, "CASOC Schools for Saudi Arabia, with attached memo by J. G. Hosmer, Folder 32," Training Department, Box 5 Mulligan Papers. Hosmer oversaw these first training projects.

56. Details courtesy of Bill Mulligan, who wrote up the history of the early days of the Jebel School in 1973 in a piece he had hoped to publish in ARAMCO's newspaper, the *Arabian Sun.* As he noted in his records, "Ismail Nawwab, for a number of understandable reasons, couldn't bring himself to run it." Folder 32, Box 5, Mulligan Papers.

57. Joy Winkie Viola, *Human Resources Development in Saudi Arabia: Multinationals and Saudization* (Boston: International Human Resources Development Corporation, 1986), 1; Norberg, "Saudi Arabs."

58. Memo, "Education and Arab Training Conference, Dhahran, Oct. 23, 1949, Arab Trade Preparatory Schools," Folder 32, Box 5, Mulligan Papers.

59. See Snyder to Sanger, Dec. 27, 1946, RG 59, 890F.42/12-2746.

60. Ibid.

61. "The Growth of Training Policy (1949–1959)," Folder 32, Box 5, Mulligan Papers.

62. Hart to State, Report 125, Weekly Summary of Events, Aug. 8 through Aug. 14, 1949, RG 59, 890F.00/8-2049.

63. Department of State Instruction, No. CA-3384, Dec. 29, 1953, "Comment on the October–Nov 1953 Strike at the Arabian American Oil Company Installations in Saudi Arabia," p. 18, RG 59, 886A.062/12-2953.

64. See John McGowan, Sheffield, England to Mulligan, Aug. 28, 1986, Folder 2, "History Project," Box 8, Mulligan Papers.

Part 2

1. TV Guide Online, Movie Database, Entry for *Island of Allah*, Richard Lyford, 1956 (one star out of five), www.tvguide.com/Movies/database, accessed May 3, 2003.

2. Holly Edwards, "The Near East and the Wild West," in *Noble Dreams, Wicked Pleasures: Orientalism in America, 1870–1930* (Princeton, NJ: Princeton University Press, 2000), 44. In 1904 *Garden of Allah* was one of the most successful novels of the early twentieth century, reprinted forty-four times and turned into a Broadway play in 1907, which was revived in 1911. That revival fueled a remarkable campaign of tie-ins—perfume, lamps, hotel and department store promotions, and Maxfield Parrish's famous print. There was also a silent film version made by the Edison studio, a remake in 1926 by Metro Golden Mayer, starring Alice Terry and Ivan Petrovich, and a second United Artists' remake in 1936, with Charles Boyer and Marlene Dietrich in her first Technicolor feature.

3. William E. Mulligan, "The Arabian Peninsula in Living Color," column from the *Arabian Sun*, undated (but from the mid-1970s presumably), Box 6, Folder 15, "Arabian Sun Articles," Mulligan Papers.

4. One of the king's sons, Prince Muhammad bin Abd al-Aziz, told a U.S. embassy officer who appeared in the film (when still in ARAMCO's employ), "I enjoyed your acting, but from the point of view of history it was a hodge-podge of surmises and imagination." See Jidda to State, Dispatch 26, 20 July 1959, Prince Mohammed Deplores Arab Disunity, Attacks the Arab Press, and Extols Former King Abdul Aziz, RG 59, 786A.00/7-2059.

5. Industrial or business filmmaking began around the same time as company magazines, but the lyrical documentary style of *Island of the Arabs/Allah* was an innovation of the 1940s associated with Standard Oil's (NJ) production *Louisiana Story* (1948) by the legendary prospector-turned-documentary filmmaker Robert Flaherty (1884–1951). See Anthony Slide, *Before Video: A History of the Non-Theatrical Film* (New York: Greenwood, 1992), 102–3.

6. "The Story of a Modern American Family," *ARAMCO World* (Apr. 1955): 13–16, but see "Injun Talk," about a Chevron educational film made in 1946 and reported in the same issue as "Partners in Oil and Progress," *Bulletin* (Autumn 1946): 3–15.

7. See Dhahran to State, 118, June 21, 1951, Aramco's Utilization of Publicity Posters, RG 59, 886A.2553/6-2151.

8. William Eddy to Mary Eddy, Washington, Aug. 17, 1948, Box 6, Folder 3, William A. Eddy, 1948–49, Eddy Papers.

9. Spingarn to Dulles, Feb. 18, 1953, and enclosures, RG 59, 50/7/2 886A.2553/2-1853; also Beirut to State, 592, Mar. 18, 1954, RG 59, 886A.2553/3-1854.

10. Hoye to Ellender, Feb. 6, 1968, Folder 23, Correspondence, 1961–69, Box 11, Mulligan Papers.

Chapter 5

1. Hoskins, Confidential Extracts from Statement Regarding My Recent Trip to the Middle East, dated Aug. 16, 1950, Folder 8, Near East College Association, 1947–1957, Box 42, Allen W. Dulles Papers, Seeley G. Mudd Manuscript Library, Princeton University, Princeton, New Jersey.

2. David E. Long, *The United States and Saudi Arabia: Ambivalent Allies* (Boulder, CO: Westview, 1985), 34–35, 106–7; Brown, *Oil, God, and Gold*, 198. The correspondence on the grandfather's clock begins with State to Jidda, 121, Mar. 3, 1950, RG 59, 786A.11/3-350.

3. Anderson, *Aramco*, 189–97; Ronald Ferrier, "The Anglo-Iranian Dispute: A Triangular Relationship," in James Bill and Roger Louis, eds., *Mussadiq, Iranian Nationalism and Oil* (Austin: University of Texas Press, 1988): 172–79.

4. See 'Abd al-'Aziz Muhammad al-Dakhil, "'Abdallah Tariki, Oil, and the Nation," *al-Mustaqbal al-'arabi* 20, 226 (Dec. 1997): 38–52 (in Arabic). Eddy provided the account of the king's senility and the finance minister's incapacitation. Cairo to State, 354, Aug.

14, 1950, Conversation with Col. Eddy, Political Advisor to ARAMCO, Regarding Conditions in Saudi Arabia, RG 59, 886A.2553/8-1450.

5. London to State, 3350, Jan. 17, 1951, A British Comment on ARAMCO'S Labor Problems, RG 59, 886A.2553/1-1751. The quote is a close but not exact transcription.

6. Mary Eddy, "Continued from Dad's Letter," Feb. 27, 1950, Dhahran, Folder 6, Mary Garvin Eddy, 1950, Box 3, Eddy Papers.

7. Mary Eddy to Family, Feb. 17, 1950, Dhahran Guest House, Folder 6, Box 3, Eddy Papers.

8. Jidda to State, 186, Dec. 18, 1950, Saudi Arabian Government Hotel in Dammam, RG 59, 5886A.02/12-1850; Jidda to State, 282, Mar. 17, 1952, RG 59, 886A.00/3-1752.

9. Jidda to State, 64, Aug. 20, 1952, Survey Trip of Embassy Officer to Riyadh and Eastern Arabia, RG 59, 886A.00/8-2052; ARAMCO, "Translation Manual," May 15, 1954, Section titled "Saudi Arabs Prominent in Commercial, Cultural, and Similar Activities."

10. See Jidda to State, 64, Aug. 20, 1952, Survey Trip of Embassy Officer to Riyadh and Eastern Arabia, RG 59, 886A.00/8-2052.

11. See memorandum titled "Arabian Research Division," Aug. 1, 1953, Folder 17, ART, AAD, Etc., 1953, Mulligan Papers. Mine is a close paraphrase but not exact transcription.

12. From Abdelrahman Munif's tribute to Tariki, *al-Safir*, September 18, 1997. "Death slipped past in silence, in obscurity, and snatched away one of the most prominent sons of the Arabian Peninsula."

13. The only scholarly article on Tariki, by Stephen Duguid, "A Biographical Approach to the Study of Social Change in the Middle East: Abdullah Tariki as a New Man," *International Journal of Middle East Studies* 1, 1970: 195–220, calls him a protégé of Faisal. Later books and articles repeat the claim, but Duguid's work is both poorly sourced and underspecified.

14. Ibid.; "Biographic Report: Abdallah ibn Hamud al-TARIQI, Former Minister of Petroleum and Mineral Wealth" [dated Feb. 26, 1970], *CIA Research Reports, Middle East, 1946–1976* (Ann Arbor: University Publications of America, 1983), microfilm, reel 3; Edward Webb, "Tariki's Life and Thought—Research Note," 2003, unpublished paper. My discussion in this section draws on Webb's invaluable research assistance.

15. Munif, *al-Safir*, September 18, 1997.

16. Barger's notes, Lecture for Society of Petroleum Engineers, Folder 5, Barger, Thomas C., Biographical Info and Obits, Box 1, Mulligan Papers; *CIA Research Reports*.

17. Vassiliev, *History of Saudi Arabia*, 332; Blair, *Control of Oil*, 50; al-Dakhil, "'Abdallah Tariki, Oil, and the Nation," 38–52.

18. Wanda Jablonski, Interview with Tariki, *Petroleum Week*, Feb. 22, 1957: 22–23.

19. "'Ila 'aina nahnu masuqun?" *al-Yamamah* 1, 12 (1954): 30–31.

20. See *al-Riyadh*, September 18, 1997.

21. "Libyan Oil Politics," *Arab Oil and Gas Journal* 2 (Oct. 1965), reprinted in Walid

Khadduri, ed., *The Complete Works of Abdallah Tariki* (Beirut: Markaz dirasat al-wihdah al-ʿarabiyah, 1999): 87–88. In Arabic.

22. Metz interview with Webb, August 11, 2001.

23. Some called him "anti-American." "His dedicated Arab nationalism is reportedly deepened by painful memories of having been confused with Mexicans in Texas," *Time* magazine reported in 1958. Anthony Cave Brown insists dramatically that his mistreatment led directly to Tariki's "return to Riyadh" and a putative call for "expropriation of ARAMCO." Brown, *Oil, God, and Gold*, 152. Tariki did no such thing. This undocumented assertion is by no means the only reason not to trust what is probably the poorest-researched and most error-filled book I have read in ten years of working on the subject. Brown seems incapable of getting a single date right.

24. *Time*, Apr. 27, 1959: 22.

25. Barger's notes, Lecture for Society of Petroleum Engineers, Folder 5, Barger, Thomas C., Biographical Info and Obits, Box 1, Mulligan Papers, emphasis mine; *CIA Research Reports*. None of the observers of the time seemed as interested in Tariki's second marriage, to Maha Jumblatt, in Lebanon, where he settled until the mid-1970s.

26. Notes for a talk dated Aug. 13, 1952, Box 15, Folder 19, "Addresses: Impact of an American Industry," Eddy Papers. My notes are an exact copy of his notes but I have rendered these as full sentences for readability. Emphasis is in the original.

27. Letter, Quint, Dhahran to Hoye, Aramco Overseas Company, Beirut, Oct. 3, 1970, Folder 1, Mulligan Papers.

28. Dhahran to State, Incoming Telegram 83, Aug. 15, 1951, RG 59, 886A.2553/8-1551.

29. William Eddy to Mary Eddy, Aug. 18, 1951, Box 6, Folder 4, William A. Eddy, Correspondence, 1950–1951, Eddy Papers.

30. Dhahran to State, Incoming Telegram 83, Aug. 15, 1951, RG 59, 886A.2553/8-1551.

31. Dhahran to State, 58, Jan. 3, 1952, 50/9 886A.452/1-352. The unsent telegram is found in White House to John Simmons, Chief of Protocol, State, Mar. 29, 1952, RG 59, 786A.11/3-2952.

32. Cairo to State, 354, Aug. 14, 1950, Conversation with Colonel Eddy, Political Advisor to ARAMCO, Regarding Conditions in Saudi Arabia, RG 59, 886A.2553/8-1450.

33. Awalt's Memo to File, Aug. 21, 1951, RG 59, 886A.2553/8-2151.

34. Awalt, Memorandum of Conversation, Mar. 31, 1952, with Duce and others, RG 59, 986A.512/3-3152.

35. Cairo to State, 2787, May 24, 1951, New Developments in ARAMCO Operations, RG 59, 886A.2553/5-2451. On the turnover rates, see as well Tehran to State, 138, Aug. 1, 1951, Labor Attaché's Comments on Dhahran, Saudi Arabia, RG 59, 886A.2553/8-151.

36. Department of State Instruction, No. CA-3384, Dec. 29, 1953, RG 59, 886A.062/12-2953. Unfortunately, I have not been able to find additional materials on the strikes and slowdowns in the early 1950s.

37. Tehran to State, 138, Aug. 1, 1951, Labor Attaché's Comments on Dhahran, Saudi

Arabia, RG 59, 886A.2553/8-151.

38. Ibid.

39. Eakens, Memorandum of Conversation, Oct. 24, 1951, General Discussion of Saudi Arabian Oil Problems, RG 59, 886A.2553/10-2451

40. Tehran to State, 138, Aug. 1, 1951, Labor Attaché's Comments on Dhahran, Saudi Arabia, RG 59, 886A.2553/8-151.

41. Department of State Instruction, No. CA-3384, Dec. 29, 1953, "Comment on the October–Nov 1953 Strike at the Arabian American Oil Company Installations in Saudi Arabia," p. 18, RG 59, 886A.062/12-2953.

42. Memorandum of Conversation, Sept. 27, 1951, "Point IV Organization in Saudi Arabia," RG 59, 886A.00-TA/9-2751. The Saudi hands at State had to break the news to the clueless TCA official that "Saudi Arabia may not be ready for a grass-roots program at present."

43. Commanding Officer, Dhahran Airfield to Secretary of Defense, Telegram Ch-58, Aug. 16, 1951, RG 59, 786A.5-MAP/8-1651.

44. Self-description of ARAMCO from a draft memo dated Feb. 1, 1951, "Why a Research Division," Folder 16, ART, AAD, Etc., 1951, Box 2, Mulligan Papers.

45. Biographies in Box 2, Mulligan Papers; Metz's interviews with Edward Webb, Aug. 11, 2001, confirming biographical material in Mulligan papers; Julius Mader and Mohamed Abdelnabi, *Who's Who in CIA: A Biographical Reference Work on 3,000 Officers of the Civil and Military Branches of Secret Services of the USA in 120 Countries* (Berlin: Julius Mader, 1968), for Alter; interview by author with George Lane, former U.S. ambassador to Yemen (on Yassin's recruitment). Headley may not have been recruited until leaving ARAMCO in 1963. I am not sure.

46. Memorandum, Dhahran, Saudi Arabia, Oct. 22, 1947, Personal and Confidential, Report of Conversation during Audiences with King Abdul Aziz al Saud, at Riyadh, Oct. 20–21 1947, Folder 6, Box 2, Eddy Papers. It was widely believed in the court nonetheless that Eddy was a spy.

47. Eddy to Reverend William A. Eddy, Jr., Jan. 12, 1956, Folder 8, Box 8, General Correspondence, Eddy Papers.

48. See memo draft dated 2-1-51 titled "Why a Research Division," Folder 16 ART, AAD, Etc., 1951, Box 2, Mulligan Papers.

49. Department of State Instruction, No. CA-3384, Dec. 29, 1953, RG 59, 886A.062/12-2953.

50. On Abu Sunayd, see "Enclosure A" to Dhahran to State, 61, Nov. 4, 1953, Labor Disturbances in Eastern Saudi Arabia, RG 59, 886A.062/11-453. For criticism of and comparison to the school of commerce TCA was building in Jidda, see Jidda to State, 40, June 8, 1953, TCA Monthly Program Summary May 1953, RG 59, 886A.00-TA/6-853.

51. *Al-bilad al-saudiya*, Mar. 3, 1953, translation enclosed in Jidda to State, 279, Mar. 22, 1953, "Reforms of Crown Prince in al-Hassa," RG 59, 886A.00/3-2253; also Thomas

Pledge, *Saudi ARAMCO and Its People: A History of Training* (Dhahran: Saudi Arabian Oil Company, 1998), 48.

52. *Umm al-qura*, Mar. 6, 1953, translation enclosed in RG 59, 886A.00/3-2253; on the police mission and Point Four initiative, see Jidda to State, 34, Apr. 7, 1953, Technical Cooperation Administration Monthly Program Summary, Mar. 1953, RG 59, 886A.00-TA/4-753.

53. These are the transliterations used in the company and embassy documents. Salih Sad al-Zaid appears in the list of original scholarship students to AUB. See Page, *Saudi ARAMCO*, 46.

54. Department of State Instruction, No. CA-3384, Dec. 29, 1953, RG 59, 886A.062/12-2953, one of the long retrospective chronologies and commentaries I have relied on, but see the comment on sources below.

55. See Jidda to State, 393, June 27, 1953, "Further Measures of S[audi] A[rabian] G[overnment] Discrimination against ARAMCO," RG 59, 886A.2553/6-2753.

56. For the "agitator" quote, Dhahran to State, 61, Nov. 4, 1953, 886A.062/11-453; for the "Christian dogs" quote, Dhahran to State, 52, Oct. 13, 1953, 886A.06/10-1353.

57. This account is based primarily on the first comprehensive narrative of the strike and the events leading up to by the U.S. consulate in Dhahran to State, 61, Nov. 4, 1953, RG 59, 886A.062/11-453.

58. Ibid. Close but not exact transcription of the source text.

59. Quoted in al-Rasheed, *History of Saudi Arabia*, 98. A possible problem in this recollection is that the archival documents make it appear that the workers' committees themselves were created after the 1953 strike.

60. Dhahran to State, 70, Oct. 19, 1953, RG 59, 886A.062/10-1953; for the description of bin Issa, Dhahran to State, 78, Nov. 3, 1953, RG 59, 886A.062/11-353.

61. Dhahran to State, 71, Oct. 20, 1953, RG 59, 886A.062/10-2053.

62. Department of State Instruction, No. CA-3384, Dec. 29, 1953, RG 59, 886A.062/12-2953.

63. Dhahran to State, 72, Oct. 22, 1953, RG 59, 886A.062/10-2153.

64. Dhahran to State, 61, Nov. 4, 1953, Labor Disturbances in Eastern Saudi Arabia, RG 59, 886A.062/11-453.

65. Dhahran to State, 52, Oct. 13, 1953, Labor Agitation in Eastern Saudi Arabia, RG 59, 886A.06/10-1353; and Dhahran to State, 61, Nov. 4, 1953, 886A.062/11-453, which notes the lack of evidence of communist influence and the likelihood instead that Sunayd and the others were driven by a desire for power.

66. Dhahran to State, 78, Nov. 3, 1953, RG 59, 886A.062/11-353.

67. Eddy quote found in Department of State Instruction, No. CA-3384, Dec. 29, 1953, RG 59, 886A.062/12-2953.

68. See Dhahran to State, 61, Nov. 4, 1953, RG 59, 886A.062/11-453.

69. Phil McConnell, Diary entry, Oct. 27, 1953, 8, Box 3, McConnell Papers.

70. Department of State Instruction, No. CA-3384, Dec. 29, 1953, RG 59, 886A.062/12-2953.

71. For the deportations and new round of firings, see Dhahran to State, 61, Nov. 4, 1953, RG 59, 886A.062/11-453; on Said's termination in particular, see Dhahran to State, 174, Apr. 17, 1956, ARAMCO Discharges Labor Agitator, Nasir Al-Said, RG 59, 886A.06/4-1756. Said wrote that he was exiled, bound, to Hail in early November, and eventually confronted the new king, demanding a parliament, the release of the workers, the end of slavery, and other reforms. See Nasir al-Sa'id, *History of the House of Saud* (Ittihad sha'b al-jazirah al-'arabiyah, 1982?), 110–21. In Arabic. Note, no place of publication is indicated in this copy and edition.

72. Jidda to State, 198, Nov. 28, 1953, RG 59, 886A.062/11-2853; Dhahran to State, 90, Jan. 25, 1954, RG 59, 786A.11/1-2554 (the king's speech); Dhahran to State, 102, Feb. 8, 1954, ARAMCO Concessions in Labor Dispute with Employees, RG 59, 886A.062/2-854.

73. Rome to State, 1949, Dec. 21, 1953, RG 59, 886A.2553/12-2153.

74. Dhahran to State, 105, Feb. 24, 1954, Reappearance of Exiled Strike Leaders in Dhahran Area, RG 59, 886A.062/2-2454.

75. Dhahran to State, 12, Aug. 25, 1954, Distribution of Subversive Leaflet in al-Khobar, RG 59, 50/1/2 786A.00/8-2554.

Chapter 6

1. The most comprehensive accounts I found are Dhahran to State, 212, June 20, 1956, Disturbances and Partial Strike by ARAMCO Saudi Workers, RG 59, 886A.062/6-2056; and Dhahran to State, 7, July 14, 1956, Further Information on Aramco Labor Unrest and Government Action June 9–21, RG 59, 886A.062/7-1456. In this latter dispatch, the reporting officer commends the USAF OSI report as the most detailed account available. The OSI was created in 1948 and modeled on the Federal Bureau of Investigation.

2. Al-Sa'id, *Tarikh Al Sa'ud*, 119.

3. See Jidda to State, 13, Further Information on Anti-Subversive Measures by SAG, RG 59, 786A.52/7-2655. I should note that many dispatches and the like in the U.S. State Department records from this period dealing with military security, countersubversion, and the royal family itself remain classified. Consider this account reported by Pakistani ambassador Khwaja Shahabuddin in this light.

4. See Jidda to State, 260, May 23, 1959, Saudi Reluctance to Discuss USMTM Problems May Result from a Disinclination to Strengthen the Regular Army, RG 59, 786A.5/5-2359.

5. See Dhahran to State, 89, May 25, 1955, Aramco Labor Dissatisfaction, RG 59, 786A.00/5-2555, for the views on the arrests of PPS members; and Dhahran to State, 93, June 1, 1955, Weekly Report of Political and Economic Developments in the Dhahran Consular District, RG 59, 786A.00/6-155, for the arrests of suspected communists.

6. A source at ARAMCO provided the embassy with an initial account of Muammar's arrest on May 3 in his room at the government hotel at the Dhahran Airport. He was seized by Colonel Muahmmad al-Dib and two members of the king's bodyguard who took him by plane back to Riyadh. See Dhahran to State, Telegram 157, May 4, 1955, RG 59, 886A.2553/5-455. Later in May, a top labor relations manager for the Texas company updated the embassy. Dhahran to State, 91, May 25, 1955, Conversation with an Executive of ARAMCO Industrial Relations Concerning Labor Dissatisfaction, RG 59, 886A.2553/5-2555.

7. For the description of Muammar's alleged role as pamphleteer and the most detailed outline I have found of his various positions in the 1950s, see Dhahran to State, 62, Abd Al Aziz Ibn Mu'ammar offered Position as Deputy Minister of Finance for National Economy by SAG, RG 59, 786A.13/8-1559.

8. Dhahran to State, 129, Abd al-Aziz Mu'ammar to be Released, Aug., 2, 1956, RG 59, 786A.52/2-856.

9. I have been unable to determine the fate of the other two men arrested with Muammar, according to U.S. embassy records. One was Khalif al-Hussaini, an Egyptian who worked as a legal advisor to the Council of Ministers, and the other was Muhiyadd-din Namaani, a Lebanese national who owned an auto repair shop in Jidda. See Jidda to State, 187, Saudi Government Combats Subversion, May 31, 1955, RG 59, 786A.52/5-3155.

10. *Department of State Bulletin* 36, 917, Jan. 21, 1957, 83–87.

11. Winton Beaven, ICPA, to Eisenhower, Feb. 7, 1957, RG 59, 786A.11/2-757; Ferris Fitzpatrick, Lansing, MI, to Dulles, Feb. 24, 1957, RG 59, 786A.11/2-2457; Telegram, Michael J. Quill, New York CIO Council, to Dulles, Jan. 25, 1957, RG 59, 786A.11/1-2557; *New York Times*, Feb. 1, 1957: 3.

12. Mayor Wagner's stance was backed by the city council's Democratic majority leader, Joseph Sharkey, "1,000 per cent. When we entertain half these monkeys who come over here, it's on orders of the State Department." Sharkey later "regretted his use of the word 'monkeys.' He had meant to say 'characters.'" Charles Bennett, "Mayor Bars Fete for Saud," *New York Times*, Jan. 30, 1957: 1.

13. Almost one-half century later, in the fall of 2001, Mayor Rudolph Giuliani returned a $10 million donation from Prince Alwaleed Ibn Talal, the son of the man who in 1961–62 would try to lead a constitutional revolution in the kingdom, before fleeing to Cairo.

14. See "Report For ARAMCO by A. J. Toynbee," n.d., copy enclosed in Jidda to State, 217, May 6, 1958, Toynbee on ARAMCO and Saudi Arabia, RG 59, 886A.2553/5-658.

15. Wadsworth to Allen, Dec. 15, 1954, RG 59, 886A.512/1-3155.

16. See Robert Vitalis, *When Capitalists Collide: Business Conflict and the End of Empire in Egypt* (Berkeley: University of California Press, 1995), for the American ambassador's turn from champion of the Egyptian Wafd Party to critic of its failed domestic policies after the party acted in ways counter to American preferences. Diplomats all

seemed to have learned the Claude Rains part in *Casablanca* (1942) where Captain Louis Renault professes his shock at learning that gambling is going on at Rick's place, orders the bar closed, but not before pocketing his night's winnings.

17. Excepts from the testimony of Fred Davies, Chairman of the Board of ARAMCO, Accompanied by James Terry Duce, Vice President, and Douglas Erskine, Tax Counsel, Before the Subcommittee on Antitrust and Monopoly of the Senate Common Judiciary and Subcommittee on Public Lands of the Senate Committee on Interior and Insular Affairs, Washington, D.C., Mar. 20–21, 1957, pp. 8–9, Box 2, McConnell Papers.

18. Jidda to State, 63, Nov. 17, 1955, Review of Events for Oct. 1955, RG 59, 786A.00/11-1755, for more of what ARAMCO Americans loved to say was really just anti-Zionism.

19. Jidda to State, Telegram 527, June 28, 1954, RG 59, 886A.00-TA/6-2754 (for the king's decision on Point Four).

20. My summary is based on the superb recounting in Nathan Citino, "Defending the 'Postwar Petroleum Order': The U.S., Britain and the 1954 Saudi-Onassis Tanker Deal," *Diplomacy & Statecraft* 11, 2 (July 2000): 137–60. I am citing from a manuscript copy of the article so have omitted page references to avoid confusion.

21. See the memorandum from Dorsey to Murphy, Oct. 4, 1954, Recent Development Regarding Onassis Agreement, RG 59, 786A.2553/10-454.

22. The contradiction was fully spelled out in an editorial "Getting into Step" in the London *Times* (Jan. 12, 1956) which noted how loans by ARAMCO translated into Saudi "mischief" in the form of subsides to Arab parties aligned against the West's more reliable Hashemite clients in Jordan and Iraq. ARAMCO's Terry Duce objected, saying it was New York bankers that were advancing the funds, not ARAMCO (but they did so in close coordination with the oil company, based on the old formula of royalties owed by the firm to the kingdom but not yet paid). State Department, Memorandum of Conversation, James Terry Duce and David Newsom, Jan. 13, 1956, ARAMCO Payments to Saudi Arabia, RG 59, 786A.00/1-1356.

23. Citino, "Defending the 'Postwar Petroleum Order.'"

24. Jidda to State, 78, Sept. 15, 1959, Professor George Lenczowski Discusses His Interview with Crown Prince Faisal, RG 59, 786A.11/9-1959. "He had spoken with Shaykh Abdullah Tariqi in the past few days, and while he admired Tariki as a sincere Arab nationalist, he felt that his oil policies, particularly his preference for 'integration' of oil operations, even at the expense of possible additional revenues, was not favorable to progress for the kingdom. Tariqi although determined and emotionally sensitive on the issue, was at least a reasonable man with whom matters could be discussed frankly, unlike the radical and emotionally unbalanced Arab nationalist, Abdul Aziz Muammar."

25. Jidda to State, 227, Apr. 27, 1959, Head of Foreign Office Gives Personal Views of Difficulties of Making Saudi Arabia a Modern State, RG 59, 786A.00/4-2759.

26. USIS Jidda, 52, Sept. 4, 1958, Education in Saudi Arabia, RG 59, 886A.432/9-458. Emphasis mine.

27. See for example Philby's first article on the kingdom following his exile (London *Sunday Times*, Oct. 23, 1955), where, with Sulayman's replacement by Muhammad Surur, he insisted (although it was hardly true) "there is nobody now able to resist the fantastic and unproductive drain on its resources." Précis in London to State, 947, Oct. 25, 1955, RG 59, 786A.00/10-2655.

28. Reporting on these matters took place regularly. For example, see Jidda to State, 41, Economic Summary for Saudi Arabia, First Two Quarters 1956, RG 59, 886A.00/8-1856.

29. See Jidda to State, 160, Jan. 31, 1957, Progress of Transfer of Ministries from Jidda to Riyadh, RG 59, 786A.13/1-3157.

30. Jidda to State, 140, Mar. 21, 1956, Fortnightly Review of the Economy in Saudi Arabia, Mar. 1–15, RG 59, 786A.00/3-2156.

31. See Jidda to State, 143, Economic and Financial Review: A Survey, Mar. 24, 1956, RG 59, 886A.00/3-2456.

32. State Department, Memorandum of Conversation, Nov. 8, 1957, U.S.–Saudi Arabian Relations, Abdul Rahman Azzam Pasha, Special Representative of King Saud, and David Newsom, RG 59, 611.86A/11-857.

33. Dhahran to State, Telegram 157, May 4, 1955, RG 59, 886A.2553/5-455. Copies of the workers' original and follow-up letters to ARAMCO are enclosed in Dhahran to State, 105, June 17, 1955, Secret Labor Group's Letters to ARAMCO Board Chairman Fred A. Davies, RG 59, 886A.2553/6-1755.

34. Dhahran to State, 88, May 19, 1955, Agitation by ARAMCO Employees [missing file number and enclosure's title].

35. Dhahran to State, 93, June 1 1955, Weekly Report of Political and Economic Developments in the Dhahran Consular District, RG 59, 786A.00/6-155.

36. Dhahran to State, 23, Aug. 24, 1955, Weekly Report, RG 59, 786A.00/8-2455; Dhahran to State, 28, Aug. 31, 1955, RG 59, 786A.00/8-3155.

37. William Eddy to Family, Ainab (Lebanon), Sept. 25, 1955, Box 3, Folder 10, Eddy Papers. He only returned to Beirut as the Americans began their occupation, during which he aided the military command's senior political advisor, Deputy Undersecretary of State Robert Murphy, in making contact with Lebanese political factions—pretty impressive for a consultant to ARAMCO on industrial relations.

38. Mary Eddy to Family, Dhahran, Saudi Arabia, Dec. 19, 1955, Box 3, Folder 10, Eddy Papers.

39. Dhahran to State, 101, Dec. 22, 1955, RG 59, 786A.00/12-2255.

40. Dhahran to State, 115, Jan. 14, 1956, ARAMCO Labor Demonstrations against New Fences around Camps, RG 59, 886A.2553/1-1456. The direct quotation in the next paragraph is from this long and detailed dispatch.

41. See letter, Barger to Duce, Dhahran, Jan. 30, 1956, and Barger's Memorandum to File, Dhahran, Jan. 21, 1956, Fencing General Camps, both enclosed in RG 59, 886A.O6/1-3056. Needless to say, this is ARAMCO's version of events as recorded in a long memo

that Barger wrote later in January for Ohliger's use in a meeting with the king about the latest disturbances.

42. Quotes not attributed specifically to Barger are from the account in Dhahran to State, 115, ARAMCO Labor Demonstration, RG 59, 886A.2553/1-1456.

43. Barger Memorandum, in RG 59, 886A.O6/1-3056, emphasis mine. The embassy's report said the appeal had failed and that the firm appealed to the governor, Ibn Jiluwi, who phoned the amir and instructed him to cancel the rally.

44. Barger Memorandum, in RG 59, 886A.06/1-3056; Dhahran to State, 112, Jan. 14, 1956, Weekly Report of Political and Economic Developments in Eastern Saudi Arabia, RG 59, 786A.00/1-1456; Dhahran to State, 115, Jan. 14, 1956, ARAMCO Labor Demonstrations against New Fences around Camps, RG 59, 886A.2553/1-1456; Dhahran to State, 116, Jan. 21, 1956, Weekly Report, 786A.00/1-2156.

45. See Dhahran to State, 133, Feb. 16, 1956, Biweekly Report of Political and Economic Developments in Eastern Saudi Arabia, RG 59, 786A.00/2-1656.

46. Dhahran to State, 125, Feb. 2, 1956, Biweekly Report of Political and Economic Developments, RG 59, 786A.00/2-256.

47. Dhahran to State, 150, Mar. 15, 1956, Biweekly Report of Political and Economic Developments in Eastern Saudi Arabia, RG 59, 786A.00/3-1556.

48. Barger Memorandum in RG 59, 886A.O6/1-3056.

49. Dhahran to State, 143, Mar. 1, 1956, Biweekly Report of Political and Economic Developments in Eastern Saudi Arabia, RG 59, 786A.00/3-156.

50. Dhahran to State, 160, Mar. 31, 1956, Biweekly Report, RG 59, 786A.00/3-3156. The direct quotation below is from the same dispatch.

51. Dhahran to State, 145, Mar. 8, 1956, Labor Petition Presented to King Saud by ARAMCO Employee, RG 59, 886A.06/3-856.

52. See "Demands of Saudi Workmen in the Arabian American Oil Company Stated in a Petition to His Majesty the King and Dictated by the Permanent Royal Committee on Labor Affairs to Sayyid Muhammad Jamil al-Shawwa for Transmittal to the Company," 24 Rabi' II 1375, Dec. 10, 1955, Translation from Arabic with Interposed Point-by-Point Explanations or Comments by ARAMCO Management, enclosed in Dhahran to State, 145, Mar. 8, 1956, Labor Petition Presented to King Saud by ARAMCO Employee, RG 59, 886A.06/3-856.

53. Dhahran to State, 173, King Saud Enunciates Policy of Cooperation with ARAMCO, RG 59, 786A.11/4-1756.

54. Dhahran to State, 174, Apr. 17, 1956, ARAMCO Discharges Labor Agitator, Nasir Al-Said, RG 59, 886A.06/4-1756.

55. Dhahran to State, 182, Apr. 28, 1956, Biweekly Report, RG 59, 786A.00/4-2856; Dhahran to State, 195, May 12, 1956, Biweekly Report, RG 59, 786A.00/5-1256.

56. Dhahran to State, 154, Mar. 24, 1956, Biweekly Report of Political and Economic Developments in Bahrain, Qatar, and Oman, RG 59, 786A.00/3-2456.

57. Dhahran to State, 199, May 26, 1956, Biweekly Report of Political and Economic Developments in Eastern Saudi Arabia, RG 59, 786A.00/5-2656.

58. Dhahran to State, 91, May 25, 1955, Conversation with an Executive of ARAMCO Industrial Relations Concerning Labor Dissatisfaction, RG 59, 886A.2553/5-2555

59. Dhahran to State, 182, Apr. 28, 1956, Biweekly Report, RG 59, 786A.00/4-2856; and Dhahran to State, 183, Apr. 30, 1956, Hints of Saudi Officials' Interest in Labor Unions, RG 59, 886A.062/4-3058. The latter dispatch includes the translation of al-Rayyis's piece, "Industrial Relations and Social Progress," from al-Yamamah, no. 29, Apr. 8, 1956, which I have excerpted here.

60. Dhahran to State, 200, May 29, 1956, New Royal Commission Investigating Labor and Related Problems in the Eastern Province, RG 59, 886A.06/5-2956

61. Dhahran to State, 210, Organized Demonstrators Disturb King Saud on Way to Banquet at ARAMCO, RG 59, 786A.11/6-1356.

62. Dhahran to State, 208, June 9, 1956, Biweekly Report, RG 59, 786A.00/6-956.

63. Dhahran to State, 212, June 20 1956, Disturbances and Partial Strike by ARAM-CO Saudi Workers, 886A.062/6-2056.

64. Dhahran to State, 215, June 25, 1956, Biweekly Report, 786A.00/6-2556; Dhahran to State, 7, July 14, 1956, Further Information on ARAMCO Labor Unrest and Government Action, June 9–21, RG 59, 886A.062/7-1456. The account of events below is drawn from these two dispatches. I have not tried to find a copy of the USAF intelligence report, but it is clear that it includes details, for instance of names of arrestees, that are not available elsewhere.

65. Dhahran to State, 215, June 25, 1956, Biweekly Report, RG 59, 786A.00/6-2556; Dhahran to State, 7, July 14, 1956, Further Information on ARAMCO Labor Unrest, RG 59, 886A.062/7-1456.

66. Jidda to State, 107, Nov. 8, 1958, ARAMCO Estimate of Labor Unrest in the Eastern Province, RG 59, 886A.06/11-1558.

67. Dhahran to State, 12, July 24, 1956, Armed Forces Television Station at Dhahran Airfield, 986A.50/7-2456.

68. Dhahran to State, 74, Mar. 11, 1957, ARAMCO's Television Station in Dhahran, RG 59, 986A.50/3-1157; and Dhahran to State, 175, Nov. 1, 1958, King Saud Shows Interest in TV Station for Riyadh, 986A.50/11-158.

69. State to Dhahran, Telegram 264, Jan. 9, 1957, RG 59, 786A.11/1-957; E. W. Kenworthy, "President to See Saud at Airfield," New York Times, Jan. 26, 1957: 9.

70. "Worried Oil Monarch," New York Times, Jan. 30, 1957: 2.

71. Dana Adams Schmidt, "Saud's Visit Seen as a Mideast Key," New York Times, Jan. 31, 1957: 1; Schmidt, "Saud Is Eager for Serious Talks, King's Aides Say," New York Times, Jan. 29, 1957: 5 (for "father of the Arabs" quote); David Newsom, Memorandum of Conversation, Jan. 15, 1957, RG 59, 786A.11/1-1557.

72. "Worried Oil Monarch," New York Times, Jan. 30, 1957: 2.

73. Memo dated Jan. 31, 1957, Box 41, Saudi Arabia, Ann Whitman Files, President Dwight D. Eisenhower's Office files, 1953–61, Part 2: International Series, Eisenhower Library, Abilene, Kansas. For the exchange of views with Dulles, see State Department, Memorandum of Conversation, Feb. 1, 1957, White House, Whitman Files, Box 41, Saudi Arabia. Note too that an unnamed Saudi official was revealing all the same things publicly to the *New York Times* a few days later.

74. For an outline of the terms, see Gordon Grey, Assistant Secretary of Defense, to Robert Murphy, Deputy Undersecretary of State, Feb. 4, 1957, RG 59, 786A.11/2-457. Also Dana Adams Schmidt, "U.S.–Saudi Talks Reach an Accord on Base and Aid," *New York Times*, Feb. 8, 1957: 1.

75. See the account of the impromptu press conference in Dana Adams Schmidt, "President Hails Saud Visit, King Praises Mideast Plan," *New York Times*, Feb. 7, 1957: 1. The memorandum from the meeting has been heavily redacted. All of the king's requests are still secrets of state except one. Might the communiqué specify that the target of American development support was "the country as a whole and not mention restricting economic development to the Dhahran area only"? Eisenhower replied that this was "put in for public relations reasons since he had to tie economic help to mutual defense. He suggested however that the communiqué state that the U.S. agreed to provide economic facilities that would serve to augment the mutual security of both countries." See Department of State, Memorandum of Conversation, Feb. 8, 1957, White House, Box 41, Saudi Arabia, Ann Whitman Files, President Dwight D. Eisenhower's Office files, 1953–61, Part 2: International Series, Eisenhower Papers.

76. "Eisenhower-Saud Text," *New York Times*, Feb. 9, 1957: 2.

77. Jidda to State, 272, Fortnightly Review of Events in Saudi Arabia, May 1–15, RG 59, 7786A.00/5-2257; Memorandum, Rountree to Murphy, Saudi Arabia: Request from King Saud for Help in Replacing Egyptian Military Personnel, Apr. 23, 1957, RG 59, 786A.5-MSP/4-2357.

78. Jidda to State, 46, Aug. 28, 1958, Conversation with Ghalib Rifa'i, RG 59, 786A.00/8-2858.

79. Editorial, "Struggle for Arabia," *New York Times*, June 15, 1957: 16. "In the struggle for the soul of the Arab peoples now under way between the pro-Soviet and imperialist elements led by President Nasser and the anti-Communist forces backing the principles of the Eisenhower Doctrine, King Saud of oil-rich Saudi Arabia is obviously emerging as the key figure." Also see the next day's story by the staunchly anticommunist Cairo correspondent, Osgood Caruthers, "King Saud Now Plays Key Role in Mideast," *New York Times*, June 16, 1957: 3.

80. Jidda to State, 289, Fortnightly Review of Events in Saudi Arabia, May 16–31, 1957, RG 59, 786A.00/6-1157.

81. Jidda to State, 305, Fortnightly Review of Events in Saudi Arabia, June 1–15, 1957, RG 59, 786A.00/6-2757.

82. Jidda to State, 8, Fortnightly Review of Events in Saudi Arabia, June 16–30, 1957, RG 59, 786A.00/7-1157.

83. Jidda to State, 81, Oct. 17, 1957, Documents Concerning ARAMCO–Saudi Arabian Government Income Tax Issue, RG 59, 886A.112/10-1757 (reviewing the course of negotiations, June–August 1957); "A Super Aramco Could Lift Profits, Saudi Aide Asserts," *New York Times*, Sept. 17, 1957: 47; and Sam Pope Brewer, "Saudi Oil Terms for U.S. to Stand," *New York Times*, Dec. 17, 1957: 5 (threatening to press for revisions of the ARAMCO fifty–fifty norm).

84. For full details, see the careful account by Douglas Little, "Cold War and Covert Action: The U.S. and Syria, 1945–1958, *Middle East Journal* 48, 1 (1990): 51–75.

85. Jidda to State, 55, Fortnightly Review of Events in Saudi Arabia, Sept. 1–15, 1957, RG 59, 786A.00/9-1857; Jidda to State, 72, Fortnightly Review, Sept. 16–30, 1957, RG 59, 786A.00/10-957.

86. Jidda to State, Telegram 535, RG 59, 786A.5-MSP/12-1057.

87. Jidda to State, 72, Fortnightly Review, Sept. 16–30, 1957, RG 59, 786A.00/10-957.

88. Memorandum by Robert Murphy to Eisenhower, Sept. 20, 1957, Subject: Meeting with Crown Prince Faisal of Saudi Arabia, White House, Box 41, Saudi Arabia, Ann Whitman Files, President Dwight D. Eisenhower's Office files, 1953–61, Part 2: International Series, Eisenhower Papers.

89. Jidda to State, 112, Fortnightly Review, Nov. 1–15, 1957, RG 59, 786A.00/11-2057; Jidda to State, Fortnightly Review, Dec. 16–31, 1957, RG 59, 786A.00/1-958; and Jidda to State, 151, Fortnightly Review, Jan. 1–15, 1958, RG 59, 786A.00/1-2158.

90. Jidda to State, 167, Fortnightly Review, Feb. 1–15, 1958, RG 59, 786A.00/2-1958.

91. Osgood Caruthers, "Nasser Is Pressing His Attack on Saud," *New York Times*, Mar. 9, 1958: 1.

92. Dhahran to State, Telegram 537, Mar. 26, 1958, RG 59, 786A.00/3-2658.

93. Jidda to State, Telegrams 897, RG 59, 786A.00/3-1558; 929, RG 59, 786A.11/3-2158; and 943, RG 59, 786A.00/3-2358.

Chapter 7

1. Telegram, Graham to Stegner, July 7, 1955, Folder 5, Correspondence, Ray Graham 1955; Letters, Stegner to Phillips, Aug. 3, 13, 22 (muckraking quote), and Oct. 2, 1955, Folder 17, Correspondence, T. O. Phillips, 55–56, 1968; contract dated Sept. 19, 1955, Folder 27, Contract for Services of Consultant, 1955, 1958; Box 29, Discovery Correspondence, Stegner Papers. "'Island of Allah' Bows," *New York Times*, June 27, 1955: 35.

2. Stegner to H. O. Thompson, Draft, Dhahran, Nov. 14, 1955, Box 29, Discovery Correspondence, Folder 19, Correspondence, H. O. Thompson, 1955–56, Stegner Papers.

3. Stegner to Sinclair, Mar. 7, 1956, Folder 18, Correspondence, Angus Sinclair, 1955–56, Box 29, Discovery Correspondence, Stegner Papers.

4. Stegner to Thompson, Jan. 5, 1957; Thompson to Stegner, Apr. 2, 1957; Stegner to Thompson, Apr. 6, 1957; Stegner to Thompson, Dec. 6, 1957; Thompson to Stegner, Dec. 9, 1957; and Stegner to Thompson, Dec. 16, 1957, Folder 19, Correspondence, H. O. Thompson, 1955–58, Box 29, Discovery Correspondence, Stegner Papers.

5. Stegner to Thompson, Jan. 24, 1958, Folder 19, Correspondence, H. O. Thompson, 1955–58, Box 29, Discovery Correspondence, Stegner Papers.

6. Stegner to Thompson, Jan. 24, 1958, Folder 19, Correspondence, H. O. Thompson, 1955–58, Box 29, Discovery Correspondence, Stegner Papers.

7. See the annotated manuscript chapters in Folder 2, First Draft II, The Negotiators, Box 26, Discovery, First Draft, Reader's Copy; and Folder 3, III, The Wildcatters; Folder 7, VII, pp. 302–12; and Folder 9, The Wildcatters, Box 27, Discovery, Stegner Papers.

8. Ibid.

9. Folder 2, First Draft II, The Negotiators, and Folder 7, First Draft, VII, Commercial Production, Box 26 Discovery, First Draft, Reader's Copy, Stegner Papers.

10. Tore Tingvold Peterson, "Anglo-American Rivalry in the Middle East: The Struggle for the Buraimi Oasis, 1952–1957," *International History Review* 14 (1992): 71–91, "greatest obstacle" quote on 87; Nathan Citino, *From Arab Nationalism to OPEC: Eisenhower, King Sa'ud and the Making of U.S.-Saudi Relations* (Bloomington: University of Indiana Press, 2002), 82–85, 90–91; Brown, *Oil, God, and Gold*, 212. Homer Mueller, the ARAMCO Arabist who met secretly with Egyptian and Saudi intelligence officers to promote a rebellion against the British in southeastern Arabia, was himself an undercover CIA officer.

11. Stegner to Thompson, Jan. 24, 1958, Folder 19, Correspondence, H. O. Thompson, 1955–58, Box 29, Discovery Correspondence, Stegner Papers.

12. Ibid.

13. Copy of Contract between Stegner and ARAMCO in Folder 27, Contract for Services of Consultant, 1955, 1958, Box 29, Discovery Correspondence, Stegner Papers. Stegner was to receive an additional flat fee of $9,250, just under $57,000 today, to produce a final version, including an introductory chapter that would situate the pioneer story and bring it up to date.

14. Hamilton to Stegner, Oct. 2, 1958, Folder 6, Correspondence, Gordon C. Hamilton, 1957–58, Box 29, Discovery Correspondence, Stegner Papers.

15. Al-Rasheed, *History of Saudi Arabia*, 117–21.

16. The historian Salim Yaqub is correct that we ought to view the final year or so of the Eisenhower administration as an effort to turn in the direction that the new Kennedy administration would go in 1961 when it sought closer ties with the United Arab Republic. See the key chapter "Death of a Doctrine, July 1958–December 1960," in *Containing Arab Nationalism: The Eisenhower Doctrine and the Middle East* (Chapel Hill: University of North Carolina Press, 2004).

17. Frank Hendryx, "A Sovereign Nation's Legal Ability to Make and Abide by a

Petroleum Concession Contract," in Secretariat General of the League of Arab States, Proceedings of the First Arab Petroleum Congress, Cairo, Apr. 1960. The copy in my possession is from Box 2, Folder 3, Ibn Muammar Papers, Los Angeles. For ARAMCO's initial reaction, see Jidda to State, 884, Apr. 22, 1959, RG 59, 886A.2553/4-2259.

18. Bill Mulligan, "Darwish Building," Box 7, Folder 22, "Dammam Building," Mulligan Papers.

19. ARD, Chronological Files, Mar. 1954, Box 2, Folder 38, Mulligan Papers.

20. See the account of a "Statement by a Saudi Clandestine Group," the Union of Sons of the Peninsula, published in the Beirut weekly *al-Huriyah*, May 8, 1960, and found in Box 3, Folder 7, AAD, Chronological File, May–June 1961, Mulligan Papers.

21. See James Knight to Garry Owen, May 17, 1960, and enclosure, draft, dated May 16, 1960, "Why Should Aramco Handle Its Relations with the Government through a Company Department Specially Designated for That Purpose," Box 5, Folder 13, Mulligan Papers; and Confidential Memorandum to Files, Special Committee, Dec. 30, 1957, Box 2, Folder 4, Mulligan Papers.

22. See the delayed response to a request for a description of the Arabian Research Division during the time of the October 1953 strike, Barger to Ohliger, Apr. 22, 1954, "The Arabian Research Division of the Arabian American Oil Company," Box 5, Folder 13, Mulligan Papers.

23. Confidential Memorandum to Files, Special Committee, Dec. 30, 1957, Box 2, Folder 57, Mulligan Papers. My transcription is a close but not exact copy of the original text.

24. See Khalil Semaan, Harpur College, Binghamton, NY, to R. G. Von Peursen, ARAMCO, Shoreham Bldg., Washington, D.C., Nov. 22, 1965, Box 5, Folder 1, Mulligan Papers. Semaan had been the translator.

25. See Mulligan's Biography of Rentz, Box 1, Folder 53, Rentz, George S.—Correspondence re, 1963–87, Mulligan Papers. For the research topics, see John Pendleton to R. A. Eeds, et al., July 20, 1957, enclosing memorandum, "Operational Considerations Arabian Research Division," Box 2, Folder 57, ARD, Chronological Files, July–Dec. 1957, Mulligan Papers.

26. Confidential Memorandum to Files, Special Committee, Dec. 30, 1957, Box 2, Folder 57, Mulligan Papers.

27. Memorandum, Confidential, by Pendleton, Special Committee, First Meeting, Jan. 15, 1958, Box 2, Folder 58, Mulligan Papers.

28. Memorandum, Confidential, Pendleton, ARAMCO Objectives, Feb. 24, 1958, Box 2, Folder 58, Mulligan Papers.

29. See the document labeled Confidential, Dhahran, Feb. 11, 1962, GRO Presentation, EMC/Budget Committee, Feb. 12, 1962, Box 5, Folder 13, Mulligan Papers.

30. See "The Growth of Training Policy (1949–1959)," Box 5, Folder 32, Mulligan Papers.

31. See Confidential, ARAMCO, Donations, Contributions, and Assistance to Saudi Arabia, 1933–70, Box 5, Folder 10, Mulligan Papers.

32. See B. C. Nelson, Manager, Employee Relations Department, "Employee Relations—Then and Now," Box 5, Folder 16, ARAMCO Management Development Seminar, Mar. 6, 1965, Mulligan Papers. For a short account at the time of the change in policy see, Memorandum by Nelson, Coordinator, Labor Relations Division to W. E. Squires, Assistant Director, Industrial Relations Department, Dhahran, Dec. 30, 1958, Special Labor Relations Report, The Boys in the Back Room, Box 6, Folder 22, Mulligan Papers.

33. Salman headed the Arab League's petroleum bureau. He planned the conference originally for March 1957. My account is based on the documentation and analysis that Rentz and others in Arabian Affairs prepared. See the binder-full of material, First Arab Petroleum Congress, Cairo, UAR, 16–23, Apr. 1959, Box 4, Folder 13, Mulligan Papers.

34. I have my doubts, but it is true nonetheless that Bustani's death just weeks before the fourth congress in 1963 left ARAMCO officials lamenting that there was "no third party present to appeal to logic and commercial reasoning or inject other issues to harass or humor the delegates. . . . Industry [thus] took a much worse beating in this one." ARAMCO, Fourth Arab Petroleum Congress, Beirut, Lebanon, Nov. 5–12, 1963, report dated Jan. 26, 1964, Box 4, Folder 16, Mulligan Papers.

35. For Tariki's response to the trading company proposal, see Jidda to State, 157, Dec. 19, 1959, Fortnightly Economic Review, Dec. 1–15, 1959, RG 59, 886A.00/12-1959; for the development foundation proposal, see Jidda to State, 296, Armco Proposal to Establish an Industrial Foundation in Saudi Arabia, June 25, 1959, RG 59, 886A.19/6-2559.

36. Jidda to State, 90, Fortnightly Economic Review, Sept. 1–15, 1959, RG 59, 786A.00/9-1759.

37. See Jidda to State, 242, RG 59, 786A.11/10-1459.

38. Jidda to State, 116 Fortnightly Economic Review, Nov. 16–31, 1958, RG 59, 786A.00/12-458.

39. Richard Sanger to Gordon Arneson, State, INR, Continued Tension within Saudi Royal Family, Oct. 1, 1958, RG 59, 786A.11/10-158; Jidda to State, 841, Apr. 2, 1959, RG 59, 786A.13/4-259; Jidda to State [from the military attaché], 349, Nov. 27, 1959, 786A.00/11-2759.

40. Jidda to State, 78, Sept. 1, 1959, Professor George Lenczowski Discusses His Interview with Crown Prince Faisal, RG 59, 786A.11/9-159.

41. See Beirut to State, 597, Feb. 15, 1961, RG 59, 886A.2553/2-1561.

42. Jidda to State, 75, Aug. 26, 1959, Prince Faisal's Policy of Cultivating Young Intellectuals, RG 59, 786A.11/8-2659.

43. Dhahran to State, 62, Aug. 15, 1950, RG 59, 786A.13/8-1559. I am grateful to Haifaa Muammar and her family members for the information about her father's time in Austria.

44. Jidda to State, 111, Nov. 15, 1958, Economic Summary, Second and Third Quarters

of 1958, RG 59, 886A.00/11-1558.

45. He was endorsing Lenczowski's view. Jidda to State, 78, Sept. 1, 1959, Professor George Lenczowski Discusses His Interview with Crown Prince Faisal, RG 59, 786A.11/9-159.

46. For Heath's frank exchange with the king and his comparison of the two positions on oil, see Jidda to State, 348, RG 59, 886A.00/11-2659; and Jidda to State, 141, Fortnightly Review, Nov. 16–30, 1959, RG 59, 786A.00/12-559. On the change in the willingness to loan the king cash, see Jidda to State, 1268, RG 59, 786A.11/6-2358; Dhahran to State, 606, RG 59, 786A.11/4-2959; and Dhahran to State, 617, RG 59, 786A.11/5-459.

47. Confidential, Dhahran, Feb. 11, 1962, GRO Presentation, EMC/Budget Committee, Feb. 12, 1962, Box 5, Folder 13, Mulligan Papers.

48. The papers were turned over to St. Antony's at Oxford more than a decade later, but the correspondence in the Mulligan Papers makes clear that the firm understood its first priority to be to find and keep out of other people's hands any materials thought "embarrassing" to either the firm or to the royal family. See, e.g., Mulligan, Manager, Research and Services, to General Manager, Gov. Relations, Dhahran, Feb. 6, 1972, Disposition of the Philby Papers, Box 1, Folder 45, Philby, H. St. John, Correspondence re 1971, Mulligan Papers. The firm also vetted Elizabeth Monroe's biography of Philby, which drew on the papers.

49. See report of conversation with Carl Barnet, assistant Aramco representative, Riyadh, Oct. 23, 1960, enclosed in Jidda to State, 155, Nov. 2, 1960, Reform in Saudi Arabia Both a Royal Pawn and a Reformist End, RG 59, 886A.00/11-260.

50. Ibid.

51. For Heath's views, see Jidda to State, 264, Apr. 4, 1960, RG 59, 886A.00/4-460. The wording here is a close but not exact transcription.

52. For Anwar Ali, see the confidential report on Ali's conversation with Heath, enclosed in Jidda to State, 284, Apr. 18, 1960, Fortnightly Economic Review, Apr. 1–15, 1960, RG 59, 886A.00/4-1860. For the view of Harold Folk, who headed the World Bank mission, see Jidda to State, G 80, June 8, 1960, RG 59, 886A.00/6-860.

53. For St. John (Abdallah) Philby's deeply cynical perspective, see Jidda to State, 217, Feb. 6, 1960, Philby Says Money is the Root of Saudi Arabia's Evil, RG 59, 786A.00/2-660.

54. Jidda to State, 589, Mar. 22, 1960, RG 59, 886A.2553/3-2260.

55. Jidda to State, 442, Jan. 10, 1960, RG 59, 786A.00/1-1060.

56. Jidda to State, 860, June 24, RG 59, 786A.00/6-2460.

57. For *al-Nahar*, see Jidda to State, 359, June 19, 1960, Revolutionary Group in Saudi Arabian Army, RG 59, 786A.00/6-1960. For *al-Jumhuriyya*, see Jidda to State, 818, June 9, 1960, RG 59, 786A.00/6-960. Date of the interview was not reported.

58. Jidda to State, 832, June 12, 1960, RG 59, 786A.00/6-1260.

59. Jidda to State, 838, June 16, 1960, RG 59, 786A.00/6-1660.

60. See Jidda to State, 359, June 19, 1960, Revolutionary Group in Saudi Arabian Army, RG 59, 786A.00/6-1960; Jidda to State, 36, July 20, 1960, RG 59, 786A.00/7-2060; Dhahran to State, 62, July 27, 1960, RG 59, 786A.00/7-2760; and Jidda to State, 76, Aug. 1, 1960, RG 59, 786A.00/8-1660. Readers should keep in mind that much remains classified in the file series I am drawing on in these pages.

61. Jidda to State, 249, Dec. 19, 1960, Periodic Assessment of Political-Military Trends and Foreign Affairs in Saudi Arabia, Mar. 15–Dec. 15, 1960, RG 59, 786A.00/12-1960.

62. For Faisal's growing anger at the firms' pricing decision, see Jidda to State, 71, Aug. 17, 1960, Biweekly Economic Review, Aug. 3–16, 1960, RG 59, 886A.00/8-1760; for Tariki's campaigning and ARAMCO's understanding of the crisis, see Jidda to State, 87, Aug. 31, 1960, Biweekly Economic Review, Aug. 17–31, 1960, RG 59, 5/886A.00/8-3160.

63. See Dhahran to State, 282, Apr. 6, 1960, Annual Petroleum Report for Saudi Arabia, RG 59, 886A.2553/4-660.

64. See Jidda to State, 175, Nov. 21, 1960, Aramco Representative Explains Cleavage between Abdullah Tariki and Abdul Aziz bin Mu'ammar, RG 59, 886A.2553/11-2160.

65. Dhahran to State, 349, Dec. 26, 1960, RG 59, 886A.2553/12-2660.

66. Jidda to State, G-95, Feb. 6, 1961, RG 59, 786A.00/1-2661.

67. See Jidda to State, 416, Feb. 14, 1961, RG 59, 786A.00/2-1460, for the quotations contained in this and the following paragraph. Additional details from the meeting with Metz and others are provided in Jidda to State, 275, Feb. 25, 1961, RG 59, 786A.00/2-2561.

68. Beirut to State, 757, Feb. 16, 1961, RG 59, 786A.00/2-1661, reporting views of unnamed ARAMCO officials, one of whom was clearly Duce, carrying on once more about Ibn Muammar being a menace and consorting with communists in Vienna, a charge for which he admitted he had no evidence. Also see the *Economist*, Mar. 18, 1961, "Arabia's New Men."

69. Jidda to State, 275, Feb. 25, 1961, RG 59, 786A.00/2-2561. The travel writer James (now Jan) Morris was one. She brutally dissected the "regal rearguard" fought by "flabby and threadbare" Saud, "an aristocrat gone to seed. . . . A strange man, with a sick look to his eyes." Morris was apparently not much impressed with the Saudi Arabian revolution—"one of the most politically stagnant of all states. Scarcely a breath of liberal democracy has yet rustled the trappings of the royal house." James Morris, "Saud—Master of the Rich Desert," *New York Times Magazine*, Feb. 19, 1961: 32.

70. Dhahran to State, 441, Feb. 15, 1961, RG 59, 786A.00/2-156.

71. Jidda to State, 275, Feb. 25, 1961, Conversations in Riyadh with Mu'ammar, Turaiqi, Wahba (Jr.), Bihaijan and an Aramco Official about the New Outlook in the Saudi Arabian Government, RG 59, 786A.00/2-2561.

72. See Dhahran to State, 358, Dec. 31, 1960, RG 59, 886A.2553/12-3160.

73. See Jidda to State, 263, Feb. 7, 1961, Tariki Press Interview Alleged Need for Changing Agreement with ARAMCO, RG 59, 886A.2553/2-761.

74. See Dhahran to State, 17, July 17, 1961, ARAMCO Voices Concern to King Saud over Recent Press Interviews of Shaikh Abdullah Tariki and Prince Talal, RG 59, 886A.2553/7-1761; and for the public relations countercampaign, Dhahran to State, 269, Apr. 25, 1961, Annual Petroleum Report for Saudi Arabia, RG 59, 886A.2553/4-2561.

75. Jidda to State, 369, Jan. 10, 1961, RG 59, 786A.00/1-1061.

76. Dhahran to State 441, Feb. 15, 1961, RG 59, 786A.00/2-1561; Jidda to State, 238, Jan. 19, 1961, Apprehensions of Pakistani Ambassador over Leftist Influence on New Saudi Arabian Government, RG 59, 786A.00/1-1961; Jidda to State, 416, Feb. 14, 1961, RG 59, 786A.00/2-1461.

77. Dhahran to State 441, Feb. 15, 1961, RG 59, 786A.00/2-1561.

78. Jidda to State, 433, Feb. 19, 1961, RG 59, 786A.13/2-1961. The file is also the source for the al-Saqqaf quote that follows.

79. Heath continued, "It would almost seem he is disintegrating before the pressure of the opposing members of the royal family and clutching at straw." See Jidda to State, 479, Mar. 13, 1961, RG 59, 611.86A7/3-1361; and State to Jidda, 366, Mar. 13, 1961, RG 59, 611.86A7/3-1361.

80. Jidda to State, 544, Apr. 10, 1961, RG 59, 786A.11/4-1161; and State to Jidda, 402, Apr. 11, 1961, RG 59, 786A.11/4-1161.

81. Dana Adams Schmidt, "King Saud's Foes Step Up Activity," *New York Times*, May 23, 1961: 4.

82. See Dhahran to State, 148, Dec. 27, 1960, Discrimination in the Eastern Province of Saudi Arabia against the Shia and Their Increasing Discontent, RG 59, 886A.413/12-2760; and Dhahran to State, 71, Sept. 9, 1961, Visit to Hofuf, RG 59, 886A.00/9-961.

83. See Jidda to State, A-68, Sept. 27, 1961, RG 59, 786A.13/9-2761.

84. See Jidda to State, 88, Sept. 25, 1961, Enclosure, Memorandum of Conversation, Political Situation in Saudi Arabia, RG 59, 786A.00/9-2561.

Chapter 8

1. Amr would console himself with an Algerian actress turned mistress and with hashish. See Kirk Beattie, *Egypt during the Nasser Years: Ideology, Politics and Civil Society* (Boulder, CO: Westview, 1994), 160.

2. Ibid. The plot was uncovered in January 1962.

3. See Dhahran to State, 170, Jan. 18, 1962, RG 59, 786A.00/1-2362, enclosing summaries and partial transcripts of Voice of the Arabs broadcasts on November, 25, December 5, and during the week of December 6–19, 1961.

4. See State Department, Memorandum of Conversation, Current Developments in Saudi Arabia, Jan. 23, 1962, RG 59, 786A.00/1-2362.

5. See Jidda to State, [telegram number not obvious], Jan. 3, 1962, 686A.86B/1-362.

6. See Jidda to State, A-142, Mar. 23, 1962, RG 59, 686A.86B/3-2462.

7. Jidda to State, A-143, Apr. 4, 1962, RG 59, 686A.86B/4-462.

8. See New York to State, 1667, Nov. 16, 1961, RG 59, 786A.11/11-1661; Dhahran to State, 153, Nov. 18, 1961, RG 59, 786A.11/11-1861; State to Jidda, 175, Nov. 27, 1961, RG 59, 786A.11/11-2761; and Chronology, compiler not listed, no date [1962], "Saudi Arabian News Chronology," Box 3, Folder 15, Mulligan Papers.

9. New York to State, 1982, Dec. 6, 1961, RG 59, 786A.11/12-661.

10. For a broadly revisionist but carefully argued account of the Eisenhower administration's late shift toward seeking "common ground," in effect, paving the way for the Kennedy advisors' initiative, see Yaqub, *Containing Arab Nationalism*. For JFK's efforts at constructive engagement, see Douglas Little, "New Frontier on the Nile, *Journal of American History* 25 (1988): 501–27; and chapters 2 and 3 of the superbly crafted Warren Bass, *Support Any Friend: Kennedy's Middle East and the Making of the U.S.-Israel Alliance* (New York: Oxford University Press, 2003), on which I have relied here.

11. See Objectives of the Mission to the United Arab Republic of Dr. Edward Mason, enclosed in Memorandum from Battle to Bundy, Washington, Feb. 27, 1962, *Foreign Relations of the United States, 1961–63*, Vol. 17, Near East, 1961–62, No. 201, www.state. gov/r/pa/ho/frus/kennedyjf/xvii/17713.htm, accessed on Dec. 4, 2004.

12. See Bass, *Support Any Friend*, 72–143 for details.

13. Ibid., 76.

14. See State Department Memorandum, Strong (NE) to Talbot (NEA), Jan. 31, 1962, RG 59, 786A.11/1-3162.

15. State Department, Memorandum of Conversation, by Seelye, President–King Saud Meeting, Feb. 13, 1962, *Foreign Relations of the United States, 1961–63*, Vol. 17, Near East, 1961–62, No. 191, www.state.gov/r/pa/ho/frus/kennedyjf/xvii/17713.htm, accessed on Dec. 5, 2004.

16. See Jidda to State, 151, Nov. 29, 1961, RG 59, 886A.2553/11-2961.

17. Jidda to State, 219, Feb. 8, 1962, Conversation with Saudi Petroleum Minister Abd Allah Tariki, RG 59, 886A.2553/2-862.

18. See Kuwait to State, A-134, Apr. 10, 1962, RG 59, 886A.2553/4-1062; Jidda to State, 637, Apr. 15, 1962, RG 59, 886A.2553/4-1562.

19. ARAMCO's Washington representative, Garry Owen, thought the report of Tariki's musings on regime change disturbing enough to alert the State Department. See 886A.2553/11-562. The source was an unnamed, "reliable" contact who had seen Tariki in Cairo.

20. The journalist Robert Lacey, who was living in the kingdom in the late 1970s while researching his book *The Kingdom: Arabia and the House of Saud* (New York: Harcourt Brace Jovanovich, 1981), had met Tariki in Riyadh (340). Although Lacey's account of this era is based mainly on interviews, the now available archival evidence shows his to be one of the most reliable (if under-referenced) books about politics in the 1950s and 1960s.

21. See Dhahran to State, 265, June 16, 1962, Eastern Province Public Security Arrests, RG 59, 786A.00/6-1662; Dhahran to State, A-25, July 14, 1962, RG 59, 786A.00/7-1462.

22. Baghdad to State, 533, Feb. 17, 1962, Transmitting Newspaper Report of Clandestine Organization in Saudi Arabia, RG 59, 786A.00/2-1762.

23. See Dhahran to State, 267, Mar. 30, 1962, RG 59, 986A.61/3-3062; Jidda to State, A-144, Mar. 31, 1962, RG 59, 986A.61/3-3162; and Jidda to State, A-148, Apr. 14, 1962, RG 59, 986A.61/4-462.

24. Jidda to State, 86, Aug. 3, 1962, RG 59, 786A.00/8-362; Jidda to State, 107, Aug. 16, 1962, RG 59, 786A.00/8-1662; "Kinsman of Saud Asks for Reform," *New York Times*, August 16, 1962: 4.

25. Cairo to State, 302, Aug. 20, 1962, RG 59, 786A.00/8-2062.

26. Dana Adams Schmidt, "Saudi Oil Money Put to New Uses," *New York Times*, May 12, 1962: 5; and "Saudis Expunge a Popular Image," *New York Times* May 13, 1962. See also, Jidda to State, 124, Aug. 25, 1962, RG 59, 786A.00/8-2562, p. 21

27. State Department, memcon, Aug. 28, 1962, Political Situation in Saudi Arabia, RG 59, 786A.00/8-2862.

28. Quotation is from Jidda to State, 99, Aug. 30, 1962, RG 59, 786A.00/8-3062.

29. Dhahran to State, A-41, Listeners to Voice of the Arabs Arrested, July 18, 1962, RG 59, 786A.00/7-1862; Dhahran to State, A-52, Aug. 4, 1962, RG 59, 786A.00/8-462; Dhahran to State, A-57, Arrests of Teen-Agers Reported, Aug. 11, 1962, RG 59, 786A.00/8-1162.

30. Jidda to State, 214, Sept. 29, 1962, RG 59, 786A.00/9-2962.

31. See Cairo to State, 491, Sept. 24, 1962, RG 59, 786A.00/9-2462.

32. See State Department memcon, Aug. 8, 1962, ARAMCO–Saudi Arabia Negotiations—OPEC Resolutions, with Owen, George Ballou, Manager, Eastern Hemisphere Operations of SOCAL, and others, RG 59, 886A.2553/8-862.

33. Dhahran to State, A-4, July 7, 1962, RG 59, 786A.00/7-762; Dhahran to State, A-50, July 25, 1962, RG 59, 786A.00/7-2562; State to Damascus, 31, Aug. 29, 1962, RG 59, 886A.2553/8-2962.

34. Jidda to State, A-58, Aug. 2, 1962, Memorandum of Conversation with Kamal Adham, RG 59, 786A.00/8-262.

35. Paul Dresch, *A History of Modern Yemen* (Cambridge, UK: Cambridge University Press, 2000), 93.

36. Dhahran to State, 77, Oct. 3, 1962, RG 59, 786A.00/10-362; Cairo to State, 559, Oct. 3, 1962, RG 59, 786A.00/10-362; Dhahran to State, 78, Oct. 3, 1962, RG 59, 786A.00/10-362.

37. Cairo to State, 564, Oct. 4, 1962, RG 59, 786A.00/10-462.

38. Jidda to State, 292, Oct. 15, 1962, RG 59, 786A.00/10-1562, reporting a two-day observation trip to Riyadh, Oct. 9–10.

39. Jidda to State, A-161, Oct. 6, 1962, Saudi Reaction to Defection of RSAF Officers, RG 59, 786A.00/10-662.

40. See Jidda to State, 296, Oct. 16, 1962, RG 59, 686A.86H/10-1662; Jidda to State, A-

183, Oct. 25, 1962, RG 59, 686A.86H/10-2562; and Memorandum from the Joint Chiefs of Staff to Secretary of Defense McNamara, JCSM-875-62, Washington, Nov. 9, 1962, Possible US Military Support to the Saudi Arabian Regime, National Archives and Records Administration, RG 218, JCS Records, 1962 Files, 9180/3100 (Oct. 19, 1962). Secret, *Foreign Relations of the United States, 1961-1963*, Vol. 18, Near East, 1961-62, Document 92, www.state.gov/r/pa/ho/frus/kennedyjf/xviii/26200.htm, accessed Dec. 26, 2004.

41. Cairo to State, 606, Oct. 11, 1962, RG 59, 786A.00/10-1162. Also see the account of Dana Adams Schmidt, the *New York Times* reporter, of his discussions with Talal in the book he would publish a few years later, *Yemen: The Unknown War* (London: Bodley Head, 1968), 51.

42. Cairo to State, 585, Oct. 8, 1962, RG 59, 786A.11/10-862.

43. Beirut to State, A-480, Nov. 19, 1962, RG 59, 786A.00/11-1962. The al-Said quote is from an interview he gave while in Yemen to the Lebanese weekly *al-Hawadith* (Nov. 23, 1962) and reported to Washington by the embassy. Beirut to State, 517, Nov. 23, 1962, RG 59, 686A.86B/11-2362.

44. Beirut to State, 567, Dec. 19, 1962, RG 59, 786A.00/12-1962. The quotation above is an expanded and thus altered version of a telegram's typical clipped prose.

45. "5 Exiled Kinsmen of King Saud Ask Permission to Return Home," *New York Times*, August 24, 1963: 4. The subtitle reads "Prince Talal and 4 Comrades Abandon Their Policy of Opposition to Monarch."

46. Memorandum from Robert W. Komer of the National Security Council Staff to President Kennedy, Washington, Oct. 4, 1962, Kennedy Library, National Security Files, Countries Series, Saudi Arabia, 10/62, *Foreign Relations of the United States, 1961–63*, Vol. 18, Near East, 1961–62, Document 68, www.state.gov/r/pa/ho/frus/kennedyjf/xviii/26200.htm, accessed Dec. 26, 2004.

47. Jidda to State A-161, Oct. 6, 1962, RG 59, 786A.00/10-662.

48. See, for instance, Bass's fast-paced reconstruction of the meeting, drawn from State Department and presidential library documents, in *Support Any Friend*, 104. He did not read U.S. ambassador Parker Hart's memoir, apparently.

49. For instance, compare Madawi al-Rasheed's account in *History of Saudi Arabia*, 120–23 (which emphasizes the mythmaking element in the Faisal-as-modernizer narrative) with Vassiliev, *History of Saudi Arabia*, 364–66, Safran, *Saudi Arabia*, 97, Lacey, *The Kingdom*, 345, among others. Even Schmidt, *Yemen*, 53, leads with the banning of slavery.

50. Hart, *Saudi Arabia and the United States*. The book appeared the year after his death.

51. Ibid., 114.

52. Peres's prediction at the time of the Yemen crisis, when he was Israel's deputy defense minister, is quoted in Bass, *Support Any Friend*, 118.

53. Dhahran to State, 95, Oct. 11, 1962, RG 59, 786A.00/10-1162.

54. See Jidda to State, A-212, Nov. 15, 1962, RG 59, 886A.40/11-1562.

55. David Holden, *Farewell to Arabia* (New York: Walker, 1966), 134–36.

56. Gary Samore, "Royal Family Politics in Saudi Arabia (1953–1982)," Ph.D. dissertation, Harvard University, 1983, 177; Safran, *Saudi Arabia*, 95.

57. It is worse. See Lippman, *Inside the Mirage*. He gets the year and the details about the reforms wrong, but he is nonetheless sure that credit belongs to Faisal and those he promoted, and blunter than most in his caricature of a hapless king. He writes that Faisal's reforms were launched within days of his becoming king in 1964 (112) thereby fulfilling his pledge to Kennedy (141–44). One problem is thus his fragile grasp of the record. One part of his brief for Faisal versus Saud is the alleged failure of the latter to make use of the good offices of the Ford Foundation inside the kingdom. Nonetheless, his own account shows that the foundation had been reluctant to fund work in the kingdom when Saud was requesting aid (since Saudi Arabia could pay); and that, smart enough or no, Lippman's own sources show it was King Saud and not Faisal who sought Ford Foundation support in 1963, unless he is misquoting them. Later in the discussion he notes that the first Ford Foundation project actually began almost two years earlier, and had nothing to do with Faisal, but everything to do with the progressives who established the Institute of Public Administration, about which he knows only what the Ford Foundation documents tell him. The institute was a UN-supported reform launched in 1960, under Saud, although he allegedly did not even comprehend the idea of development. Unfortunately, Lippman did not use the only historical study to date on developments in the Saud-Faisal era based on what were then newly available archival sources, published by Sarah Yizraeli in 1997, years earlier than his. Her book goes far to explode the precious myths about Faisal.

58. See Dhahran to State, 95, Oct. 11, 1962, RG 59, 786A.00/10-1162.

59. Yizraeli, *Remaking of Saudi Arabia*, 141.

60. Spencer Mawbry, "The Clandestine Defence of Empire: British Special Operations in Yemen, 1951–64," *Intelligence and National Security* 17 (Fall 2002): 105–30.

61. Faisal had to give up the dream of restoring the imamate, but he could absorb the cost. "The Yemeni army and security services were purged of leftists. Many royalists returned from Saudi Arabia; some of them were appointed to the supreme administrative bodies." Faisal's government recognized the now right-tilting republic in July 1970 and looked forward to his and his successors wielding increased influence over the north, prefigured in the defense pact Yemen signed with the Saudis in March 1971. Vassiliev, *History of Saudi Arabia*, 378.

62. As the historian Warren Bass puts it, "the attempt to woo Nasser became an attempt to deter him from waging war on Saudi Arabia." Bass, *Support Any Friend*, 115.

63. Hart, *Saudi Arabia*, 250.

64. Herman Eilts, review of Parker Hart, *Saudi Arabia and the United States*, in *Middle East Policy* 6, 4 (1999) n.p. in electronic edition, www.mepc.org/public_asp/journal_vol6/eilts.html, accessed Dec. 30, 2005.

65. "Saudi Arabia Bars Nationalization," *New York Times*, Nov. 9, 1963: 47. Tariki had called for nationalization of the industries in all the producer states.

66. See Dhahran to State, 128, Nov. 14, 1962, RG 59, 886A.2553/11-1462.

67. See the remarkable Jidda to State, 309, Jan. 17, 1963, RG 59, 686A.861/1-1763 XR 786A.13. I leave it to others to reconcile this account with Lacey's, *The Kingdom*, 347, "Faisal held war rallies in Riyadh which he addressed with a vigour and eloquence no one has suspected. He was a natural and inspiring orator."

68. One of those arrested as a communist and later released was an ARAMCO employee turned writer, Ibrahim al-Nasir (possibly Ibrahim al-Nasir Humaydan?), who published a book of short stories about work in the TAPLINE province, *Ard bila Matar* (Land without Rain), in 1967. I have not been able to locate a copy. I note this for others who, in some better world, might want to document the fate of those who lived under Faisal's reign.

69. See John Jones to GRO, Mar. 3, 1965, Enclosing report by Quint, untitled, confidential, reporting on the Ford Foundation mission, Box 3, Folder 23, AAD, Chronological Files, Jan.–Apr. 1965, Mulligan Papers.

70. See Quint, Memo to file, July 15, 1965, Trip to Riyadh, Box 3, Folder 24, AAD, Mulligan Papers.

71. Hart, *Saudi Arabia*, 265 n5; Lacey, *The Kingdom*, 381. The journalist unfortunately never explains how he knows what Faisal said. I leave the quotation marks in because he uses them, but take it literally at your peril.

72. Haifaa Muammar to Vitalis, Feb. 9, 2005. Published sources are unclear and contradictory about the date of Ibn Muammar's arrest. His daughter has her mother's calendar from the year of the arrest, with the date marked, "so that we would never forget."

73. Harry McDonald, letter excerpt dated July 13, 1963, recipient unspecified but presumably Mulligan, Box X, Folder 13, "McDonald Letters," Mulligan Papers.

74. Ludovic Kennedy, *Very Lovely People: A Personal Look at Some Americans Living Abroad* (New York: Simon Schuster, 1969), 321. Kennedy was a BBC personality who visited Dhahran in 1965. "It bears little relation to the American way of life it is said to represent. The essence of small town American life is government by democracy, but government by ARAMCO is benevolent dictatorship. . . . To live as an ARAMCO employee is to be a citizen of a superbly well-organized welfare state." He commented too on women's isolation and often alienation, "with perhaps the bottle or the Church or the occasional affair to help them along the road," 311.

75. Nora Johnson, *You Can Go Home Again* (New York: Doubleday, 1982), 42. She says she wrote the paragraph I quoted "just after I left there, in 1958, and know I can never tell better—awkwardness and all—what is was like there then."

76. Joy Wilson, "Raising Children in an American Oil Camp in Saudi Arabia," *Radcliffe Quarterly* (Mar. 1977): 8–9.

77. Speers to Mulligan, Apr. 5–6, 1977, Box 10, Mulligan Papers. On his and Ellen's background, see Eddy to Duce, memo, LWN-142, ARAMCO, Washington, D.C., Mar. 25, 1949, "Report and Recommendations Regarding Special Employee-Trainees 1949/1950," Folder 11, Box 2, Mulligan Papers; and Brown, *Oil, God, and Gold*, 142–45.

78. Biographical material on Kimball is from the Inventory of the Solon Toothaker Kimball Papers, Newberry Library, Special Collections, Chicago, www.newberry.org/collections/FindingAids/kimball/kimball.html, accessed on Feb. 16, 2005. Brown, *Oil, God, and Gold*, gets Kimball's story ("Barger called in experts to make a study of the town for the review of the senior staff") and the citation for the paper ("presented to the U.S. Anthropology Section of Oceanography and Meteorology"—-what can this possibly mean?) wrong, spectacularly, as usual, 140 and 383 n26. See instead Solon T. Kimball, "American Culture in Saudi Arabia," *Transactions of the New York Academy of Sciences*, Series 2 (18, 5), 1956: 469–84.

79. Johnson, *You Can Go Home Again*, 45–46. For examples from ARAMCO archives of the easy racism, parties, and pranks, including another Banana Nose incident, see Vitalis, "Aramco World."

80. See Confidential, Memo to File, Dhahran, Feb. 21, 1967, Miss Lorraine A. Landry, 90899, Folder 15; and his typescripts on Catherine Brown and Lovelind Hoel, Folder 18, Box 9, Mulligan Papers.

81. It is unclear why O'Dea was hired. He had spent 1955–56 at Stanford's Center for Advanced Study in the Behavioral Sciences, where Stegner was then completing his manuscript. For O'Dea's biography, see Robert Michaelsen et al., 1976, University of California: In Memoriam, Thomas F. O'Dea, Sociology; Religious Studies: Santa Barbara, 1915–1974, http://dynaweb.oac.cdlib.org:8088/dynaweb/uchist/public/inmemoriam/inmemoriam1976/@Generic__BookTextView/1896;pt=1962, accessed Feb. 13, 2005.

82. See Mulligan to Barger, Memo, Oct. 2, 1968, Social Changes in Saudi Arabia: Five Years after "the O'Dea Report," AAD Chronological Files, July–Dec. 1968, Folder 31, Box 3, Mulligan Papers.

83. Thomas F. O'Dea, "Social Change in Saudi Arabia: Problems and Prospects," Report Prepared from Research Conducted by the Special Study Group, [George Maranjian, Salma Najjar, O'Dea, and Federico S. Vidal], Feb.–Aug. 1963, unpublished, copy in J. Willard Marriott Library, Special Collections, University of Utah, Salt Lake City, 1–2, emphasis mine.

84. Ibid., 161.

85. Ibid., 19–20.

86. Ibid., 33–34.

87. See Malcolm Quint, Memo to file, Mar. 25, 1964, Conversation with Employee of Minypet Jiddah, Folder 20, Box 3, Mulligan Papers.

88. O'Dea, "Social Change," 34.

89. Ibid., 51–52.

90. Ibid., 137–38.

91. Ibid., 146–47.

92. Ibid., 163–64.

93. Ibid., 165–67.

94. Ibid., 168–75.

95. Mary Eddy, excerpts of condolences, typescript, Folder 5, Condolences, 1962, Box 1, Eddy Papers.

96. Parker T. Hart to Mary Eddy, Jidda, Saudi Arabia, Oct. 14, 1963, Folder 4, Box 4, Eddy Papers.

97. Mary Eddy to Zelia [?], from the Barger house, Dhahran, Saudi Arabia, June 3, 1964, Folder 5, Box 4, Eddy Papers.

98. "Malcolm Says He Is Backed Abroad," *New York Times*, May 22, 1964: 22.

99. See Jack and Marion White, "Comments on Draft Code of Behavior," enclosed in White to Mulligan, n.d., Folder 16, P.T.A. at ARAMCO Schools, Box 16, Mulligan Papers.

100. See Quint, Memo to files, July 20, 1964, Employee Reaction to Boycott of Company Facilities, Folder 18, Box 5, Mulligan Papers.

101. ARAMCO translator's (A. H. Kamal) summary, Dec. 19, 1966, *Shuara al-rass al-nabitiyun* [The Vernacular Poets of al-Rass], Collected and edited by Fahd al-Rushayd, volume 1, al-Hashimiyah Press, Damascus, 1965, AAD, Chronological Files, June–Dec. 1966, Folder 26, Box 3, Mulligan Papers.

Afterword

1. See the two Memoranda of Conversations, May 17, 1945, Saudi Arabian Finance, and May 28, 1945, Financial Assistance to Saudi Arabia, RG 59, 890F.51/5-1745 and 890F.51/5-2845.

2. See these tired tropes of generations of company historians pressed into service yet again in Schwantes, *Vision and Enterprise*, preface, xix.

3. Crawford, *Building the Workingman's Paradise*, 37–45, 101–25, 174–95.

4. Roland Marchand, *Creating the Corporate Soul: The Rise of Public Relations and Corporate Imagery in American Big Business* (Berkeley: University of California Press, 1998), 108–10.

5. Cairo to State, A-470, Dec. 10, 1962, RG 59, 886A.2553/12-1062. I have made the clipped text of the telegram readable and corrected a transcription error found in the original.

6. For the covert joint ventures, see Rachel Bronson, *Thicker Than Oil: America's Uneasy Partnership with Saudi Arabia* (Oxford: Oxford University Press, 2006); for military modernization, see Joshua Teitelbaum, "Civil-Military Relations in Saudi Arabia," paper presented at the Conference on Saudi Futures, International Institute for the Study of Islam in the Modern World, Leiden, Feb. 19–21, 2004; for petrodollar recycling,

see David Spero, *The Hidden Hand of American Hegemony: Petrodollar Recycling and International Markets* (Ithaca, NY: Cornell University Press, 1999); for arms sales and oil pricing, see Vitalis, "The Closing of the Arabian Oil Frontier"; and for imagining how oil and hegemony matter to one another in the period after ARAMCO's nationalization, see Simon Bromley, "The United States and World Oil," *Government and Opposition* 40, 2 (2005): 225–55.

7. Michael Watts, "Resource Curse? Governmentality, Oil and Power in the Niger Delta, Nigeria," *Geopolitics* 9, 1 (2004): 50–80.

8. Gregory Llewelyn, "Developers Grab Death Warrant for Saudi Camp, Town with a Soul," *Arab News*, July 23, 1979, Box 6, Folder 19, ARAMCO, Mulligan Papers.

BIBLIOGRAPHY

Government Archives

Dwight D. Eisenhower Presidential Library, Abilene, KS.
    President Dwight D. Eisenhower's Office files, 1953–61, Part 2: International Series.

National Archives and Record Administration, College Park, MD.
    Record Group 59, General Records of the Department of State.
    Record Group 218, Records of the Joint Chiefs of Staff.

Private Manuscript Collections

'Abd al-'Aziz Ibn Mu'ammar Papers, Los Angeles, California

American Heritage Center, University of Wyoming, Laramie
    Frank Ayer Collection
    John G. Douglas Collection
    James Terry Duce Collection
    Harold B. Hoskins Collection
    Richard H. Sanger Collection
    C. Stribling Snodgrass Collection

Hoover Institution Library and Archives, Stanford University, Palo Alto, California
    Philip C. McConnell Papers

Lauinger Library, Special Collections, Georgetown University, Washington, D.C.
    John DeNovo Papers
    William E. Mulligan Papers

J. Willard Marriott Library, Special Collections, University of Utah, Salt Lake City
    Wallace Earle Stegner Papers

Seeley G. Mudd Manuscript Library, Princeton University, Princeton, New Jersey
Allen W. Dulles Papers
William Alfred Eddy Papers
Karl Twitchell Papers

Secondary Works Cited

Adas, Michael. "From Settler Colony to Global Hegemon: Integrating the Exceptionalist Narrative of the American Experience into World History." *American Historical Review* 16, 5 (2001): 1692–1720.

Anderson, Irvine. *Aramco, The United States and Saudi Arabia: A Study of the Dynamics of Foreign Oil Policy 1933–1950.* Princeton, NJ: Princeton University Press, 1981.

Ansell, Martin R. *Oil Baron of the Southwest: Edward L. Doheny and the Development of the Petroleum Industry in California and Mexico.* Columbus: Ohio State University Press, 1998.

Bamberg, J. H. *The History of the British Petroleum Company. Volume 2. The Anglo-Iranian Years, 1928–1954.* Cambridge, UK: Cambridge University Press, 1994.

Barger, Thomas C. *Out in the Blue: Letters from Arabia, 1937 to 1940.* Vista, CA: Selwa Press, 2000.

Barnes, Larry. *Looking Back Over My Shoulders.* Peterborough: Private edition, 1979.

Bass, Warren. *Support Any Friend: Kennedy's Middle East and the Making of the U.S.-Israel Alliance.* New York: Oxford University Press, 2003.

Beattie, Kirk. *Egypt during the Nasser Years: Ideology, Politics and Civil Society.* Boulder, CO: Westview, 1994.

Becker, David, et al. *Postimperialism: International Capitalism and Development in the Late Twentieth Century.* Boulder, CO: Lynne Reinner, 1987.

Blair, John. *The Control of Oil.* New York: Pantheon, 1976.

Blatt, Jessica. "'To Bring Out the Best That Is in Their Blood': Race, Reform, and Civilization in the *Journal of Race Development* (1910–1919)." *Ethnic and Racial Studies* 27 (2004): 691–709.

Borstelmann, Thomas. *The Cold War and the Color Line.* Cambridge, MA: Harvard University Press, 2003.

Bromell, Henry. *Little America.* New York: Knopf, 2001.

Bromley, Simon. "The United States and World Oil." *Government and Opposition* 40 (2005): 225–55.

Bronson, Rachel. *Thicker Than Oil: America's Uneasy Partnership with Saudi Arabia.* Oxford: Oxford University Press, 2006.

Brown, Anthony Cave. *Oil, God, and Gold: The Story of Aramco and the Saudi Kings.* Boston: Houghton Mifflin, 1999.

Brown, Jonathan C. *Oil and Revolution in Mexico.* Berkeley: University of California Press, 1993.

Byrkit, James. *Forging the Copper Collar: Arizona's Labor-Management War of 1901–1921.* Tucson: University of Arizona Press, 1982.

Cheng, Charles. "The Cold War: Its Impact on the Black Liberation Struggle within the United States." In Ernest Kaiser, ed. *A Freedomways Reader: Afro-America in the Seventies.* New York: International Publishers, 1977.

Citino, Nathan. "Defending the 'Postwar Petroleum Order': The US, Britain and the 1954 Saudi-Onassis Tanker Deal." *Diplomacy & Statecraft* 11 (2000): 137–60.

———. *From Arab Nationalism to OPEC: Eisenhower, King Saʿud and the Making of U.S.-Saudi Relations.* Bloomington: University of Indiana Press, 2002.

Cleland, Robert Glass. *A History of Phelps Dodge.* New York: Alfred Knopf, 1954.

Conniff, Michael L. *Black Labor on a White Canal: Panama, 1904–1981.* Pittsburgh: University of Pittsburgh Press, 1985.

Coon, Carleton. "Operation Bultiste: Promoting Industrial Development in Saudi Arabia." In H. Teaf and P. Franck, eds. *Hands Across Frontiers: Case Studies in Technical Cooperation.* Ithaca, NY: Cornell University Press, 1955.

Crawford, Margaret. *Building the Workingman's Paradise: The Design of American Company Towns.* London: Verso, 1995.

Crinson, Mark. "Abadan: Planning and Architecture under the Anglo-Iranian Oil Company." *Planning Perspectives* 12 (1997): 341–59.

Davis, Margaret Leslie. *Dark Side of Fortune: Triumph and Scandal in the Life of Oil Tycoon Edward L. Doheny.* Berkeley: University of California Press, 1998.

Davis, Simon. "Keeping the Americans in Line: Britain, The United States, and Saudi Arabia, 1939–1945: Inter-Allied Rivalry in the Middle East Revisited." *Diplomacy and Statecraft* 8 (1997): 96–136.

Dilworth, Leah. *Imagining Indians in the Southwest: Persistent Visions of a Primitive Past.* Washington, D.C.: Smithsonian Institution Press, 1996.

Dodge, Phyllis B. *Tales of the Phelps-Dodge Family: A Chronicle of Five Generations.* New York: New York Historical Society, 1987.

Dresch, Paul. *A History of Modern Yemen.* Cambridge, UK: Cambridge University Press, 2000.

Du Bois, William E. B. "Education in Africa." *Crisis* (June, 1926): 86–89.

Dudziak, Mary. *Cold War Civil Rights: Race and the Image of American Democracy.* Princeton, NJ: Princeton University Press, 2000.

———. "Desegregation as a Cold War Imperative." *Stanford Law Review* 41 (1988): 61–120.

Duguid, Stephen. "A Biographical Approach to the Study of Social Change in the Middle East: Abullah Tariki as a New Man." *International Journal of Middle East Studies* 1, 1970: 195–220.

Edwards, Holly. "The Near East and the Wild West." In *Noble Dreams, Wicked Pleasures: Orientalism in America, 1870–1930.* Princeton, NJ: Princeton University Press, 2000.

Ferrier, Ronald. "The Anglo-Iranian Dispute: A Triangular Relationship." In James Bill and Roger Louis, eds. *Mussadiq, Iranian Nationalism and Oil*. Austin: University of Texas Press, 1988.

Field, Michael. *Merchants: The Big Business Families of Saudi Arabia and the Gulf States*. Woodstock, NY: Overlook Press, 1985.

Foner, Eric, and Jon Wiener. "Fighting for the West." *The Nation* (July 29, 1991): 163–66.

Frieden, Jeffry A. "Oil and the Evolution of U.S. Policy Towards the Developing Areas, 1900–1950: An Essay in Interpretation." In R. W. Ferrier and A. Fursenko, eds. *Oil in the World Economy*. London: Routledge, 1989.

Füredi, Frank. *The Silent War: Imperialism and the Changing Perception of Race*. London: Pluto Press, 1998.

Gause, F. Gregory. *Oil Monarchies: Domestic and Security Challenges in the Arab Gulf States*. New York: Council on Foreign Relations, 1994.

Gerstle, Gary. *American Crucible: Race and Nation in the Twentieth Century*. Princeton, NJ: Princeton University Press, 2001.

Godfried, Nathan. *Bridging the Gap between Rich and Poor: American Economic Development Policy toward the Arab East, 1942–1949*. New York: Greenwood, 1987.

Graham, Don. "Frank Norris." In *A Literary History of the American West*. Fort Worth: Texas Christian University Press, 1987: 370–80.

Halliday, Fred. "A Curious and Close Liaison: Saudi Arabia's Relations with the United States." In Tim Niblock, ed. *State, Society and Economy in Saudi Arabia*. London: Croom Helm, 1992: 125–47.

Hanrahan, Gene Z. *The Bad Yankee / El Peligro Yankee*. Vol. 1. Chapel Hill, NC: Documentary Publications, 1985.

Harik, Iliya. "The Origins of the Arab State System." In Giacomo Luciani, ed. *The Arab State*. Berkeley: University of California Press, 1990: 1–28.

Hart, Parker. *Saudi Arabia and the United States: Birth of a Security Partnership*. Bloomington: Indiana University Press, 1998.

Hildebrand, George, and Garth Mangum. *Capital and Labor in American Copper, 1845–1990*. Cambridge, MA: Harvard University Press, 1992.

Holden, David. *Farewell to Arabia*. New York: Walker, 1966.

Ikenberry, G. John. *After Victory: Institutions, Strategic Restraint, and the Rebuilding of Order after Major Wars*. Princeton, NJ: Princeton University Press, 2001.

Indonesia Project. Center for International Studies, MIT. *Stanvac in Indonesia*. Sixth Case Study in a National Planning Association Series on United States Business Performance Abroad. Washington, D.C.: NPA, 1957.

Ingersoll, Ralph. *In and Under Mexico*. New York: Century Co., 1924.

Jackson, Robert H. *Quasi-States: Sovereignty, International Relations and the Third World*. Cambridge, UK: Cambridge University Press, 1990.

Jacoby, Sanford M. *Modern Manors: Welfare Capitalism since the New Deal.* Princeton, NJ: Princeton University Press, 1997.

Johnson, Nora. *You Can Go Home Again.* New York: Doubleday, 1982.

Kaplan, Robert. *Arabists: The Romance of an American Elite.* New York: Free Press, 1993.

Katz, Friedrich. *The Secret War in Mexico: Europe, the United States, and the Mexican Revolution.* Chicago: University of Chicago Press, 1981.

Kelly, Robin D. G. "'But a Local Phase of a World Problem': Black History's Global Vision, 1883–1950." *Journal of American History* 86 (1999): 1045–77.

Kennedy, Ludovic. *Very Lovely People: A Personal Look at Some Americans Living Abroad.* New York: Simon Schuster, 1969.

Khadduri, Walid, ed. *'Abdallah al-Tariki al-a'mal al-kamilah.* Beirut: Markaz dirasat al-wihdah al-'arabiyah, 1999.

Killgore, Andrew I. "David Stuart Dodge." *Washington Report on Middle East Affairs.* September 9, 1985.

Kimball, Solon T. "American Culture in Saudi Arabia." *Transactions of the New York Academy of Sciences.* Series 2. 18, 5 (1956): 469–84.

King, Carl B., and Howard W. Risher, Jr. *The Negro in the Petroleum Industry.* Report No. 5 in the series the Racial Policies of American Industry. University of Pennsylvania. Wharton School of Finance and Commerce. Industrial Research Unit, 1969.

King, Kenneth James. *Pan-Africanism and Education: A Study of Race Philanthropy and Education in the Southern States of America and East Africa.* Oxford: Clarendon Press, 1971.

Kluger, James R. *The Clifton-Morenci Strike: Labor Difficulty in Arizona, 1915–1916.* Tucson: University of Arizona Press, 1970.

Kostiner, Joseph. *The Making of Saudi Arabia, 1916–1936: From Chieftaincy to Monarchical State.* New York: Oxford University Press, 1993.

Lacey, Robert. *The Kingdom: Arabia and the House of Saud.* New York: Harcourt Brace Jovanovich, 1981.

Lackner, Helen. *A House Built on Sand: A Political Economy of Saudi Arabia.* London: Ithaca Press, 1978.

LaFeber, Walter. *The Panama Canal: The Crisis in Historical Perspective.* Expanded edition. New York: Oxford University Press, 1979.

Larson, Henrietta, Evelyn Knowlton, and Charles Popple. *New Horizons, 1927–1950.* New York: Harper and Row, 1971.

Layton, Azza Salama. *International Politics and Civil Rights Policies in the United States 1941–1960.* Cambridge, UK: Cambridge University Press, 2000.

LeBotz, Dan. *Edward L. Doheny: Petroleum, Power, and Politics in the United States and Mexico.* New York: Praeger, 1991.

Lemann, Nicholas. "The Way They Were." *New Yorker,* April 15, 2002.

Lewis, David Levering. *W. E. B. Du Bois: Biography of a Race, 1868–1919*. New York: Henry Holt, 1993.

———. *W. E. B. Du Bois: The Fight for Equality and the American Century, 1919–1963*. New York: Henry Holt, 2000.

Lieuwen, Edwin. *Petroleum in Venezuela: A History*. University of California Publications in History. Vol. 47. Berkeley: University of California Press, 1954.

Limerick, Patricia. *The Legacy of Conquest: The Unbroken Past of the American West*. New Haven, CT: Yale University Press, 1987.

Lincoln, C. Eric. "The Race Problem and International Relations." In George Shepherd, Jr., ed. *Racial Influences on American Foreign Policy*. New York: Basic Books, 1970.

Lindsey, Donal F. *Indians at Hampton Institute, 1877–1923*. Urbana: University of Illinois Press, 1995.

Lippman, Thomas W. *Inside the Mirage: America's Fragile Partnership with Saudi Arabia*. Boulder, CO: Westview, 2004.

Lipsky, George A. *Saudi Arabia: Its People, Its Society, Its Culture*. New Haven, CT: Human Relations Area Files Press, 1959.

Little, Douglas. *American Orientalism: The United States and the Middle East since 1945*. Chapel Hill: University of North Carolina Press, 2002.

———. "Cold War and Covert Action: The U.S. and Syria, 1945–1958." *Middle East Journal* 48 (1990): 51–75.

———. "New Frontier on the Nile." *Journal of American History* 25 (1988): 501–27.

———. "Pipeline Politics: America, TAPLINE, and the Arabs." *Business History Review* 64 (1990): 255–85.

Long, David E. *The United States and Saudi Arabia: Ambivalent Allies*. Boulder, CO: Westview, 1985.

Louis, Wm. Roger. *The British Empire in the Middle East, 1945–1951: Arab Nationalism, the United States, and Postwar Imperialism*. New York: Oxford University Press, 1984.

Mader, Julius, and Mohamed Abdelnabi. *Who's Who in CIA: A Biographical Reference Work on 3,000 Officers of the Civil and Military Branches of Secret Services of the USA in 120 Countries*. Berlin: Julius Mader, 1968.

Marchand, Roland. *Creating the Corporate Soul: The Rise of Public Relations and Corporate Imagery in American Big Business*. Berkeley: University of California Press, 1998.

Mawbry, Spencer. "The Clandestine Defence of Empire: British Special Operations in Yemen, 1951–64." *Intelligence and National Security* 17 (2002): 105–30.

Mellinger, Philip J. *Race and Labor in Western Copper*. Tucson: University of Arizona Press, 1995.

Meyer, Lorenzo. *Mexico and the United States in the Oil Controversy, 1917–1942*. Austin: University of Texas Press, 1977.

Miller, Aaron David. *Search for Security: Saudi Arabian Oil and American Foreign Policy, 1939–1949*. Chapel Hill: University of North Carolina Press, 1980.

Mohamedi, Fareed. "Oil Prices and Regime Resilience in the Gulf." *Middle East Report* 232 (2004): 36–39.

Morrison, Tony. *Playing in the Dark: Whiteness and the Literary Imagination.* New York: Random House, 1993.

———. "Unspeakable Things Unspoken: The Afro-American Presences in American Literature." In Henry B. Wonham, ed. *Criticism and the Color Line: Desegregating American Literary Studies.* New Brunswick: Rutgers University Press, 1996.

Norberg, Robert. "Saudi Arabs, Americans, and Oil," *Saudi-American Forum*, Essay Series 10 (March 20, 2003), www.saudi-american-forum.org/Newsletters/SAF_Essay_10.htm, accessed on April 16, 2003.

Nore, Petter, and Terisa Turner, eds. *Oil and Class Struggle.* London: Zed Press, 1981.

Nowell, Gregory. *Mercantile States and the World Oil Cartel, 1900–1939.* Ithaca, NY: Cornell University Press, 1994.

O'Dea, Thomas F. "Social Change in Saudi Arabia: Problems and Prospects." Report Prepared from Research Conducted by the Special Study Group. [George Maranjian, Salma Najjar, O'Dea, and Federico S. Vidal]. February–August 1963.

Olien, Roger, and Diana David Olien. *Life in the Oil Fields.* Austin: Texas Monthly Press, 1986.

Painter, David. *Oil and the American Century: The Political Economy of U.S. Foreign Oil Policy, 1941–1954.* Baltimore: Johns Hopkins University Press, 1986.

Peterson, Tore Tingvold. "Anglo-American Rivalry in the Middle East: The Struggle for the Buraimi Oasis, 1952–1957." *International History Review* 14 (1992): 71–91.

Philby, Harry St. John Bridger. *Arabian Oil Ventures.* Washington, D.C.: Middle East Institute, 1964.

Pledge, Thomas. *Saudi ARAMCO and Its People: A History of Training.* Dhahran: Saudi Arabian Oil Company, 1998.

Plummer, Brenda. *Rising Wind: Black Americans and U.S. Foreign Affairs, 1935–1960.* Chapel Hill: University of North Carolina Press, 1996.

Quam-Wickham, Nancy Lynn. "Petroleocrats and Proletarians: Work, Class, and Politics in the California Oil Industry, 1917–1925." Ph.D. dissertation, University of California, Berkeley, 1994.

Rabe, Stephen G. *The Road to Opec: United States Relations with Venezuela, 1919–1976.* Austin: University of Texas Press, 1982.

Randall, Laura. *The Political Economy of Venezuelan Oil.* New York: Praeger, 1987.

Rasheed, Madawi al-. *History of Saudi Arabia.* Cambridge, UK: Cambridge University Press, 2002.

———, and Robert Vitalis. *Counternarratives: History, Society and Politics in Saudi Arabia and Yemen.* New York: Palgrave / St. Martins Press, 2003.

Reed, Adolph, Jr. *W. E. B. Du Bois and American Political Thought: Fabianism and the Color Line.* New York: Oxford University Press, 1996.

Rivas, Darlene. "Like Boxing with Joe Louis: Nelson Rockefeller in Venezuela, 1945–1948." In Peter Hahn and Mary Ann Heiss, eds. *Empire and Revolution: The United States and the Third World since 1945*. Columbus: Ohio State University Press, 2001: 217–41.

Robbins, William G. *Colony and Empire: The Capitalist Transformation of the American West*. Lawrence: University of Kansas Press, 1994.

———. "The 'Plundered Province' Thesis and the Recent Historiography of the American West." *Pacific Historical Review* 55 (1986): 577–97.

Roberts, David. *Once They Moved Like the Wind: Cochise, Geronimo, and the Apache Wars*. New York: Simon and Schuster, 1993.

Rodgers, Daniel. "Exceptionalism." In Anthony Mohlo and Gordon Wood, eds. *Imagined Histories: American Historians Interpret the Past*. Princeton, NJ: Princeton University Press, 1998: 21–40.

Ross, Dorothy. "The New and Newer Histories: Social Theory and Historiography in an American Key." In Anthony Mohlo and Gordon Wood, eds. *Imagined Histories: American Historians Interpret the Past*. Princeton, NJ: Princeton University Press, 1998: 85–206.

Ruiz, Ramón. *The People of Sonora and Yankee Capitalists*. Tucson: University of Arizona Press, 1988.

Safran, Nadav. *Saudi Arabia: The Ceaseless Quest for Security*. Cambridge, MA: Harvard University Press, 1985.

Sa'id, Nasir al-. *Tarikh Al Sa'ud*. Ittihad sha'b al-jazirah al-'arabiyah, 1982.

Samore, Gary. "Royal Family Politics in Saudi Arabia (1953–1982)." Ph.D. dissertation, Harvard University, 1983.

Sanger, Richard. *Arabian Peninsula*. Ithaca, NY: Cornell University Press, 1954.

Schmidt, Dana Adams. *Yemen: The Unknown War*. London: Bodley Head, 1968.

Schwantes, Carlos A., ed. *Bisbee: Urban Outpost on the Frontier*. Tucson: University of Arizona Press, 1992.

———. *Vision & Enterprise: Exploring the History of Phelps Dodge Corporation*. Tucson: University of Arizona Press, 2000.

Scott, William R. *The Americans in Panama*. New York: Statler, 1912.

Seccombe, Ian. "A Disgrace to American Enterprise: Italian Labor and the Arabian American Oil Company in Saudi Arabia, 1944–56." *Immigrants and Minorities* 5 (1986): 233–57.

Sellars, Nigel Anthony. *Oil, Wheat and Wobblies: The Industrial Workers of the World in Oklahoma, 1905–1930*. Norman: University of Oklahoma Press, 1998.

Slide, Anthony. *Before Video: A History of the Non-Theatrical Film*. New York: Greenwood, 1992.

Slotkin, Richard. "Buffalo Bill's 'Wild West.'" In Amy Kaplan and Donald E. Pease, eds. *Cultures of United States Imperialism*. Durham, NC: Duke University Press, 1993.

Snyder, Harry Roscoe. "Community College Education for Saudi Arabia: A Report of a Type A Project." Ph.D. dissertation, Teachers College, Columbia University, 1963.

Spero, David. *The Hidden Hand of American Hegemony: Petrodollar Recycling and International Markets.* Ithaca, NY: Cornell University Press, 1999.

Stanfield, John H. *Philanthropy and Jim Crow in American Social Science.* Westport, CT: Greenwood, 1985.

Stegner, Wallace. *Discovery! The Search for Arabian Oil.* Beirut: Middle East Export Press, 1971.

Sundquist, Eric J., ed. *The Oxford W. E. B. Du Bois Reader.* New York: Oxford University Press, 1996.

Taylor, Wayne, and John Lindemann. *The Creole Petroleum Corporation in Venezuela.* United States Business Performance Abroad. Washington, D.C.: National Planning Association, 1955.

Teitelbaum, Joshua. "Civil-Military Relations in Saudi Arabia." Paper presented at the Conference on Saudi Futures. International Institute for the Study of Islam in the Modern World. Leiden, Feb. 19–21, 2004.

———. *The Rise and Fall of the Hashimite Kingdom of Arabia.* New York: New York University Press. 2001.

Tinker Salas, Miguel. *In the Shadow of the Eagle: Sonora and the Transformation of the Border during the Porfiriato.* Berkeley: University of California Press, 1997.

———. "Venezuelans, West Indians, and Asians: The Politics of Race in Venezuelan Oil Fields, 1920–1940." In Vincent Peloso, ed. *Work, Protest and Identity in Twentieth Century Latin America.* Wilmington, DE: Scholarly Resources, 2003.

Truettner, William H., ed. *The West as America: Reinterpreting Images of the Frontier, 1820–1920.* Washington, D.C.: Smithsonian Institution Press, 1991.

Vassiliev, Alexei. *History of Saudi Arabia.* New York: New York University Press, 2000.

Vernon, Raymond. *Storm over the Multinationals: The Real Issues.* Cambridge, UK: Harvard University Press, 1977.

Viola, Joy Winkie. *Human Resources Development in Saudi Arabia: Multinationals and Saudization.* Boston: International Human Resources Development Corporation, 1986.

Vitalis, Robert. "Aramco World: Business and Culture on the Arabian Oil Frontier." In Karen Merrill, ed. *The Modern Worlds of Business and Industry: Cultures, Technology, Labor.* Turnhout, Belgium: Brepols, 1999: 3–28.

———. "Birth of a Discipline." In Brian Schmidt and David Long, eds. *Imperialism and Internationalism in the Discipline of International Relations.* Albany: State University of New York Press, 2005: 159–82.

———. "The Closing of the Arabian Oil Frontier and the Future of Saudi-American Relations." *Middle East Report* 204 (1997): 15–21.

———. *When Capitalists Collide: Business Conflict and the End of Empire in Egypt.* Berkeley: University of California Press, 1995.

Watts, Michael. "Resource Curse? Governmentality, Oil and Power in the Niger Delta, Nigeria." *Geopolitics* 9 (2004): 50–80.

White, Gerald T. *Formative Years in the Far West: A History of Standard Oil Company of California and Predecessors through 1919.* New York: Appleton-Century-Crofts, 1962.

White, Richard. *Roots of Dependency: Subsistence, Environment and Social Change among the Choctaws, Pawnees and Navajos.* Lincoln: University of Nebraska Press, 1983.

Wilentz, Sean. "America Made Easy." *New Republic* (July 2, 2001): 3.

Wilkins, Mira. *The Emergence of Multinational Enterprise: American Business Abroad from the Colonial Era to 1914.* Cambridge, MA: Harvard University Press, 1970.

Wilson, Joy. "Raising Children in an American Oil Camp in Saudi Arabia." *Radcliffe Quarterly* (March 1977): 8–9.

Winant, Howard. *The World is a Ghetto.* New York: Basic Books, 2002.

Yaqub, Salim. *Containing Arab Nationalism: The Eisenhower Doctrine and the Middle East.* Chapel Hill: University of North Carolina Press, 2004.

Yergin, Daniel. *The Prize: The Epic Quest for Oil, Money, and Power.* New York: Simon and Schuster, 1991.

Yizraeli, Sarah. *The Remaking of Saudi Arabia: The Struggle Between King Sa'ud and Crown Prince Faysal, 1953–1962.* Tel Aviv: Moshe Dayan Center, 1997.

INDEX

Anglo-Iranian Oil Company (AIOC), 24, 107, 113, 120, 129, 130.137, 140, 142–143, 299n40
Aoueini, Hussein, 295n25
Arab League, 185, 191, 317n33
Arab Liberation Front, 236, 239–240
Arab Petroleum Conference
  Cairo (1959), 197, 199, 208–210
  Beirut (1963), 248
Arabian Affairs Division (AAD), 31, 69, 95, 99, 145, 154, 196–197, 200–204, 252
Arabian American Oil Company (ARAMCO), x, xi–xii. *See also* Government Relations Organization; more specific entries, e.g. public relations efforts of ARAMCO
  bust of early 1960s, effects of, 218–219
  corporate biographies of Saudi figures of influence, 204
  defensiveness of, 251–252, 255
  documents and records, xvi–xvii
  early American corporations, development and organization compared to, 31–37, 53–61
  exceptionalism, myth of, xii–xiii, 9–12, 68, 69, 76–79, 118
  Heritage Center, 273–274
  historiographical issues regarding, xii–xviii, 11–18, 22–26
  internal reform efforts, 200–208
  loyalty claims of, 97
  Marshall or Point Four program, self-view as private version of, 35, 70, 120, 218, 252, 267
  maximum exploitation/minimum outlay approach, 201, 205
  naming of, 79, 281n11
  nationalization, 25, 69, 140, 206, 248, 266, 272
  O'Dea report on future of, 256–262
  origins of, 31–32, 52–61
  postwar transition, 79–86
  "special relationship" between U.S. and Saudi Arabia attributed to importance of, 164
Arabian Frontiers (Sanger), 34–35, 37, 116
Arizona, history of mining industry in. *See* Phelps Dodge Company; Western frontier corporations
Armenian genocide, 53
Arnold, José, 72
As-Sufayyan, Abdul Aziz, 147
Aswan Dam, 185
Atlanta Compromise, 286–287n22

Austria, Ibn Muammar in, 212, 319n68
Azzam Pasha, Abd al-Rahman, 106, 171

Baathists, 10, 128, 161, 162, 199, 211
Badr Ibn Abd al-Aziz, 211, 227, 235
al-Badr, Muhammad, 238
Baghdad Pact, 165–166, 171
al-Bahijan, Abd al-Rahman, 147, 148, 156, 172, 173, 222, 223
Bahrain, xvii, xix, 53, 55, 56–57, 60, 64, 66, 78, 92, 98, 103, 119, 130, 138, 154, 177, 180, 183, 189, 226
Bailey, Waldo, 101
Bandar Ibn Sultan Ibn Abd al-Aziz Al Saud, 3–4, 5
barasti housing provided for local workers, 56, 89, 90, 101, 102, 112, 255, *plate c*
Barger, Kathleen, 58, 59, 68, 263
Barger, T[h]eresa, 263, 275
Barger, Tom, xii.xxiv
  challenges of postwar world and nationalist populism, strategies for meeting, 145, 151, 153
  on Egypt-Saudi conflicts, 230, 231, 237
  formation of policy on ARAMCO camp life and labor negotiations, 92, 93, 96, 100, 105, 108.109, 113
  internal reforms at ARAMCO in response to rise of populist nationalism in Middle East, 200, 201, 203, 204, 205, 206, 207
  labor unrest of 1955–1956, 176, 177, 178, 179, 183
  popular feeling against, 264
  racist terms, use of, 58–60
  on Stegner's *Discovery!*, 195, 196, 197
  struggle between Saud and Faisal (1959–1961), 220, 222
  on Tariki, 134, 136
  in WWII, 64, 68–71, 73
Barnet, Carl, 219
Bass, Warren, 235, 324n62
Beard, Charles, 8
Bechtel Company and Bechtel brothers, 32, 61, 74–76, 85, 101, 131, 137, 248, 298n27
S.S. *Bellows*, 83, 92, 296n41
Benedict, Ruth, 119
bin Adwan, Abdallah, 140, 148, 149
bin Ali al-Khalifah, Isa, 57
bin Farraj, Juman, 172
bin Laden family, 133
bin Laden, Muhammad, xxii, 74, 169, 170, 201

Stanford Studies in Middle Eastern and Islamic Societies and Cultures

2006

Jessica Winegar, *Creative Reckonings: The Politics of Art and Culture in Contemporary Egypt*

Joel Beinin and Rebecca L. Stein, *The Struggle for Sovereignty: Palestine and Israel, 1993–2005*